Lecture Notes in Mathematics 2045

Editors:
J.-M. Morel, Cachan
B. Teissier, Paris

T0236671

For further volumes:
http://www.springer.com/series/304

FONDAZIONE CIME
ROBERTO CONTI
CENTRO INTERNAZIONALE MATEMATICO ESTIVO
INTERNATIONAL MATHEMATICAL SUMMER CENTER

Fondazione C.I.M.E., Firenze

C.I.M.E. stands for *Centro Internazionale Matematico Estivo*, that is, International Mathematical Summer Centre. Conceived in the early fifties, it was born in 1954 in Florence, Italy, and welcomed by the world mathematical community: it continues successfully, year for year, to this day.

Many mathematicians from all over the world have been involved in a way or another in C.I.M.E.'s activities over the years. The main purpose and mode of functioning of the Centre may be summarised as follows: every year, during the summer, sessions on different themes from pure and applied mathematics are offered by application to mathematicians from all countries. A Session is generally based on three or four main courses given by specialists of international renown, plus a certain number of seminars, and is held in an attractive rural location in Italy.

The aim of a C.I.M.E. session is to bring to the attention of younger researchers the origins, development, and perspectives of some very active branch of mathematical research. The topics of the courses are generally of international resonance. The full immersion atmosphere of the courses and the daily exchange among participants are thus an initiation to international collaboration in mathematical research.

C.I.M.E. Director
Pietro ZECCA
Dipartimento di Energetica "S. Stecco"
Università di Firenze
Via S. Marta, 3
50139 Florence
Italy
e-mail: zecca@unifi.it

C.I.M.E. Secretary
Elvira MASCOLO
Dipartimento di Matematica "U. Dini"
Università di Firenze
viale G.B. Morgagni 67/A
50134 Florence
Italy
e-mail: mascolo@math.unifi.it

For more information see CIME's homepage: http://www.cime.unifi.it

CIME activity is carried out with the collaboration and financial support of:

- INdAM (Istituto Nazionale di Alta Matematica)

- MIUR (Ministero dell'Universita' e della Ricerca)

John Lewis • Peter Lindqvist
Juan J. Manfredi • Sandro Salsa

Regularity Estimates for Nonlinear Elliptic and Parabolic Problems

Cetraro, Italy 2009

Editors:
Ugo Gianazza
John Lewis

 Springer

FONDAZIONE
CIME
ROBERTO CONTI

John Lewis
University of Kentucky
Lexington, KY
USA

Juan J. Manfredi
University of Pittsburgh
PA, USA

Sandro Salsa
Politecnico di Milano
Italy

Peter Lindqvist
Norwegian University of
Science and Technology
Trondheim, Norway

ISBN 978-3-642-27144-1   e-ISBN 978-3-642-27145-8
DOI 10.1007/978-3-642-27145-8
Springer Heidelberg Dordrecht London New York

Lecture Notes in Mathematics ISSN print edition: 0075-8434
                                    ISSN electronic edition: 1617-9692

Library of Congress Control Number: 2012933109

Mathematics Subject Classification (2010): 5J70, 35J75, 35J92, 35K65, 35K67, 35K86, 35K92, 35Q91,
                          35R11, 35R35, 49K20, 49N60

Printed on acid-free paper

Springer is part of Springer Science+Business Media (www.springer.com)

Preface

This volume collects the notes of the CIME course **Regularity Estimates for Nonlinear Elliptic and Parabolic Problems** held in Cetraro (Italy) on June 22–27, 2009. The school consisted in five series of lectures, delivered by

Emmanuele DiBenedetto (Vanderbilt University, Nashville, USA)
John Lewis (University of Kentucky, Lexington, USA)
Peter Lindqvist (Norwegian University of Science and Technology, Trondheim, Norway)
Juan J. Manfredi (University of Pittsburgh, Pittsburgh, USA)
Sandro Salsa (Politecnico di Milano, Milano, Italy).

The issue of regularity has obviously played a central role in the theory of Partial Differential Equations, almost since its inception, and despite the tremendous development, it still remains a very fruitful research field.

In particular regularity estimates for degenerate and singular elliptic and parabolic equations have developed considerably in the last years, in many unexpected and challenging directions.

Because of all these recent results, it seemed timely to trace an overview that would highlight emerging trends and issues of this fascinating research topic in a proper and effective way.

The course aimed at showing the deep connections among all these topics and at opening new research directions, through the contribution of leading experts in all these fields.

Emmanuele DiBenedetto gave a course on
Introduction to Regularity Theory
for Degenerate Parabolic Equations in Divergence Form
discussing some techniques recently introduced to investigate the local and global behavior of solutions to degenerate parabolic equations when their principal part fails to be coercive. The equations have to be regarded in their own intrinsic geometry, and the solutions have a limited degree of regularity. DiBenedetto showed how identifying regularity classes as functions

of the degenerate and/or singular structure of the equation is part of an emerging theory which promises to yield an understanding of degeneracy and/or singularity in Partial Differential Equations. Unfortunately there are no notes of this course.

The course of *John Lewis* on
Applications of Boundary Harnack Inequalities
for p-Harmonic Functions and Related Topics
discussed applications of recent work and techniques concerning the boundary behavior of positive p-harmonic functions vanishing on a portion of the boundary of Lipschitz, chord arc, and Reifenberg flat domains. At first fundamental properties of p-harmonic functions and elliptic measure were presented. Then the dimension of p-harmonic measure was dealt with. The final part of the course first considered boundary Harnack inequalities and the Martin boundary problem in Reifenberg flat and Lipschitz domains, and at the end uniqueness and regularity both in free boundary and inverse type problems.

Peter Lindqvist presented in his course
Regularity of Supersolutions
a general theory for supersolutions of the p-Laplace Equation. Indeed the regularity theory for *solutions* to the parabolic p-laplacian is a well-developed topic, but when it comes to (semicontinuous) *supersolutions* and *subsolutions* a lot remains to be done. Supersolutions are often auxiliary tools as in the celebrated Perron method, for example, but they are also interesting in their own right. Therefore, the lectures were entirely focused on this important issue.

Juan J. Manfredi delivered a series of lectures on
Introduction to random Tug-of-War games and PDEs
providing an introduction to the connection between the theory of stochastic tug-of-war games and nonlinear equations of p-Laplacian type in the Euclidean and discrete cases. The fundamental contributions of Kolmogorov, Ito, Kakutani, Doob, Hunt, Lévy, and many others have shown the profound and powerful connection between classical linear potential theory and probability theory. The idea behind the classical interplay is that harmonic functions and martingales share a common cancellation property that can be expressed by using mean value properties. In his lectures, Manfredi showed how this approach turns out to be very useful in the nonlinear theory as well.

Sandro Salsa taught a course on
The Problems of the Obstacle in Lower Dimension
and for the Fractional Laplacian
giving a somewhat self-contained presentation of the results concerning the analysis of the solution and the free boundary of the thin obstacle problem and more generally of the obstacle for the fractional Laplacian. He started from the thin obstacle problem, considering the case of *zero obstacle*. In

this case, the main ideas and tools were clearly seen and developed without too many technicalities and in a somewhat self-contained fashion. Later, he extended the results on the optimal regularity and the analysis of the *regular part* of the free boundary to the general case for $(-\Delta)^s$.

This series of lectures attracted approximately 50 participants, largely PhD students or post-docs, and also senior researchers; we are sure that this CIME course was rich of useful suggestions and ideas for inspiring new developments, and opening new research prospects in the near future.

We wish to thank all the lecturers for their active participation and their valuable contribution, and the CIME foundation, in particular the director Prof. Pietro Zecca and the secretary Prof. Elvira Mascolo, for their helpful support and for the organization of such a remarkable event in Cetraro.

Pavia, Italy   *Ugo Gianazza*
Lexington, KY   *John Lewis*

Contents

Applications of Boundary Harnack Inequalities for p Harmonic Functions and Related Topics 1

J. Lewis

1   Outline of the Course 1

   1.1   Ode to the p Laplacian 1

   1.2   My Introduction to p Harmonic Functions 2

2   Basic Estimates for the p Laplacian 2

   2.1 p Harmonic Functions in NTA Domains........................ 4

   2.2   The p Laplacian and Elliptic PDE............................. 6

   2.3   Degenerate Elliptic Equations 7

3   p Harmonic Measure 9

   3.1 p Harmonic Measure in Simply Connected Domains 15

   3.2   Preliminary Reductions for Theorem 2.6 15

   3.3   Proof of Theorem 2.8 16

   3.4   The Final Proof 19

   3.5 p Harmonic Measure in Space 21

   3.6   Open Problems for p Harmonic Measure...................... 22

4   Boundary Harnack Inequalities and the Martin Boundary Problem for p Harmonic Functions 23

   4.1   History of Theorem 3.1 24

   4.2   Proof of Step 1 26

   4.3   Proof of Step 2 27

   4.4   Proof of Step 3 30

   4.5   Proof of Step 4 and Theorem 3.1 33

   4.6   More on Boundary Harnack Inequalities 37

   4.7   The Martin Boundary Problem 38

   4.8   Proof of Theorem 3.9 42

   4.9   Further Remarks 46

5   Uniqueness and Regularity in Free Boundary:
    Inverse Type Problems 46
    5.1   History of Theorem 4.1 46
    5.2   Proof of Theorem 4.1 49
    5.3   Further Uniqueness Results 50
    5.4   Boundary Regularity of p Harmonic Functions 51
    5.5   Proof of Theorem 4.3 52
    5.6   Proof of Theorem 4.4 55
    5.7   Proof of Theorem 4.5 57
    5.8   Regularity in a Lipschitz Free Boundary Problem 59
    5.9   History of Theorem 4.11 60
    5.10 Proof of Theorem 4.11 60
    5.11 Enlargement of the Cone of Monotonicity in the Interior 61
    5.12 Enlargement of the Cone of Monotonicity
          at the Free Boundary 61
    5.13 An Application of Theorem 4.11 63
    5.14 Proof of (161) 65
    5.15 Closing Remarks 68
References 69

Regularity of Supersolutions 73
Peter Lindqvist
1   Introduction 73
2   The Stationary Equation 78
3   The Evolutionary Equation 91
    3.1   Definitions 92
    3.2   Bounded Supersolutions 94
    3.3   Unbounded Supersolutions 102
    3.4   Reduction to Zero Boundary Values 109
4   Weak Supersolutions are Semicontinuous 111
5   The Equation with Measure Data 122
6   Pointwise Behaviour 123
    6.1   The Stationary Equation 123
    6.2   The Evolutionary Equation 125
References 130

Introduction to Random Tug-of-War Games and PDEs 133
Juan J. Manfredi
1   Introduction 133
2   Probability Background 133
3   The p-Laplacian Gambling House 141
4   p-harmonious Functions 144
5   Directed Trees 147
6   Epilogue 150
References 151

**The Problems of the Obstacle in Lower
Dimension and for the Fractional Laplacian** 153
Sandro Salsa
1   Introduction.. 153
2   The Zero Obstacle Problem 160
    2.1   Setting of the Problem... 160
    2.2   Lipschitz Continuity and Semiconvexity....................... 162
    2.3   Local $C^{1,\alpha}$ Estimate... 166
    2.4   Optimal Regularity for Tangentially Convex
          Global Solutions 170
    2.5   Almgren's Frequency Formula................................... 173
    2.6   Asymptotic Profiles and Optimal Regularity.................. 177
    2.7   Lipschitz Continuity of the Free Boundary
          at Stable Points 180
    2.8   Boundary Harnack Principles and $C^{1,\alpha}$ Regularity
          of the Free Boundary at Stable Points........................ 183
    2.9   Structure of the Singular Set 188
3   Obstacle Problem for the Fractional Laplacian 201
    3.1   Construction of the Solution and Basic Properties 202
    3.2   Lipschitz Continuity, Semiconvexity and $C^{1,\alpha}$ Estimates 203
    3.3   Thin Obstacle for the Operator L_a: Local $C^{1,\alpha}$ Estimates.... 204
    3.4   Minimizers of the Weighted Rayleigh Quotient
          and a Monotonicity Formula 206
    3.5   Optimal Regularity for Tangentially Convex
          Global Solutions 207
    3.6   Frequency Formula 211
    3.7   Blow-up Sequences and Optimal Regularity 217
    3.8   Nondegenerate Case: Lipschitz Continuity
          of the Free Boundary 225
    3.9   Boundary Harnack Principles and $C^{1,\alpha}$ Regularity
          of the Free Boundary 227
Appendix A: The Fractional Laplacian................................... 231
    Definition and Basic Properties 231
    Supersolutions and comparison....................................... 232
Appendix B: The Operator L_a 234
    Definition and Preliminary Facts..................................... 234
    Harnack inequality, Liouville theorem and mean value property 236
    Poincaré inequalities 240
Appendix C: Relation between $(-\Delta)^s$ and L_a 240
References.. 243

List of Participants... 245

Applications of Boundary Harnack Inequalities for p Harmonic Functions and Related Topics

J. Lewis

1 Outline of the Course

This course will be concerned with applications of recent work—techniques concerning the boundary behavior of positive p harmonic functions vanishing on a portion of the boundary of Lipschitz, chord arc, and Reifenberg flat domains. An optimistic outline follows:

1. Fundamental properties of p harmonic functions and elliptic measure.
2. The dimension of p harmonic measure.
3. Boundary Harnack inequalities and the Martin boundary problem in Reifenberg flat and Lipschitz domains.
4. Uniqueness and regularity in free boundary—inverse type problems.

The lectures concerning 2 will be drawn from [6, 8, 48, 63]. Lectures involving 3 will be based on [49, 50, 52, 62]. Lectures on 4 will be concerned with [55–60] and [49–54].

1.1 *Ode to the p Laplacian*

I used to be in love with the Laplacian so worked hard to please her with beautiful theorems. However she often scorned me for the likes of Albert Baernstein, Björn Dahlberg, Carlos Kenig, and Thomas Wolff. Gradually I became interested in her sister the p Laplacian, $1 < p < \infty, p \neq 2$. I did not find her as pretty as the Laplacian and she was often difficult to handle

---

J. Lewis (✉)

Mathematics Department, University of Kentucky, Lexington, KY 40506, USA

e-mail: john@ms.uky.edu

J. Lewis et al., *Regularity Estimates for Nonlinear Elliptic and Parabolic Problems*, Lecture Notes in Mathematics 2045, DOI 10.1007/978-3-642-27145-8_1,
© Springer-Verlag Berlin Heidelberg 2012

because of her nonlinearity. However over many years I took a shine to her and eventually developed an understanding of her disposition. Today she is my girl and the Laplacian pales in comparison to her.

1.2   *My Introduction to p Harmonic Functions*

I was trained in function theory—subharmonic functions by my advisor, Maurice Heins and postdoctoral advisor Matts Essén. My first paper on elliptic PDE and p harmonic functions (see [47]) was entitled 'Capacitary Functions in Convex Rings.' The catalyst for this paper was a problem in [31] which read as follows:

'If D is a convex domain in space of 3 or more dimensions can we assert any inequalities for the Green's function $g(P,Q)$ which generalize the results for two dimensions, that follow from the classical inequalities for schlicht functions. Gabriel [29] has shown that the level surfaces of $g(P,Q) = \lambda$ are convex but his proof is long. It would be interesting to find a simpler proof.'

I tried to find a simpler proof of Gabriel's result but failed so eventually read his paper. In contrast to the author of the problem, I found Gabriel's proof easy to follow and quite ingenious. Thus instead of finding a different proof I found a different PDE, the p Laplacian, to use Gabriel's technique on. Moreover in writing the above paper I was forced to learn some classical PDE techniques (Möser iteration, Schauder techniques) in order to deal with this degenerate nonlinear divergence form elliptic PDE.

2   Basic Estimates for the p Laplacian

We shall be working in Euclidean n space $\mathbf{R}^n$. Points will be denoted by $x = (x_1, \ldots, x_n)$ or (x', x_n) where $x' = (x_1, \ldots, x_{n-1}) \in \mathbf{R}^{n-1}$. Let $\bar{E}, \partial E$, be the usual closure, boundary, of the set E and $d(y, E) =$ the distance from y to E. $\langle \cdot, \cdot \rangle$ denotes the standard inner product on $\mathbf{R}^n$ and $|x| = \langle x, x \rangle^{1/2}$ is the Euclidean norm of x. $B(x, r) = \{y \in \mathbf{R}^n : |x - y| < r\}$ and dx denotes Lebesgue n measure on $\mathbf{R}^n$. Let e_i be the i unit coordinate vector. If $O \subset \mathbf{R}^n$ is open and $1 \leq q \leq \infty$, let $W^{1,q}(O)$, denote the usual Sobolev space of equivalence classes of functions f with distributional gradient $\nabla f = (f_{x_1}, \ldots, f_{x_n})$, both of which are qth power integrable on O with Sobolev norm, $\|f\|_{1,q} = \|f\|_q + \| |\nabla f| \|_q$. Next let $C_0^\infty(O)$ be infinitely differentiable functions with compact support in O and let $W_0^{1,q}(O)$ be the closure of $C_0^\infty(O)$ in the norm of $W^{1,q}(O)$.

Given G a bounded domain (i.e., a connected open set) and $1 < p < \infty$, we say that u is p harmonic in G provided $u \in W^{1,p}(G)$ and

$$\int |\nabla u|^{p-2} \langle \nabla u, \nabla \theta \rangle \, dx = 0 \tag{1}$$

whenever $\theta \in W_0^{1,p}(G)$. Observe that if u is smooth and $\nabla u \neq 0$ in G, then

$$\nabla \cdot (|\nabla u|^{p-2} \nabla u) \equiv 0 \text{ in } G \tag{2}$$

so u is a classical solution in G to the p Laplace partial differential equation. Equation (1) arises in the study of the following Dirichlet problem: Given $g \in W^{1,p}(\mathbf{R}^n)$ let $\mathcal{F} = \{h : h - g \in W_0^{1,p}(G)\}$. Find

$$\inf_{h \in \mathcal{F}} \int_G |\nabla h|^p \, dx. \tag{3}$$

it is well known that the infimum in (3) occurs for a unique function u with $u - g \in W_0^{1,p}(G)$. Moreover u satisfies (1) as follows from the fact that u is a minimum and the usual calculus of variations type argument.

v is said to be subpharmonic (superpharmonic) in G if $v \in W^{1,p}(G)$ and whenever $\theta \geq 0 \in W_0^{1,p}(G)$,

$$\int |\nabla v|^{p-2} \langle \nabla v, \nabla \theta \rangle \, dx \leq 0 \, (\geq 0) \tag{4}$$

Lemma 1.1. *(Boundary Maximum Principle) If v is subpharmonic in G, while w is superpharmonic in G with $\min\{v - w, 0\} \in W_0^{1,p}(G)$, then $v - w \leq 0$ a.e in G.*

Lemma 1.2. *(Interior Estimates for u) Given $p, 1 < p < \infty$, let u be a positive p harmonic function in $B(w, 2r)$. Then*

(i) Caccioppoli Inequality:

$$r^{p-n} \int_{B(w,r/2)} |\nabla u|^p \, dx \leq c \, (\max_{B(w,r)} u)^p,$$

(ii) Harnack's Inequality:

$$\max_{B(w,r)} u \leq c \min_{B(w,r)} u.$$

Furthermore, there exists $\alpha = \alpha(p, n) \in (0, 1)$ such that if $x, y \in B(w, r/2)$ then

(iii) Hölder Continuity:

$$|u(x) - u(y)| \leq c \left(\frac{|x-y|}{r} \right)^\alpha \max_{B(w,r)} u.$$

Lemma 1.3. *(Interior Estimates for ∇u) Let $1 < p < \infty$ and suppose u is p harmonic in $B(w, 2r)$. Then u has a representative in $W^{1,p}(B(w, 2r))$ with Hölder continuous partial derivatives in $B(w, 2r)$. In particular there exists $\sigma \in (0, 1]$, depending only on p, n, such that if $x, y \in B(w, r/2)$, then for some $c = c(p, n)$,*

$$c^{-1} \left| \nabla u(x) - \nabla u(y) \right| \leq (|x - y|/r)^\sigma \max_{B(w,r)} |\nabla u|$$

$$\leq c r^{-1} (|x - y|/r)^\sigma \max_{B(w,2r)} |u|.$$

Also if $\nabla u(x) \neq 0$, then u is C^∞ near x.

For a proof of Lemmas 1.1 and 1.2, see [69]. Numerous proofs have been given of Hölder continuity of ∇u in Lemma 1.3. Perhaps the first was due to Ural'tseva for $p > 2$ while DiBenedetto, myself, and Tolksdorff all gave proofs independently and nearly at the same time (1983, 1984) for $1 < p < \infty$. A proof which even applies to the parabolic p Laplacian and systems can be found in [19].

If $p > 2$ it is known that u (as above) need not be C^2 locally. For $1 < p < 2$ this question is still open in $\mathbf{R}^n, n \geq 3$. In two dimensions, Iwaniec and Manfredi [36] showed that solutions are C^k where $k = k(p) \geq 2$ when $1 < p < 2$ and $k \to \infty$ as $p \to 1$.

2.1 p Harmonic Functions in NTA Domains

Definition A. *A domain Ω is called non tangentially accessible (NTA), if there exist $M \geq 2$ and $0 < r_0 \leq \infty$ such that the following are fulfilled,*

(i) *Corkscrew condition: For any $w \in \partial\Omega, 0 < r < r_0$, there exists $a_r(w) \in \Omega$ satisfying $M^{-1}r < |a_r(w) - w| < r, d(a_r(w), \partial\Omega) > M^{-1}r$,*
(ii) *$\mathbf{R}^n \setminus \bar{\Omega}$ satisfies the corkscrew condition,*
(iii) *Uniform condition: If $w \in \partial\Omega$, and $w_1, w_2 \in B(w, r_0) \cap \Omega$, then there is a rectifiable curve $\gamma : [0, 1] \to \Omega$ with $\gamma(0) = w_1, \gamma(1) = w_2$, and*

(a) *$H^1(\gamma) \leq M |w_1 - w_2|$,*
(b) *$\min\{H^1(\gamma([0, t])), H^1(\gamma([t, 1]))\} \leq M\, d(\gamma(t), \partial\Omega)$.*

In Definition A, H^1 denotes length or Hausdorff one measure. Often in our applications Ω will at least be an NTA domain with constants M, r_0 while p is fixed, $1 < p < \infty$. Also, $c \geq 1$ will be a positive constant which may only depend on p, n, M unless otherwise stated. Let $w \in \partial\Omega, 0 < r < r_0$, and suppose $u > 0$ is p harmonic in $\Omega \cap B(w, 2r)$ with $u = 0$ on $\partial\Omega \cap B(w, 2r)$ in the usual Sobolev sense. We extend u to $B(w, 2r)$ by putting $u = 0$ on $B(w, 2r) \setminus \Omega$. Under this scenario we state

Lemma 1.4. *Let u, p, w, Ω be as above. Then $u \in W^{1,p}(B(w, 2r))$ and*

$$r^{p-n} \int\limits_{B(w,r/2)} |\nabla u|^p \, dx \leq c \, (\max_{B(w,r)} u)^p.$$

Moreover there exists $\beta = \beta(p, n, M) \in (0, 1)$ such that u has a Hölder continuous representative in $B(w, 2r)$ with

$$|u(x) - u(y)| \leq c \, (|x - y|/r)^\beta \max_{B(w,r)} u$$

whenever $x, y \in B(w, r)$.

Lemma 1.5. *Let u, p, w, Ω, r, be as in Lemma 1.4. There exists c such that if $\hat{r} = r/c$, then*

$$\max_{B(w,\hat{r})} u \leq c u(a_{\hat{r}}(w)).$$

2.1.1   Outline of Proofs

The first display in Lemma 1.4 is a standard subsolution type inequality (use u times a cutoff as a test function in (1)). As for the last display in Lemma 1.4 if $p > n$, this display is a consequence of Morrey's Theorem and the first display. If $1 < p \leq n$, then from the interior estimates in Lemma 1.2, we deduce that it suffices to consider only the case when $y \in \partial\Omega \cap B(w, r)$. One then shows for some $\theta = \theta(p, n, M), 0 < \theta < 1$, that

$$\max_{B(z,\rho/2)} u \leq \theta \max_{B(z,\rho)} u \tag{5}$$

whenever $0 < \rho < r/4$ and $z \in \partial\Omega \cap B(w, r)$. Equation (5) can then be iterated to get Hölder continuity in Lemma 1.4 for x, y as above. To prove (5) one uses the fact that $(\mathbf{R}^n \setminus \Omega) \cap B(z, \rho/2)$ and $B(z, \rho/2)$ have comparable p capacities, as well as estimates for subsolutions to elliptic partial differential equations of p Laplacian type. These estimates are due to [68] for the p Laplacian (see also [30]). Lemma 1.5 for harmonic functions is often called Carleson's lemma although apparently it could be due to Domar. This lemma for uniformly elliptic PDE in divergence form is usually attributed to [14]. All proofs use only Harnack's inequality and Hölder continuity near the boundary. Thus Lemma 1.5 is also valid for solutions to many PDE's including the p Laplacian.   □

In our study of p harmonic measure we shall outline a proof of a similar inequality when the geometry is considerably more complicated. That is when $\Omega \subset \mathbf{R}^2$ is only a bounded simply connected domain.

2.2   The p Laplacian and Elliptic PDE

Let u be a solution to the p Laplace equation in (2) and suppose ∇u is nonzero as well as sufficiently smooth in a neighborhood of $x \in \Omega$. Let $\eta \in \mathbf{R}^n$ with $|\eta| = 1$ and put $\zeta = \langle \nabla u, \eta \rangle$. Then differentiating $\nabla \cdot (|\nabla u|^{p-2} \nabla u) = 0$ with respect to η one gets that ζ is a strong solution at x to

$$L\zeta = \nabla \cdot [(p-2)|\nabla u|^{p-4}\langle \nabla u, \nabla \zeta \rangle \nabla u + |\nabla u|^{p-2}\nabla \zeta] = 0. \qquad (6)$$

Clearly, (7)

$$Lu = (p-1)\nabla \cdot \left[|\nabla u|^{p-2} \nabla u \right] = 0 \text{ at } x. \qquad (7)$$

Equation (6) can be rewritten in the form

$$L\zeta = \sum_{i,j=1}^{n} \frac{\partial}{\partial x_i} \left[b_{ij}(x)\zeta_{x_j}(x) \right] = 0, \qquad (8)$$

where $b_{ij}(x) = |\nabla u|^{p-4}[(p-2)u_{x_i}u_{x_j} + \delta_{ij}|\nabla u|^2](x),$ $\qquad (9)$

for $1 \le i, j \le n$, and δ_{ij} is the Kronecker δ. In many of our applications it is of fundamental importance that u, derivatives of u, both satisfy the same divergence form PDE in (6), (7). For example, in several of our papers we integrate by parts functions of $u, \nabla u$ and the bad terms drop out because both functions satisfy the same PDE. Thus we study (8), (9). We note that if $\xi \in \mathbf{R}^n$, then

$$\min\{p-1,1\}|\xi|^2 |\nabla u(x)|^{p-2} \le \sum_{i,k=1}^{n} b_{ik}\,\xi_i\xi_k$$

$$\qquad (10)$$

$$\le \max\{1, p-1\}|\nabla u(x)|^{p-2}|\xi|^2.$$

Observe from (10) that L can be degenerate elliptic if $\nabla u = 0$. Thus in many of our papers we also prove the fundamental inequality:

$$c^{-1}u(x)/d(x, \partial\Omega) \le |\nabla u(x)| \le cu(x)/d(x, \partial\Omega), \qquad (11)$$

for some constant c and x near $\partial\Omega$. Note that (10), (11), and Harnack's inequality for u imply that $(b_{ik}(x))$ are locally uniformly elliptic in Ω.

Behaviour near the boundary, such as boundary Harnack inequalities, are more involved. The easiest case for our methods to work is when $\partial\Omega$ is sufficiently flat in the sense of Reifenberg (to be defined later). In this case we will be able to show that $|\nabla u|^{p-2}$ is an A_2 weight (also to be defined). Thus we list some theorems on degenerate elliptic equations whose degeneracy is given in terms of an A_2 weight.

2.3 Degenerate Elliptic Equations

Let $w \in \mathbf{R}^n$, $0 < r$ and let λ be a real valued Lebesgue measurable function defined almost everywhere on $B(w, 2r)$. λ is said to belong to the class $A_2(B(w, r))$ if there exists a constant γ such that

$$\tilde{r}^{-2n} \int_{B(\tilde{w}, \tilde{r})} \lambda \, dx \cdot \int_{B(\tilde{w}, \tilde{r})} \lambda^{-1} dx \leq \gamma$$

whenever $\tilde{w} \in B(w, r)$ and $0 < \tilde{r} \leq r$. If $\lambda(x)$ belongs to the class $A_2(B(w, r))$ then λ is referred to as an $A_2(B(w, r))$-weight. The smallest γ such that the above display holds is referred to as the constant of the weight.

Once again let $\Omega \subset \mathbf{R}^n$ be a NTA domain with NTA-constants M, r_0. Let $w \in \partial\Omega$, $0 < r < r_0$, and consider the operator

$$\hat{L} = \sum_{i,j=1}^{n} \frac{\partial}{\partial x_i} \left(\hat{b}_{ij}(x) \frac{\partial}{\partial x_j} \right) \tag{12}$$

in $\Omega \cap B(w, 2r)$. We assume that the coefficients $\{\hat{b}_{ij}(x)\}$ are bounded, Lebesgue measurable functions defined almost everywhere on $B(w, 2r)$. Moreover, $\hat{b}_{ij} = \hat{b}_{ji}$ for all $i, j \in \{1, .., n\}$, and

$$c^{-1}\lambda(x)|\xi|^2 \leq \sum_{i,j=1}^{n} \hat{b}_{ij}(x)\xi_i\xi_j \leq c|\xi|^2\lambda(x) \tag{13}$$

for almost every $x \in B(w, r)$, where $\lambda \in A_2(B(w, r))$. If $O \subset B(w, 2r)$ is open let $\tilde{W}^{1,2}(O)$ be the weighted Sobolev space of equivalence classes of functions v with distributional gradient ∇v and norm

$$\|v\|_{1,2}^2 = \int_O v^2 \lambda dx + \int_O |\nabla v|^2 \lambda dx < \infty.$$

Let $\tilde{W}_0^{1,2}(O)$ be the closure of $C_0^\infty(O)$ in the norm $\tilde{W}^{1,2}(O)$. We say that v is a weak solution to $\hat{L}v = 0$ in O provided $v \in \tilde{W}^{1,2}(O)$ and

$$\int_O \sum_{i,j} \hat{b}_{ij} v_{x_i} \phi_{x_j} dx = 0 \tag{14}$$

whenever $\phi \in C_0^\infty(O)$.

The following three lemmas, Lemmas 1.6–1.8, are tailored to our situation and based on the results in [21–23].

Lemma 1.6. *Let $\Omega \subset \mathbf{R}^n$ be a NTA-domain with constant M, $w \in \partial\Omega$, $0 < r < r_0$, and let λ be an $A_2(B(w,r))$-weight with constant γ. Suppose that v is a positive weak solution to $\hat{L}v = 0$ in $\Omega \cap B(w, 2r)$. Then there exists a constant c, $1 \leq c < \infty$, depending only on n, M and γ, such that if $\tilde{w} \in \Omega$ and $B(\tilde{w}, 2\tilde{r}) \subset \Omega \cap B(w,r)$, then*

(i)

$$c^{-1}\tilde{r}^2 \int_{B(\tilde{w},\tilde{r})} |\nabla v|^2 \lambda dx \leq c \int_{B(\tilde{w},2\tilde{r})} |v|^2 \lambda dx,$$

(ii)

$$\max_{B(\tilde{w},\tilde{r})} v \leq c \min_{B(\tilde{w},\tilde{r})} v.$$

Furthermore, there exists $\alpha = \alpha(n, M, \gamma) \in (0,1)$ such that if $x, y \in B(\tilde{w}, \tilde{r})$ then

(iii)

$$|v(x) - v(y)| \leq c\left(\frac{|x-y|}{\tilde{r}}\right)^\alpha \max_{B(\tilde{w},2\tilde{r})} v.$$

Lemma 1.7. *Let $\Omega \subset \mathbf{R}^n$ be a NTA-domain with constant M, $w \in \partial\Omega$, $0 < r < r_0$, and let λ be an $A_2(B(w,r))$-weight with constant γ. Suppose that v is a positive weak solution to $\hat{L}v = 0$ in $\Omega \cap B(w, 2r)$ and that $v = 0$ on $\partial\Omega \cap B(w, 2r)$ in the weighted Sobolev sense. Extend v to $B(w, 2r)$ by setting $v \equiv 0$ in $B(w, 2r) \setminus \bar{\Omega}$. Then $v \in \tilde{W}^{1,2}(B(w, 2r))$ and there exists $\tilde{c} = \tilde{c}(n, M, \gamma)$, $1 \leq \tilde{c} < \infty$, such that the following holds with $\tilde{r} = r/\tilde{c}$.*

(i)

$$r^2 \int_{\Omega \cap B(w,r/2)} |\nabla v|^2 \lambda dx \leq \tilde{c} \int_{\Omega \cap B(w,r)} |v|^2 \lambda dx,$$

(ii)

$$\max_{\Omega \cap B(w,\tilde{r})} v \leq \tilde{c} v(a_{\tilde{r}}(w)).$$

Furthermore, there exists $\alpha = \alpha(n, M, \gamma) \in (0,1)$ such that if $x, y \in \Omega \cap B(w, \tilde{r})$, then

(iii)

$$|v(x) - v(y)| \leq c\left(\frac{|x-y|}{r}\right)^\alpha \max_{\Omega \cap B(w,2\tilde{r})} v.$$

Lemma 1.8. *Let $\Omega \subset \mathbf{R}^n$ be a NTA-domain with constant M, $w \in \partial\Omega$, $0 < r < r_0$, and let λ be an $A_2(B(w,r))$-weight with constant γ. Suppose that v_1 and v_2 are two positive weak solutions to $\hat{L}v = 0$ in $\Omega \cap B(w, 2r)$ and $v_1 = 0 = v_2$ on $\partial\Omega \cap B(w, 2r)$ in the weighted Sobolev sense. Then there exist*

$c = c(n, M, \gamma)$, $1 \leq c < \infty$, and $\alpha = \alpha(n, M, \gamma) \in (0, 1)$ such that if $\tilde{r} = r/c$, and $y_1, y_2 \in \Omega \cap B(w, r/c)$, then

$$\left| \frac{v_1(y_1)}{v_2(y_1)} - \frac{v_1(y_2)}{v_2(y_2)} \right| \leq c \frac{v_1(y_1)}{v_2(y_1)} \left(\frac{|y_1 - y_2|}{r} \right)^{\alpha}.$$

Note: The last display implies v_1/v_2 is Hölder continuous, as well as bounded above and below by its value at any one point in $\Omega \cap B(w, r/c)$. We refer to the last display as a boundary Harnack inequality.

3 p Harmonic Measure

If $\gamma > 0$ is a positive function on $(0, r_0)$ with $\lim_{r \to 0} \gamma(r) = 0$ define H^γ Hausdorff measure on $\mathbf{R}^n$ as follows: For fixed $0 < \delta < r_0$ and $E \subseteq \mathbf{R}^2$, let $L(\delta) = \{B(z_i, r_i)\}$ be such that $E \subseteq \bigcup B(z_i, r_i)$ and $0 < r_i < \delta$, $i = 1, 2, \ldots$ Set

$$\phi_\delta^\gamma(E) = \inf_{L(\delta)} \sum \gamma(r_i).$$

Then

$$H^\gamma(E) = \lim_{\delta \to 0} \phi_\delta^\gamma(E).$$

In case $\gamma(r) = r^k$ we write H^k for H^γ.

Next let $\Omega \subset \mathbf{R}^n, n \geq 2$, be a bounded domain, p fixed, $1 < p < \infty$, and N an open neighborhood of $\partial\Omega$. Let v be p harmonic in $\Omega \cap N$ and suppose that v is positive on $\Omega \cap N$ with boundary value zero on $\partial\Omega$, in the $W^{1,p}$ Sobolev sense. Extend v to a function in $W^{1,p}(N)$ by setting $v \equiv 0$ on $N \setminus \Omega$.

Then there exists (see [33]) a unique positive Borel measure ν on $\mathbf{R}^n$ with support $\subset \partial\Omega$, for which

$$\int |\nabla v|^{p-2} \langle \nabla v, \nabla \phi \rangle \, dx = - \int \phi \, d\nu \tag{15}$$

whenever $\phi \in C_0^\infty(N)$. In fact if $\partial\Omega, |\nabla v|$, are smooth

$$d\nu = |\nabla v|^{p-1} dH^{n-1} \text{ on } \partial\Omega.$$

Existence of ν follows if one can show for $\phi \geq 0$ as above,

$$\int |\nabla v|^{p-2} \langle \nabla v, \nabla \phi \rangle \, dx \leq 0. \tag{16}$$

Indeed, assuming (15) one can define a positive operator on the space of continuous functions and using basic Caccioppoli inequalities—the Riesz representation theorem, get the existence of ν. If v has continuous boundary value zero one can get (16) as follows. Let $\theta = [(\eta + \max[v - \epsilon, 0])^\epsilon - \eta^\epsilon] \phi$. Then one can show that θ may be used as a test function in (1). Doing this we deduce

$$\int_{\{v \geq \epsilon\} \cap N} [(\eta + \max[v - \epsilon, 0])^\epsilon - \eta^\epsilon]$$

$$\times |\nabla v|^{p-2} \langle \nabla v, \nabla \phi \rangle dx \leq 0.$$

Using dominated convergence, letting η and then $\epsilon \to 0$, we get (16).

If $p = 2$ and v is the Green's function with pole at $x_0 \in \Omega$, then $\nu = \omega(\cdot, x_0)$ is harmonic measure with respect to $x_0 \in \Omega$. Green's functions can be defined for the p Laplacian when $1 < p < \infty$, but are not very useful due to the nonlinearity of the p Laplacian when $p \neq 2$. Instead we often study the measure, μ, associated with a p capacitary function, say u, in $\Omega \setminus \bar{B}(x_0, r)$, where $B(x_0, 4r) \subset \Omega$. That is, u is p harmonic in $\Omega \setminus \bar{B}(x_0, r)$ with continuous boundary values, $u \equiv 1$ on $\partial B(x_0, r)$ and $u \equiv 0$ on $\partial \Omega$.

Remark 1. μ is different from the so called p harmonic measure introduced by Martio, which in fact is not a measure (see [65]).

Define the Hausdorff dimension of μ by

$$\text{H-dim } \mu = \inf\{k : \text{ there exists } E \text{ Borel} \subset \partial \Omega$$
$$\text{with } H^k(E) = 0 \text{ and } \mu(E) = \mu(\partial \Omega)\}.$$

Remark 2. We discuss for a fixed $p, 1 < p < \infty$, what is known about H-dim μ when μ corresponds to a positive p harmonic function u in $\Omega \cap N$ with boundary value 0 in the $W^{1,p}$ Sobolev sense. It turns out that H-dim μ is independent of u as above. Thus we often refer to H-dim μ as the dimension of p harmonic measure in Ω. For $p = 2, n = 2$, and harmonic measure, Carleson [15] used ideas from ergodic theory and boundary Harnack inequalities for harmonic functions to deduce H-dim $\omega = 1$ when $\Omega \subset \mathbf{R}^2$ is a 'snowflake' type domain and H-dim $\omega \leq 1$ when $\Omega \subset \mathbf{R}^2$ is the complement of a self similar Cantor set. He was also the first to recognize the importance of

$$\int_{\partial \Omega_n} |\nabla g_n| \log |\nabla g_n| \, dH^1$$

(g_n is Green's function for Ω_n with pole at zero and (Ω_n) is an increasing sequence of domains whose union is Ω). Wolff [72] used Carleson's ideas and brilliant ideas of his own to study the dimension of harmonic measure, ω, with respect to a point in domains bounded by 'Wolff snowflakes' $\subset \mathbf{R}^3$. He constructed snowflakes for which H-dim $\omega > 2$ and snowflakes for which H-dim $\omega < 2$.

In [61] we constructed Wolff Snowflakes, for which the harmonic measures on both sides of the snowflake were of H-dim $< n - 1$ and also a snowflake for which the harmonic measures on both sides were of H-dim $> n - 1$. Soon after we finished this paper, Björn Bennewitz became my Ph.D. student. We began studying the dimension of the measure μ as in (15) for fixed $p, 1 < p < \infty, p \neq 2$. We tried to imitate the Carleson–Wolff construction in order to produce examples of snowflakes where we could estimate H-dim μ, when $\Omega \subset \mathbf{R}^2$ is a snowflake. To indicate the difficulties involved we note that Wolff showed Carleson's integral over $\partial\Omega_n$ can be estimated at the nth step in the construction of certain snowflakes $\subset \mathbf{R}^3$. His calculations make key use of a boundary Harnack inequality for positive harmonic functions vanishing on a portion of the boundary of a NTA domain. Thus we proved $u_1/u_2 \leq c$ on $B(z, r/2) \cap \Omega$ whenever $z \in \partial\Omega$ and $0 < r \leq r_0$. Here $u_1, u_2 > 0$ are p harmonic in $B(z, r) \cap \Omega$ and vanish continuously on $B(z, r) \cap \partial\Omega$. Using our boundary Harnack inequality, we were able to deduce that μ had a certain weak mixing property and consequently, arguing as in Carleson–Wolff, we obtained an ergodic measure $\approx \mu$ on $\partial\Omega$. Applying the ergodic theorem of Birkhoff and entropy theorem of Shannon–McMillan–Breiman it followed that

$$\lim_{r \to 0} \frac{\log \mu[B(x, r)]}{\log r} = \text{H-dim } \mu \text{ for } \mu \text{ almost every } x \in \partial\Omega. \qquad (17)$$

Wolff uses Hölder continuity of the ratio and other arguments in order to make effective use of (17) in his estimates of H-dim μ. We first tried to avoid many of these estimates by a finesse type argument. However, later this argument fell through because of a calculus mistake. Finally we decided that instead of Wolff's argument we should use the divergence theorem and try to find a partial differential equation for which u is a solution and $\log |\nabla u|$ is a subsolution (supersolution) when $p > 2$ $(1 < p < 2)$. We succeeded, in fact the PDE is given in (8), (9):

$$L\zeta(x) = \sum_{i,j=1}^{n} \frac{\partial}{\partial x_i} (b_{ij}\zeta_{x_j})(x)$$

$$b_{ij}(x) = |\nabla u|^{p-4}[(p-2)u_{x_i}u_{x_j} + \delta_{ij}|\nabla u|^2](x), 1 \leq i, j \leq n.$$

Moreover for domains $\subset \mathbf{R}^2$ whose boundary is a quasi circle. we were able to prove the fundamental inequality in (11):

$$c^{-1}u(x)/d(x, \partial\Omega) \leq |\nabla u(x)| \leq cu(x)/d(x, \partial\Omega),$$

for some constant c and x near $\partial\Omega$. Thus interior estimates for uniformly elliptic non divergence form PDE could be applied to solutions of L. Armed with this knowledge we eventually proved in [6]:

Theorem 2.1. *Fix $p, 1 < p < \infty$, and let $u > 0$ be p harmonic in $\Omega \cap N \subset \mathbf{R}^2$ with $u = 0$ continuously on $\partial\Omega$. If $\partial\Omega$ is a snowflake and $1 < p < 2$, then H-dim $\mu > 1$ while if $2 < p < \infty$, then H-dim $\mu < 1$.*

Theorem 2.2. *Let p, u, μ be as in Theorem 2.1. If $\partial\Omega \subset \mathbf{R}^2$ is a self similar Cantor set and $2 < p < \infty$, then H-dim $\mu < 1$.*

Theorem 2.3. *Let p, u, μ be as in Theorem 2.2. If $\partial\Omega \subset \mathbf{R}^2$ is a k quasicircle, then H-dim $\mu \leq 1$ for $2 < p < \infty$, while H-dim $\mu \geq 1$ for $1 < p < 2$.*

To outline the proof of Theorem 2.1 we note that the boundary Harnack inequality mentioned earlier implies that all measures associated with functions in Theorems 2.1–2.3, have the same Hausdorff dimension. Also, since the p Laplacian is translation, dilation, and rotation invariant we may assume that u is the p capacitary function for $\Omega \setminus \bar{B}(0, 1)$ where $d(0, \partial\Omega) = 4$.

That is u is p harmonic in $\Omega \setminus \bar{B}(0, 1)$ with continuous boundary values, $u \equiv 1$ on $\partial B(0, 1)$ and $u \equiv 0$ on $\partial\Omega$. Let μ be the measure associated with u as in (15). Let $\Omega_n \subset \Omega$ be a sequence of approximating domains constructed in the usual way and for large n let u_n be the p capacitary function for $\Omega_n \setminus \bar{B}(0, 1)$. Then one first proves:

Lemma 2.4. *For fixed $p, 1 < p < \infty$,*

$$\eta = \lim_{n \to \infty} n^{-1} \int_{\partial\Omega_n} |\nabla u_n|^{p-1} \log |\nabla u_n| \, dH^1 x$$

exists. If $\eta > 0$ then H-dim $\mu < 1$ while if $\eta < 0$, then H-dim $\mu > 1$.

To prove Lemma 2.4 we followed Carleson–Wolff but our argument was necessarily more complicated, due to the non-linearity of the p Laplacian. $\square$

To prove Theorem 2.1 let u_n, Ω_n be as in Lemma 2.4. We note that one can show $\nabla u_n \neq 0$ in $\Omega_n \setminus \bar{B}(0, 1)$ and if $v = \log |\nabla u_n|, p \neq 2, 1 < p < \infty$, that

$$\frac{Lv}{p - 2} \geq \min(1, p - 1) \sum_{i,j=1}^{2} |\nabla u|^{p-4} (u_{x_i x_j})^2. \tag{18}$$

where L is as in (8), (9).

Next we apply the divergence theorem for large n in $O_n = \Omega_n \setminus \bar{B}(0, 1)$ to the vector field whose ith component, $i = 1, 2$, is

$$u_n \sum_{k=1}^{2} b_{ik} v_{x_k} - v \sum_{k=1}^{2} b_{ik} (u_n)_{x_k}.$$

We get

$$\int_{O_n} u_n \, Lv \, dx = \int_{\partial\Omega_n} \sum_{i,k=1}^{2} b_{ik} \xi_i [u_n v_{x_k} - v (u_n)_{x_k}] \, dH^1 x + O(1) \tag{19}$$

where ξ denotes the outer unit normal to $\partial\Omega_n$. Using the fact that $\xi = -\nabla u_n / |\nabla u_n|$ and the definition of (b_{ik}), we find that

$$\int_{\partial\Omega_n} \sum_{i,k=1}^{2} b_{ik}\xi_i \left[u_n v_{x_k} - v(u_n)_{x_k} \right] dH^1 x$$

$$(20)$$

$$= (p-1) \int_{\partial\Omega_n} |\nabla u_n|^{p-1} \log |\nabla u_n| \, dH^1 x + O(1).$$

From (18)–(20), and Lemma 2.4 we conclude that in order to prove Theorem 2.1 it suffices to show

$$\liminf_{n\to\infty} \left(n^{-1} \int_{O_n} u_n |\nabla u_n|^{p-4} \sum_{i,j=1}^{2} (u_n)^2_{x_i x_j} \, dx \, dx \right) > 0. \qquad (21)$$

To prove (21) we showed the existence of $\lambda \in (0,1)$ such that if $x \in \Omega_n \setminus B(0,2)$ and $d(x, \partial\Omega_n) \geq 3^{-n}$, then

$$c \int_{O_n \cap B(x,\lambda d(x,\partial\Omega_n))} u_n |\nabla u_n|^{p-4} (u_n)^2_{y_i y_j} \, dy \geq \mu_n(B(x, 2d(x, \partial\Omega_n))) \qquad (22)$$

where c depends on p and the k quasi-conformality of Ω. Covering $\{3^{-m-1} \leq d(x, \partial\Omega_n) \leq 3^{-m}\}$ by balls and summing over $1 \leq m \leq n-1$ we obtain first (21) and then Theorem 2.1. □

To prove Theorem 2.3 let $w(x) = \max(v - c, 0)$ when $1 < p < 2$ and $w(x) = \max(-v - c, 0)$ when $p > 2$. Here c is chosen so large that $|v| \leq c$ on $B(0,2)$. Following Makarov [66] we prove:

Lemma 2.5. *Let m be a nonnegative integer. There exists $c_+ = c_+(k,p) \geq 1$ such that for $0 < t < 1$,*

$$\int_{\{x:u(x)=t\}} |\nabla u|^{p-1} w^{2m} \, dH^1 x \leq c_+^{m+1} m! \, [\log(2/t)]^m .$$

To outline the proof of Lemma 2.5 let $\Omega(t) = \Omega \setminus \{x : u(x) \leq t\}$, whenever $0 < t < 1$. Using the fact that $Lw \leq 0$ for $1 < p < \infty$ when $w > 0$, one computes,

$$L(w^{2m})(x) \leq 2m(2m-1)p|\nabla u|^{p-2}(x)\, w^{2m-2}(x)\, |\nabla w|^2(x). \qquad (23)$$

Next we use (23) and apply the divergence theorem in $\Omega(t)$ to the vector field whose ith component for $i = 1, 2$ is

$$(u - t) \sum_{j=1}^{2} b_{ij}(w^{2m})_{x_j} - w^{2m} \sum_{j=1}^{2} b_{ij} u_{x_j}.$$

We get

$$(p-1)\int_{\{x:u(x)=t\}} |\nabla u|^{p-1}\, w^{2m}\, dH^1 x$$

$$\leq 2m(2m-1)p\int_{\Omega(t)} u\,|\nabla u|^{p-2}\, w^{2m-2}\,|\nabla w|^2\, dx. \tag{24}$$

Using interior estimates for solutions to the p Laplace equation from Section 2, the coarea formula and once again and our fundamental inequality, $|\nabla u(\cdot)| \geq u(\cdot)/d(\cdot,\partial\Omega)$, we deduce from (24) that

$$I_m(t) = \int_{\{x:u(x)=t\}} |\nabla u|^{p-1}\, w^{2m}\, dH^1 x \leq 2m(2m-1)c\int_t^1 I_{m-1}(\tau)\,\tau^{-1}\, d\tau$$

where $c = c(p,k)$. Lemma 2.5 then follows from an inductive type argument, using $I_0 \equiv$ constant on $(0,1)$. $\qquad\square$

Dividing the display in Lemma 2.5 by $(2c_+)^m\, m!\, \log^m(2/t)$ and summing we see for $0 < t < 1$ that

$$\int_{\{x:u(x)=t\}} |\nabla u|^{p-1} \exp\left[\frac{w^2}{2c_+\log(2/t)}\right] dH^1 x \leq 2c_+. \tag{25}$$

Using (25) and weak type estimates it follows that if

$$\lambda(t) = \sqrt{4\,c_+\,\log(2/t)}\,\sqrt{\log(-\log t)} \text{ for } 0 < t < e^{-2},$$

$$F(t) = \{x : u(x) = t \text{ and } w(x) \geq \lambda(t)\} \tag{26}$$

then

$$\int_{F(t)} |\nabla u|^{p-1}\, dH^1 x \leq \frac{2c^+}{\log^2(1/t)} \tag{27}$$

One can use (27) to show that if Hausdorff measure (denoted H^γ) is defined with respect to

$$\gamma(r) = \begin{cases} r\, e^{a\lambda(r)} & \text{when } 1 < p < 2 \\ r\, e^{-a\lambda(r)} & \text{when } p > 2. \end{cases} \tag{28}$$

and a is sufficiently large then

$$\mu \text{ is absolutely continuous with respect to } H^\gamma \text{ when } 1 < p < 2 \tag{29}$$

$$\mu \text{ is concentrated on a set of } \sigma \text{ finite } H^\gamma \text{ measure when } p > 2. \tag{30}$$

Clearly (29), (30) imply Theorem 2.3. $\qquad\square$

3.1   p Harmonic Measure in Simply Connected Domains

Recently in [63] we have managed to prove the following theorem.

Theorem 2.6. *Fix $p, 1 < p < \infty$, and let $u > 0$ be p harmonic in $\Omega \cap N$, where Ω is simply connected, $\partial\Omega$ is compact, and N is a neighborhood of $\partial\Omega$. Suppose u has continuous boundary value 0 on $\partial\Omega$ and let μ be the measure associated with u as in (1). If λ, γ, are as in (26), (28), then (29), (30) are valid for $a = a(p)$ sufficiently large. Hence Theorem 2.3 remains valid in simply connected domains.*

We note that Makarov in [66] proved for harmonic measure (i.e., $p = 2$) the stronger theorem:

Theorem 2.7. *Let ω be harmonic measure with respect to a point in the simply connected domain Ω. Then*

(a) ω is concentrated on a set of σ finite H^1 measure

(b) ω is absolutely continuous with respect to $H^{\hat\gamma}$ measure defined relative to $\hat\gamma(r) = r \exp[A\sqrt{\log 1/r \, \log\log\log 1/r}]$ for A sufficiently large.

The best known value of A in the definition of $\hat\gamma$ appears to be $A = 6\sqrt{\frac{\sqrt{24}-3}{5}}$ given in [32].

3.2   Preliminary Reductions for Theorem 2.6

To outline the proof of Theorem 2.6 we first claim, as in the proof of Theorem 2.5, that all measures associated with functions satisfying the hypotheses of Theorem 2.6, will have the same Hausdorff dimension. Indeed let $u_1, u_2 > 0$ be p harmonic functions in $\Omega \cap N$ and let μ_1, μ_2, be the corresponding measures as in (15). Observe from the maximum principle for p harmonic functions and continuity of u_1, u_2, that there is a neighborhood $N_1 \subset N$ of $\partial\Omega$ with

$$M^{-1}u_1 \le u_2 \le Mu_1 \text{ in } N_1 \cap \Omega. \tag{31}$$

One can also show there exists $r_0 > 0, c = c(p) < \infty$, such that whenever $w \in \partial\Omega, 0 < r \le r_0$, and $i = 1, 2$,

$$c^{-1} r^{p-2} \mu_i[B(w, r/2)] \le \max_{B(w,r)} u_i^{p-1} \le c\, r^{p-2} \mu_i[B(w, 2r)]. \tag{32}$$

Using (31), (32), and a covering argument it follows that μ_1, μ_2 are mutually absolutely continuous. Mutual absolute continuity is easily seen to imply H-dim $\mu_1 =$ H-dim μ_2. Thus we may assume, as in the proof of Theorem 2.3,

that u is the p capacitary function for $D = \Omega \setminus \bar{B}(0,1)$ and $d(0, \partial\Omega) = 4$.
The major obstacle to proving Theorem 2.6 in our earlier paper was that
we could not prove the fundamental inequality in (11). That is, in our new
paper, we prove

Theorem 2.8. *If u is the p capacitary function for D, then there exists*
$c = c(p) \geq 1$, *such that*

$$c|\nabla u|(z) \geq \frac{u(z)}{d(z, \partial\Omega)} \ whenever \ z \in D.$$

Given Theorem 2.8 we can copy the argument leading to (27) in the proof
of Theorem 2.3. However one has to work harder in order to deduce (29),
(30) from (27) as previously we used the doubling property of μ in (32) and
this property is not available in the simply connected case. Still we omit
the additional measure theoretic argument and shall regard the proof of
Theorem 2.6 as complete once we sketch the proof of Theorem 2.8.

3.3 Proof of Theorem 2.8

To prove Theorem 2.8 we assume, as we may, that $\partial\Omega$ is a Jordan curve, since
otherwise we can approximate Ω in the Hausdorff distance sense by Jordan
domains and use the fact that the constant in Theorem 2.8 depends only on p
to eventually get this theorem for Ω. We continue under this assumption and
shall use complex notation. Let $z = x + iy$, where $i = \sqrt{-1}$ and for $a, b \in D$,
let $\rho(a, b)$ denote the hyperbolic distance from $a, b \in D$ to $\partial\Omega$.

Fact A. *u is real analytic in D, $\nabla u \neq 0$ in D, and $u_z = (1/2)(u_x - iu_y)$, is*
$k = k(p)$ quasi-regular in D. Consequently, $\log|\nabla u|$ is a weak solution to a
divergence form PDE for which a Harnack inequality holds. That is, if $h \geq 0$
is a weak solution to this PDE in $B(\zeta, r) \subset D$, then $\displaystyle\max_{B(\zeta, r/2)} h \leq \tilde{c} \min_{B(\zeta, r/2)} h$,
where $\tilde{c} = \tilde{c}(p)$.

From Fact A and Lemma 1.3 one can show that

$$|\nabla u(z)| \leq cu(z)/d(z, \partial\Omega) \text{ in } D \tag{33}$$

and that Theorem 2.8 is valid in $B(0,2) \setminus \bar{B}(0,1)$. Next we use Fact A and
(33) to show that Theorem 2.8 for $x \in D \setminus B(0,2)$ is a consequence of the
following lemma.

Lemma 2.9. *There is a constant $c = c(p) \geq 1$ such that for any point*
$z_1 \in D \setminus B(0,2)$, there exists $z^\star \in D \setminus B(0,2)$ with $u(z^\star) = u(z_1)/2$ and
$\rho(z_1, z^\star) \leq c$.

Assuming Lemma 2.9 one gets Theorem 2.8 from the following argument. Let Γ be the hyperbolic geodesic connecting z_1 to z^* and suppose that $\Gamma \subset D$. From properties of ρ one sees for some $c = c(p)$ that

$$H^1(\Gamma) \leq cd(z_1, \partial\Omega) \text{ and } d(\Gamma, \partial\Omega) \geq c^{-1}d(z_1, \partial\Omega). \tag{34}$$

Thus

$$\frac{1}{2}u(z_1) \leq u(z_1) - u(z^*) \leq \int_\Gamma |\nabla u(z)||dz|$$
$$\leq cH^1(\Gamma) \max_\Gamma |\nabla u| \leq cd(z_1, \partial\Omega) \max_\Gamma |\nabla u|.$$

So for some $\zeta \in \Gamma$ and $c^* = c^*(p) \geq 1$,

$$c^*|\nabla u(\zeta)| \geq \frac{u(z_1)}{d(z_1, \partial\Omega)}. \tag{35}$$

Also from (35), we deduce the existence of Whitney balls $\{B(w_j, r_j\}$, with $w_j \in \Gamma$, $r_j \approx d(z_1, \partial\Omega)$, connecting ζ to z_1 and

$$|\nabla u(z)| \leq cu(z_1)/d(z_1, \partial\Omega) \text{ when } z \in \bigcup_j B(w_j, r_j). \tag{36}$$

From (35), (36), we see that if $c = c(p)$ is large enough and

$$h(z) = \log\left(\frac{cu(z_1)}{d(z_1, \partial\Omega)|\nabla u(z)|}\right) \text{ for } z \in \bigcup_j B(w_i, r_i)$$

then $h > 0$ in $\cup_i B(w_i, r_i)$ and $h(\zeta) \leq c$. From Fact A we see that Harnack's inequality can be applied to h in successive balls of the form $B(w_i, r_i/2)$. Doing this we obtain $h(z_1) \leq c'$ where $c' = c'(p)$. Clearly, this inequality implies Theorem 2.8.

We note that if $\partial\Omega$ is a quasicircle one can choose z^* to be a point on the line segment connecting z_1 to $w \in \partial\Omega$ where $|w - z_1| = d(z_1, \partial\Omega)$ The proof uses Hölder continuity of u near $\partial\Omega$ and the fact that for some $c = c(p, k)$, $cu(z_1) \geq \max_{B(z_1, 2d(z_1, \partial\Omega))} u$. (see Lemmas 1.4 and 1.5). This inequality need not hold in a Jordan domain and so we have to give a more complicated argument to get Lemma 2.9. To this end, we construct a Jordan arc σ : $(-1, 1) \to D$ with $\sigma(0) = z_1$, $\sigma(\pm 1) = \lim_{t \to \pm 1} \sigma(t) \in \partial\Omega$, and $\sigma(1) \neq \sigma(-1)$. Moreover, for some $c = c(p)$,

$$(\alpha) \ H^1(\sigma) \leq cd(z_1, \partial\Omega)$$
$$\tag{37}$$
$$(\beta) \ u \leq cu(z_1) \text{ on } \sigma.$$

Let Ω_1 be the component of $\Omega \setminus \sigma$ not containing $B(0,1)$. Then we also require that there is a point w_0 on $\partial\Omega \cap \partial\Omega_1$ with

$$|w_o - z_1| \leq cd(z_1, \partial\Omega) \text{ and } d(w_0, \sigma) \geq c^{-1}d(z_1, \partial\Omega). \tag{38}$$

Finally we shall show the existence of a Lipschitz curve $\tau : (0,1) \to \Omega_1$ with $\tau(0) = z_1$, $\tau(1) = w_0$, satisfying the cigar condition:

$$\min\{H^1(\tau[0,t]), H^1(\tau[t,1])\} \leq \hat{c}d(\tau(t), \partial\Omega), \tag{39}$$

for $0 < t < 1$ and some absolute constant $\hat{c}$.

To get Lemma 2.9 from (37)–(39) let $u_1 = u$ in Ω_1 and $u_1 \equiv 0$ outside of Ω_1. From PDE estimates, (37) (β), and (38) one finds $\theta > 0, c < \infty$ such that

$$\max_{B(w_0,t)} u_1 \leq cu(z_1) \left(\frac{t}{d(z_1, \partial\Omega)}\right)^\theta \text{ for } 0 < t < d(w_0, \sigma). \tag{40}$$

From (40), (39) we conclude the existence of z^* with $\rho(z_1, z^*) \leq c$ and $u(z^*) = 1/2$, which is Lemma 2.9. To construct σ, τ let f be the Riemann mapping function from the upper half plane, $\mathbb{H}$, onto Ω with $f(i) = 0$ and $f(a) = z_1$, where $a = is$ for some $s, 0 < s < 1$. Note that f has a continuous extension to $\bar{\mathbb{H}}$, since $\partial\Omega$ is a Jordan curve. Let $I(b) = [\text{ Re } b - \text{ Im } b, \text{ Re } b + \text{ Im } b]$ whenever $b \in \mathbb{H}$. We need the following lemmas.

Lemma 2.10. *There is a set $E(b) \subset I(b)$ such that for $x \in E(b)$*

$$\int_0^{\text{Im } b} |f'(x+iy)|dy \leq c^* d(f(b), \partial\Omega)$$

for some absolute constant c^*, and also

$$H^1(E(b)) \geq (1 - 10^{-100})H^1(I(b)).$$

Lemma 2.11. *Given $0 < \delta < 10^{-1000}$, there is an absolute constant $\hat{c}$ such that if $\delta_* = e^{-\hat{c}/\delta}$ then, whenever $x \in E(b)$ there is an interval $J = J(x)$ centered at x with*

$$2\delta_* \text{ Im } b \leq H^1(J) \leq c\delta^{1/2} \text{ Im } b \leq \frac{\text{Im } b}{10000}$$

(for some absolute constant c) and a subset $F = F(x) \subset J$ with $H^1(F) \geq (1 - 10^{-100})H^1(J)$ so that

$$\int_0^{\delta_* \text{ Im } b} |f'(t+iy)|dy \leq \delta d(f(b), \partial\Omega) \text{ for every } t \in F.$$

Lemma 2.12. Let $\hat{F} = \bigcup_{x \in E(b)} F(x)$. If $L \subset I(b)$ is an interval with $H^1(L) \geq \dfrac{Im\ b}{100}$, then

$$H^1(E(b) \cap \hat{F} \cap L) \geq \frac{Im\ b}{1000}.$$

Moreover, if $\{\tau_1, \tau_2, \ldots, \tau_m\}$ is a subset of $I(b)$, then there exists τ_{m+1} in $E(b) \cap \hat{F} \cap L$ with

$$|f(\tau_{m+1}) - f(\tau_j)| \geq \frac{d(f(b), \partial\Omega)}{10^{10}\ m^2} \quad whenever\ 1 \leq j \leq m.$$

To construct σ, τ from Lemma 2.12 we put $b = a = is$, and deduce for given $\delta, 0 < \delta < 10^{-1000}$, the existence of $x_1, x_2, x_3 \in E(a)$ with $-s < x_1 < -s/2, -\frac{1}{8}s < x_3 < \frac{1}{8}s$, and $\frac{1}{2}s < x_2 < s$. Moreover,

$$\int_0^{\delta_* s} |f'(x_j + iy|\, dy \leq \delta d(z_1, \partial\Omega) \text{ for } 1 \leq j \leq 3, \tag{41}$$

$$\min\{|f(x_1) - f(x_3)|, |f(x_2) - f(x_3)|\} \geq 10^{-11} d(z_1, \partial\Omega) \tag{42}$$

Let $\tilde{Q}(a)$ be the rectangle whose boundary in $\mathbb{H}, \xi$, consists of the horizontal line segment from $x_1 + is$ to $x_2 + is$, and the vertical line segments from x_j to $x_j + is$, for $j = 1, 2$. Put $\sigma = f(\xi)$ and note from (41) that (37) (α) is valid. To construct τ we put $t_0 = 0, s_0 = s, a_0 = t_0 + is_0$. Let $s_1 = \delta_* s_0, t_1 = x_3$, and $a_1 = t_1 + is_1$. By induction, suppose $a_m = s_m + it_m$ has been defined for $1 \leq m \leq k - 1$. We then choose $t_k \in E(a_{k-1})$ so that the last display in Lemma 2.11 holds with $t = t_k$. Set $s_k = \delta_* s_{k-1}$ and $a_k = s_k + it_k$.

Let λ_k be the curve consisting of the horizontal segment from a_{k-1} to $t_k + is_{k-1}$ and the vertical line segment from a_{k-1} to a_k. Put $\lambda = \bigcup \lambda_k$ and $\tau = f(\lambda)$. From our construction we deduce that τ satisfies the cigar condition in (39) for $\delta > 0$ small. Also $x_0 = \lim_{t \to 1} \lambda(t)$ exists, $|x_0| < 1/4$, and (38) holds for $w_0 = f(x_0)$, thanks to (42) and our construction.

3.4   The Final Proof

It remains to prove $u \leq cu(z_1)$ on σ which is (37) (β). The proof is by contradiction. Suppose $u > Au(z_1)$ on σ. We shall obtain a contradiction if $A = A(p)$ is suitably large. Our argument is based on the recurrence type scheme mentioned after Lemma 1.5 (often attributed to Carleson–Domar in the complex world and Caffarelli et al. in the PDE world). Given the rectangle $\tilde{Q}(a)$ we let $b_{j,1} = x_j + i\delta_*$ Im $a, j = 1, 2$, and note that $b_{j,1}, j = 1, 2$, are points on the vertical sides of $\tilde{Q}(a)$. These points will spawn two new boxes $\tilde{Q}(b_{j,1}), j = 1, 2$, which in turn will each spawn two more new boxes, and so on. Without loss of generality, we focus on $\tilde{Q}(b_{1,1})$.

This box is constructed in the same way as $\tilde{Q}(a)$ and we also construct, using Lemma 2.12 once again, a polygonal path $\lambda_{1,1}$ from $b_{1,1}$ to some point $x_{0,1} \in I(b_{1,1})$. λ_{11} is defined relative to $b_{1,1}$ in the same way that λ was defined relative to a. Also in view of Lemma 2.12 we can require that $\lambda_{1,1} \subset \{$ Re $z <$ Re $b_{1,1}\}$. $\lambda_{2,1}$ with endpoints, $b_{2,1}, x_{0,2}$ is constructed similarly to lie in $\{$ Re $z >$ Re $b_{2,1}\}$. Next let Λ be a Harnack constant such that

$$\max\{u(f(b_{1,1})), u(f(b_{2,1}))\} \leq \Lambda u(z_1). \tag{43}$$

From Harnack's inequality for u and Lemma 2.12 with δ fixed, it is clear that Λ in (43) can be chosen to depend only on p, so can also be used in further iterations.

Let $U = u \circ f$. By the maximum principle, since $A > \Lambda$ and $\lambda_{1,1}, \lambda_{2,1}$ lie outside of $\tilde{Q}(a)$, there will be a point $z \in \lambda_{1,1} \cup \lambda_{2,1}$ such that $U(z) > AU(a) = Au(z_1)$. Suppose $z \in \lambda_{1,1}$. The larger the constant A, the closer z will be to $\mathbf{R}$. More precisely, if $A > \Lambda^k$ then Im $z \leq \delta_*^k$ Im a, as follows from Harnack's inequality for u, and the construction of $\lambda_{1,1}$. In fact we can show that
$$|f(z) - f(x_{0,1})| \leq C\delta^{k-1} d(f(b_{1,1}), \partial\Omega).$$

The argument now is similar to the argument showing the existence of z^* given σ, τ. Let $\xi_{1,1}$ be the boundary of $\tilde{Q}(b_{1,1})$ which is in $\mathbb{H}$ and let $\sigma_{1,1} = f(\xi_{1,1})$. Set $w_{0,1} = f(x_{0,1})$. Then

$$B(w_{0,1}, d(w_{0,1}, \sigma_{1,1})) \subset f(\tilde{Q}(b_{1,1})).$$

and since $d(w_{0,1}, \sigma_{1,1}) \approx d(f(b_{1,1}), \partial\Omega)$ it follows from Hölder continuity of u that
$$U(z) \leq C\delta^{\theta k} \max_{\tilde{Q}(b_{1,1})} U.$$

Choose k, depending only on p, to be the least positive integer such that

$$C\delta^{\theta k} < \Lambda^{-1}.$$

This choice of k determines A (say $A = 2\Lambda^k$) which therefore also depends only on p. With this choice of A we have

$$\max_{\xi_{1,1}} U > \Lambda U(z) > \Lambda A U(a). \tag{44}$$

Since $U(b_{1,1}) \leq \Lambda U(a)$ we see from (44) that we can now repeat the above argument with $\tilde{Q}(b_{1,1})$ playing the role of $\tilde{Q}(a)$. That is, we find $b_{1,2}$ on the vertical sides of $\tilde{Q}(b_{1,1})$ with $\text{Im } b_{1,2} = \delta_*^2 \, \text{Im } a$ and a box $\tilde{Q}(b_{1,2})$ with boundary $\xi_{1,2}$ such that

$$\max_{\xi_{1,2}} U > \Lambda^2 \Lambda U(a) \geq \Lambda U(b_{1,2}).$$

Continuing by induction we get a contradiction because $U = 0$ continuously on $\partial\Omega$. The proof of (37) (β), Theorem 2.8, and Theorem 2.6 is now complete.

$\square$

3.5 p Harmonic Measure in Space

In [48] we proved

Theorem 2.13. *Let p, u, μ, be as in Theorem 2.3. There exists $k_0(p) > 0$ such that if $\partial\Omega$ is a k quasicircle, $0 < k < k_0(p)$, then*

(a) μ is concentrated on a set of σ finite H^1 measure when $p \, \mathrel{\raise.2ex\hbox{$\scriptstyle\langle$}} \, 2$.
(b) There exists $A = A(p), 0 < A(p) < \infty$, such that if $1 < p < 2$, then μ is absolutely continuous with respect to $H^{\hat\lambda}$ where $\hat\lambda(r) = r \exp[A\sqrt{\log 1/r \, \log\log\log 1/r}]$.

In [8] we prove an analogue of Theorem 2.13 when $p \geq n$. To be more specific we need a definition.

Definition B. *Let $\Omega \subset \mathbf{R}^n$ be a (δ, r_0) NTA domain and $0 < r \leq r_0$. Then Ω and $\partial\Omega$ are said to be (δ, r_0), Reifenberg flat provided that whenever $w \in \partial\Omega$, there exists a hyperplane, $P = P(w, r)$, containing w such that*

(a) $\Psi(\partial\Omega \cap B(w, r), P \cap B(w, r)) \leq \delta r$
(b) $\{x \in \Omega \cap B(w, r) : d(x, \partial\Omega) \geq 2\delta r\} \subset$ one component of $\mathbf{R}^n \setminus P$.

In Definition B, $\Psi(E, F)$ denotes the Hausdorff distance between the sets E and F defined by

$$\Psi(E, F) = \max(\sup\{d(y, E) : y \in F\}, \sup\{d(y, F) : y \in E\})$$

Theorem 2.14. *Let $\Omega \subset \mathbf{R}^n, n \geq 3$, be a (δ, r_0) Reifenberg flat domain, $w \in \partial\Omega$, and p fixed, $n \leq p < \infty$. Let $u > 0$ be p harmonic in Ω with $u = 0$ continuously on $\partial\Omega$. Let μ be the measure associated with u as in (15). There exists, $\hat\delta = \hat\delta(p, n) > 0$, such that if $0 < \delta \leq \hat\delta$, then μ is concentrated on a set of σ finite H^{n-1} measure.* To outline the proof of Theorem 2.14, we shall need the following result from [50]:

Theorem 2.15. *Let Ω be (δ, r_0) Reifenberg flat, $1 < p < \infty$, and $u > 0$, a p harmonic function in Ω with $u \equiv 0$ on $\partial\Omega$. Then there exists, $\delta_0 > 0, c_1 \geq 1$,*

*depending only on p, n, such that if $0 < \delta \leq \delta_0$ and $x \in \Omega$, then $u \in C^\infty(\Omega)$
and*

(a) $c_1^{-1}|\nabla u(x)| \leq u(x)/d(x, \partial\Omega) \leq c_1|\nabla u(x)|, x \in \Omega$,
(b) $|\nabla u|^{p-2}$ extends to an A_2 weight on $\mathbf{R}^n$ with constant $\leq c_1$.

An outline of the proof of Theorem 2.15 will be given in the next lecture. From Theorem 2.15 we see that $(b_{ik}(x))$ in (8), (9) are locally uniformly elliptic in Ω with ellipticity constants given in terms of an A_2 weight on $\mathbf{R}^n$. Thus Lemmas 1.6–1.8 can be used. To prove Theorem 2.14 we need a key lemma:

Lemma 2.16. *Let u, Ω, be as in Theorem 2.14 and $p \geq n$. Then $Lv \geq 0$ where L is as in (8), (9), and $v = \log|\nabla u|$.*

Using Lemma 2.16 and Theorem 2.15 we can essentially repeat the proof in [48] which in turn was based on the a proof in [66]. The main difficulty involves showing that if

$$\Theta = \{y \in \partial\Omega : v(x) \to -\infty \text{ as } x \to y \text{ nontangentially}\}$$

then $\mu(\Theta) = 0$. To accomplish this we use some results on elliptic PDE whose degeneracy is given in terms of an A_2 weight. □

3.6 Open Problems for p Harmonic Measure

Note. In problems (1)–(8) the surrounding space is $\mathbf{R}^2$.

1. Can Theorem 2.6 for simply connected domains be generalized to:
    (a) μ is concentrated on a set of σ finite H^1 measure whenever $p > 2$.
    (b) If $a = a(p) > 1$ is large enough and $1 < p < 2$, then μ is absolutely continuous with respect to $H^{\hat\gamma}$ measure where $\hat\gamma$ is defined in Theorem 2.7

2. Is H-dim μ concentrated on a set of σ finite H^1 measure when $p > 2$ and Ω is any planar domain. For harmonic measure this result is in [39, 71].
3. What is the exact value of H-dim μ for a given p when $\partial\Omega$ is the Van Koch snowflake and $p \neq 2$?
4. For a given p, what is the supremum ($p < 2$) or infimum ($p > 2$) of H-dim μ taken over the class of quasi-circles and/or simply connected domains?.
5. Is H-dim μ continuous and/or decreasing as a function of p when $\partial\Omega$ is the Van Koch snowflake?
    Regarding this question, the proof of Theorem 2.1 gives that H-dim $\mu = 1 + O(|p - 2|)$ as $p \to 2$ for a snowflake domain.
6. Are the p harmonic measures defined on each side of a snowflake mutually singular? The answer is yes when $p = 2$ as shown in [9].

7. Is it always true for $1 < p < \infty$ that H-dim μ < Hausdorff dimension of $\partial\Omega$ when $\partial\Omega$ is a snowflake or a self similar Cantor set? The answer is yes when $p = 2$ for the snowflake as shown in [40]. The answer is also yes for self similar Cantor sets when $p = 2$. This question and continuity questions for H-dim ω on certain four cornered Cantor sets are answered by Batakis in [3–5].

8. We noted in Remark 2 that H-dim μ was independent of the choice of u vanishing on $\partial\Omega$. However in more general scenarios we do not know whether H-dim μ is independent of u. For example, suppose $x_0 \in \partial\Omega$ and $u > 0$ is p harmonic in $\Omega \cap B(x_0, r)$ with $u = 0$ on $\partial\Omega \cap B(x_0, r)$ in the $W^{1,p}$ sense. If $\partial\Omega \cap B(x_0, r)$ has positive p capacity, then there exists a measure μ satisfying (15) with $\phi \in C_0^\infty(N)$ replaced by $\phi \in C_0^\infty(B(x_0, r))$. Is H-dim $\mu|_{B(x_0, r/2)}$ independent of u? If Ω is simply connected and $p = 2$, then I believe the answer to this question is yes. In general this problem appears to be linked with boundary Harnack inequalities.

9. Is it true for $p \geq n$ that H-dim $\mu \leq 1$ whenever $\Omega \subset \mathbf{R}^n$? If not is there a more general class of domains than Reifenberg flat domains (see Theorem 2.14) for which this inequality holds? Compare with Problem 2.

10. What can be said about the dimension of Wolff snowflakes? Regarding this question it appears that we can perturb off the $p = 2$ case (see [61]) in order to construct Wolff snowflakes for $0 < |p - 2| < \epsilon, \epsilon > 0$ small, for which the H-dim of the corresponding p harmonic measures on both sides of the snowflake are $< n - 1$ and also examples for which the H-dim of these measures are $> n - 1$.

11. What can be said for the dimension of p harmonic measure, $p > 3 - \log 4/\log 3$, or even just harmonic measure in $\Omega = \mathbf{R}^3 \setminus J$ where J is the Van Koch snowflake?

12. The existence of a measure μ, corresponding to a weak solution u with vanishing boundary values, as in (2), exists for a large class of divergence form partial differential equations. What can be said about analogues of Theorems 2.1–2.3 or Theorems 2.6, 2.8, 2.14, 2.15 for the measures corresponding to these solutions? What can be said about analogues of problems (1)–(11)?

4 Boundary Harnack Inequalities and the Martin Boundary Problem for p Harmonic Functions

Recall from Section 3 the definition of nontangentially accessible and Reifenberg flat domains. We note that a Lipschitz domain (i.e., a domain which is locally the graph of a Lipschitz function) is NTA. Also a Reifenberg flat domain need not have a rectifiable boundary or tangent planes in the geometric measure sense anywhere (e.g., the Van Koch snowflake, in two dimensions). In [50] we prove

Theorem 3.1. *Let $\Omega \subset \mathbf{R}^n$ be a (δ, r_0)-Reifenberg flat domain. Suppose that u, v are positive p-harmonic functions in $\Omega \cap B(w, 4r)$, that u, v are continuous in $\bar{\Omega} \cap B(w, 4r)$ and $u = 0 = v$ on $\partial\Omega \cap B(w, 4r)$. There exists $\tilde{\delta}, \sigma > 0$ and $c_1 \geq 1$, all depending only on p, n, such that*

$$\left| \log \frac{u(y_1)}{v(y_1)} - \log \frac{u(y_2)}{v(y_2)} \right| \leq c_1 \left(\frac{|y_1 - y_2|}{r} \right)^{\sigma}$$

whenever $y_1, y_2 \in \Omega \cap B(w, r/c_1)$. Here $w \in \partial\Omega, 0 < r < r_0$, and $0 < \delta < \tilde{\delta}$.

Observe that the last display in Theorem 3.1 is equivalent to:

$$\left| \frac{u(y_1)}{v(y_1)} - \frac{u(y_2)}{v(y_2)} \right| \leq c_1 \frac{u(y_1)}{v(y_1)} \left(\frac{|y_1 - y_2|}{r} \right)^{\sigma}$$

whenever $y_1, y_2 \in \Omega \cap B(w, r/c_1)$ so Theorem 3.1 is a boundary Harnack inequality for positive p harmonic functions vanishing on a portion of a sufficiently flat Reifenberg domain.

4.1   History of Theorem 3.1

The term boundary Harnack inequality for harmonic functions was first introduced by Kemper in [41]. He attempted to show the ratio of two positive harmonic functions vanishing on a portion of a Lipschitz domain was bounded. Unfortunately Kemper's proof was not correct, as Brelot later pointed out. This inequality for harmonic functions in Lipschitz domains was later proved independently and at about the same time in [2, 17, 73]. Jerison and Kenig in [37] proved Theorem 3.1 for NTA domains. Moreover boundary Harnack inequalities for solutions to linear divergence form uniformly elliptic PDE are proved in [14] while these inequalities for degenerate linear divergence form elliptic PDE whose degeneracy is specified in terms of an A_2 weight were proved by [21–23], as mentioned earlier. Theorem 3.1 will be proved in the following steps:

Step 1:   We prove Theorem 3.1 for

$$Q = \{x : |x_i| < 1, 1 \leq i \leq n - 1, 0 < x_n < 2\}.$$

Step 2:   (The 'fundamental inequality' for $|\nabla u|$) In this step, for u as in Theorem 3.1, we show there exist $\hat{c} = \hat{c}(p, n)$ and $\bar{\lambda} = \bar{\lambda}(p, n)$, such that if $0 < \delta \leq \delta_1$, then

$$\bar{\lambda}^{-1}\frac{u(y)}{d(y,\partial\Omega)} \leq |\nabla u(y)| \leq \bar{\lambda}\frac{u(y)}{d(y,\partial\Omega)} \qquad (45)$$

whenever $y \in \Omega \cap B(w, r/\hat{c})$.

Step 3: In this step we show that $|\nabla u|^{p-2}$ extends to an A_2-weight locally with constants depending only on p, n (provided δ is small enough).

Step 4: (Deformation of p harmonic functions). Let u, v be as in Theorem 3.1, $r^* = r/c'$, c' large and for $0 \leq \tau \leq 1$, let $\tilde{u}(\cdot, \tau)$ be the p harmonic function in $\Omega \cap B(w, 4r^*)$ with continuous boundary values,

$$\tilde{u}(y, \tau) = \tau v(y) + (1 - \tau)u(y)$$

whenever $y \in \partial(\Omega \cap B(w, 4r^*))$ and $\tau \in [0, 1]$. To simplify matters assume that

$$0 \leq u \leq v/2 \text{ and } v \leq c \text{ in } \bar{\Omega} \cap B(w, 4r^*), \qquad (46)$$

where c, as in the rest of this lecture, may depend only on p, n. Then from the maximum principle for p harmonic functions (Lemma 1.1) we have

$$0 \leq \frac{\tilde{u}(\cdot, \tau_2) - \tilde{u}(\cdot, \tau_1)}{\tau_2 - \tau_1} \leq c \qquad (47)$$

whenever $0 \leq \tau_1, \tau_2 \leq 1$. Proceeding operationally we note that if $\tilde{u}(\cdot, \tau)$ has partial derivatives with respect to τ on $B(w, 4r^*)$, then differentiating the p Laplace equation: $\nabla \cdot (|\nabla \tilde{u}(x)|^{p-2}\nabla \tilde{u}(x, \tau)) = 0$ with respect to τ one finds that $\tilde{u}_\tau(x, \tau)$ is a solution for $x \in \Omega \cap B(w, 4r^*)$ to the PDE

$$\tilde{L}\zeta = \sum_{i,j=1}^{n} \frac{\partial}{\partial x_i}\left(\tilde{b}_{ij}(x, \tau)\zeta_{x_j}(x, \tau)\right) = 0 \qquad (48)$$

where $(\tilde{b}_{ij})(\cdot, \tau)$ are defined as in (9) relative to $\tilde{u}(\cdot, \tau)$. Thus, $\tilde{u}(\cdot, \tau)$ and $\tilde{u}_\tau(\cdot, \tau)$, both satisfy the same PDE. Now

$$\log\frac{v(x)}{u(x)} = \log\frac{\tilde{u}(x, 1)}{\tilde{u}(x, 0)} = \int_0^1 \frac{\tilde{u}_\tau(x, \tau)}{\tilde{u}(x, \tau)}\,d\tau \qquad (49)$$

when $x \in \Omega \cap B(w, 4r^*)$. Observe also from (47) that $\tilde{u}_\tau \geq 0$ with continuous boundary value zero on $\partial\Omega \cap B(w, 4r^*)$. From this fact, (49), we see that to prove Theorem 3.1 it suffices to prove boundary Harnack inequalities for the PDE in (48) with constants independent of $\tau \in [0, 1]$. Moreover, from Steps 1–2, we see that $u(\cdot, \tau)$ is a solution to a uniformly elliptic PDE whose degeneracy is given in terms of an A_2 weight with A_2 constant independent of τ. Thus Lemma 1.8 can be applied to $\tilde{u}_\tau, \tilde{u}$ in order to conclude Theorem 3.1.

4.2 Proof of Step 1

Step 1 is stated formally as
Lemma 3.2. *Let*

$$Q = \{x : |x_i| < 1, 1 \le i \le n - 1, 0 < x_n < 2\}.$$

Given $p, 1 < p < \infty$, *let* $u, v > 0$ *be* p *harmonic in* Q, *continuous in* $\bar{Q}$ *with* $u \equiv v \equiv 0$ *on* $\partial Q \cap \{x : x_n = 0\}$. *Then for some* $c = c(p, n) \ge 1$,

$$\left| \log\left(\frac{u(z)}{v(z)} \right) - \log\left(\frac{u(y)}{v(y)} \right) \right| \le c|z - y|^{\sigma}$$

whenever $z, y \in Q \cap B(0, 1/16)$, *where* σ *is the exponent in Lemma 1.3.*
Proof. To begin the proof of Lemma 3.2, observe that x_n is p harmonic and vanishes when $x_n = 0$. Thus from the triangle inequality, it suffices to prove Lemma 3.2 when $v = x_n$. To prove that u/v is bounded in $Q \cap B(0, 1/2)$ we use barrier estimates. Given $x \in B(0, 1/2)$ with $x_n \le 1/100$ let $\hat{x} = (x', 1/8)$ and let f be p harmonic in $D = B(\hat{x}, 1/8) \setminus \bar{B}(\hat{x}, 1/100)$ with continuous boundary values, $f = u(e_n/4)$ on $\partial B(\hat{x}, 1/100)$ while $f \equiv 0$ on $\partial B(\hat{x}, 1/8)$. From Lemma 1.1 we see that for some a, b,

$$f(x) = \begin{cases} a|x - \hat{x}|^{(p-n)/(p-1)} + b, \ p \ne n, \\ a \ln |x - \hat{x}| + b, \ p = n, \end{cases} \tag{50}$$

Also $f \le cu$ in D thanks to Harnack's inequality in Lemma 1.2. Using these facts and (50) it follows from direct calculation that

$$cu(e_n/4) \le \frac{u(x)}{x_n} \tag{51}$$

for some $c = c(p, n)$ when $x_n \in Q \cap B(0, 1/2)$ and $x_n \le 1/100$. From Harnack's inequality we see that this inequality holds on $Q \cap B(0, 1/2)$. Next we extend u to

$$Q' = \{x : |x_i| < 1, 1 \le i \le n - 1, |x_n| < 2\}$$

by putting $u(x', x_n) = -u(x', -x_n)$ when $x_n < 0$ (Schwarz Reflection). It is easily shown that u is p harmonic in Q'. We can now use Lemmas 1.3 and 1.5 for u to deduce for $x \in Q \cap B(0, 1/8)$, that

$$|\nabla u(x)| \le c \max_{B(0, 1/4)} u. \le c^2 u(e_n/4). \tag{52}$$

From (52) and the mean value theorem we get

$$cu(e_n/4) \ge \frac{u(x)}{x_n} \tag{53}$$

when $x \in Q \cap B(0, 1/8)$. Combining (53), (51), we obtain

$$c^{-1}u(e_n/4) \le u(x)/x_n \le cu(e_n/4). \tag{54}$$

Hölder continuity of the above ratio follows from (54) and Lemma 1.3. We omit the details.   □

4.3   Proof of Step 2

In the proof of (45) we shall need the following comparison lemma.

Lemma 3.3. *Let O be an open set, $w \in \partial O, r > 0$, and suppose that $\hat{u}, \hat{v}$ are positive p harmonic functions in $O \cap B(w, 4r)$. Let $a \ge 1, x \in O$, and suppose that*

$$a^{-1}\frac{\hat{v}(x)}{d(x, \partial \Omega)} \le |\nabla \hat{v}(x)| \le a\frac{\hat{v}(x)}{d(x, \partial \Omega)}.$$

If $\tilde{\epsilon}^{-1} = (ca)^{(1+\sigma)/\sigma}$, where σ is as in Lemma 1.3, then for $c = c(p, n)$ suitably large, the following statement is true. If

$$(1 - \tilde{\epsilon})\hat{L} \le \frac{\hat{v}}{\hat{u}} \le (1 + \tilde{\epsilon})\hat{L}$$

in $B(x, \frac{1}{4}d(x, \partial O))$ for some $\hat{L}, 0 < \hat{L} < \infty$, then

$$\frac{1}{ca}\frac{\hat{u}(x)}{d(x, \partial O)} \le |\nabla \hat{u}(x)| \le ca\frac{\hat{u}(x)}{d(x, \partial O)}$$

Proof. Using Lemmas 1.2 and 1.3, we see that if $z_1, z_2, \in B(x, td(x, \partial O))$ and $0 < t \le 1/100$,

$$|\nabla \hat{u}(z_1) - \nabla \hat{u}(z_2)| \le ct^\sigma \max_{B(x, td(x, \partial O))} |\nabla \hat{u}(\cdot)|$$

$$\le c^2 t^\sigma \hat{u}(x)/d(x, \partial O). \tag{55}$$

Here c depends only on p, n. From (55) we conclude that we only have to prove bounds from below for the gradient of $\hat{u}$ at x. To do this, suppose for some small $\zeta > 0$ (to be chosen) that,

$$|\nabla \hat{u}(x)| \le \zeta \hat{u}(x)/d(x, \partial O). \tag{56}$$

From (55) with $z = z_1, x = z_2$ and (56) we deduce

$$|\nabla \hat{u}(z)| \leq [\zeta + c^2 t^\sigma] \, \hat{u}(x)/d(x, \partial O) \tag{57}$$

for $z \in B(x, td(x, \partial O))$. Integrating, it follows that if $y \in \partial B(x, td(x, \partial O))$, with $|x - y| = td(x, \partial O)$, $t = \zeta^{1/\sigma}$, then

$$|\hat{u}(y) - \hat{u}(x)| \leq c' \zeta^{1+1/\sigma} \, \hat{u}(x). \tag{58}$$

Constants in (57), (58) depend only on p, n. On the other hand (55) also holds with $\hat{u}$ replaced by $\hat{v}$. Let $\lambda = \frac{\nabla \hat{v}(x)}{|\nabla \hat{v}(x)|}$. Then from (55) for $\hat{v}$ and the non-degeneracy assumption on $|\nabla \hat{v}|$ in Lemma 3.3, we find

$$\langle \nabla \hat{v}(z), \lambda \rangle \geq (1 - c \, a\zeta)|\nabla \hat{v}(x)| \text{ in } \bar{B}(x, \zeta^{1/\sigma} d(x, \partial O)),$$

where $c = c(p, n)$. If $\zeta \leq (2ca)^{-1}$, where c is the constant in the above display, then we can integrate, to get for $y = x + \zeta^{1/\sigma} d(x, \partial O)\lambda$, that

$$c^*(\hat{v}(y) - \hat{v}(x)) \geq a^{-1} \zeta^{1/\sigma} \hat{v}(x) \tag{59}$$

with a constant c^* depending only on p, n. From (59), (58), we see that if $\tilde{\epsilon}$ is as in Lemma 3.3, then

$$(1 - \tilde{\epsilon})\hat{L} \leq \frac{\hat{u}(y)}{\hat{v}(y)} \leq \left(\frac{1 + c' \zeta^{1+1/\sigma}}{1 + \zeta^{1/\sigma}/(ac^*)} \right) \frac{\hat{u}(x)}{\hat{v}(x)}$$
$$\leq (1 + \tilde{\epsilon}) \left(\frac{1 + c' \zeta^{1+1/\sigma}}{1 + \zeta^{1/\sigma}/(ac^*)} \right) \hat{L} < (1 - \tilde{\epsilon})\hat{L} \tag{60}$$

provided $1/(a\tilde{c})^{1/\sigma} \geq \zeta^{1/\sigma} \geq a\tilde{c}\tilde{\epsilon}$ for some large $\tilde{c} = \tilde{c}(p, n)$. This inequality and (59) are satisfied if $\tilde{\epsilon}^{-1} = (\tilde{c}a)^{(1+\sigma)/\sigma}$ and $\zeta^{-1} = \tilde{c}a$. Moreover, if the hypotheses of Lemma 3.3 hold for this $\tilde{\epsilon}$, then in order to avoid the contradiction in (60) it must be true that (56) is false for this choice of ζ. Hence Lemma 3.3 is true.   □

As an application of Lemmas 3.2 and 3.3 we note that if $\hat{u}$ is p harmonic in $D = B(\zeta, \rho) \cap \{y : y_n > \zeta_n\}$ and $\hat{u}$ has continuous boundary value zero on $\partial D \cap \{y : y_n = \zeta_n\}$, then there exists $c = c(p, n) \geq 1$ such that

$$c^{-1} \frac{\hat{u}(x)}{d(x, \partial D)} \leq |\nabla \hat{u}(x)| \leq c \frac{\hat{u}(x)}{d(x, \partial D)} \tag{61}$$

in $D \cap B(\zeta, \rho/c)$. Indeed, let $\hat{v} = x_n - \zeta_n$ and put $\tilde{L} = \hat{u}(z)/\hat{v}(z)$ where z is a fixed point in $D \cap B(\zeta, \rho/c_+)$. Using Lemma 3.2 we see that

$$\left| \tilde{L} - \frac{\hat{u}(y)}{\hat{v}(y)} \right| \leq c' c_+^{-\sigma} \tilde{L}, \text{ when } y \in D \cap B(\zeta, \rho/c_+). \tag{62}$$

Choosing c_+ large enough we deduce that Lemma 3.2 applies, so (61) is true. Equation (61) could also be proved more or less directly using barrier arguments, Schwarz reflection, and Lemma 1.3.

Next we use Lemma 3.3 to get the nondegeneracy property in (45). We restate this property as,

Lemma 3.4. *Let Ω be (δ, r_0) Reifenberg flat. Let $u > 0$ be p harmonic in $\Omega \cap B(w, 2r)$, continuous in $B(w, 2r)$ with $u \equiv 0$ in $B(w, 2r) \setminus \Omega$. There exists $\delta^* = \delta^*(p, n)$, $c_1 = c_1(p, n)$, such that if $0 < \delta \leq \delta^*$ and $y \in \Omega \cap B(w, r/c_1)$, then*

$$c_1^{-1} \frac{u(y)}{d(y, \partial\Omega)} \leq |\nabla u(y)| \leq c_1 \frac{u(y)}{d(y, \partial\Omega)}.$$

Proof. Let c_* be the constant in (61) and choose $c' \geq 1000c_*$ so that if $x \in \Omega \cap B(w, r/c')$, $s = 4c_* d(x, \partial\Omega)$, and $z \in \partial\Omega$ with $!x - z| = d(x, \partial\Omega)$, then

$$\max_{B(z, 4s)} u \leq cu(x) \tag{63}$$

for some $c = c(p, n)$, which is possible thanks to Lemma 1.5. From Definition B with w, r replaced by $z, 4s$, there exists a plane $P = P(z, 4s)$ with

$$\Psi(\partial\Omega \cap B(z, 4s), P \cap B(z, 4s)) \leq 4\delta s.$$

Since the p Laplacian is invariant under rotations, and translations, we may assume that $z = 0$ and $P = \{y : y_n = 0\}$, Also, if

$$G = \{y \in B(0, 2s) : y_n > 8\delta s\}$$

then we may assume

$$G \subset \Omega \cap B(0, 2s).$$

Let v be the p harmonic function in G with boundary values in the Sobolev sense as follows: $v \leq u$ on ∂G and

$$v(y) = u(y) \text{ when } y \in \partial G \text{ and } y_n > 32\delta s,$$
$$v(y) = (\tfrac{y_n}{16\delta s} - 1) u(y) \text{ when } y \in \partial G \text{ and } 16\delta s < y_n \leq 32\delta s .$$
$$v(y) = 0 \text{ when } y \in \partial G \text{ and } y_n < 16\delta s .$$

From Lemma 1.1 (i.e., the maximum principle for p harmonic functions) we have $v \leq u$ in G. Also, since each point of ∂G where $u(x) \neq v(x)$ lies within $100\delta s$ of a point where u is zero, it follows from (63) and Lemma 1.4 that $u \leq v + c\delta^\beta u(x)$ on ∂G. Using Lemma 1.1 we conclude that

$$v \leq u \leq v + c\delta^\alpha u(x) \text{ in } G.$$

Thus,

$$1 \leq \frac{u(y)}{v(y)} \leq (1 - c\delta^\alpha)^{-1}, y \in B(x, d(x, \partial G)/4). \tag{64}$$

From (64) and (61) with $\hat{u}$ replaced by v, we conclude that the hypotheses of Lemma 3.3 are satisfied with $O = G$. Applying Lemma 3.3 and using $d(x, \partial G) \approx d(x, \partial\Omega)$, we obtain Lemma 3.4. $\qquad\square$

4.4   Proof of Step 3

In the proof of Step 3 we shall need the following lemma.

Lemma 3.5. *Let $\Omega \subset \mathbf{R}^n$ be (δ, r_0) Reifenberg flat. Let $w \in \partial\Omega$, $0 < r < r_0$, and suppose $u > 0$ is p harmonic in $\Omega \cap B(w, 2r)$, continuous in $B(w, 2r)$, with $u \equiv 0$ on $B(w, 2r) \setminus \Omega$. Given $\epsilon > 0$, there exist $\hat{\delta} = \hat{\delta}(p, n, \epsilon) > 0$ and $c = c(p, n, \epsilon)$, $1 \leq c < \infty$, such that*

$$c^{-1} \left(\frac{\hat{r}}{r} \right)^{1+\epsilon} \leq \frac{u(a_{\hat{r}}(w))}{u(a_r(w))} \leq c \left(\frac{\hat{r}}{r} \right)^{1-\epsilon}$$

whenever $0 < \delta \leq \hat{\delta}$ and $0 < \hat{r} < r/4$.

Proof. Lemma 3.5 can be proved by a barrier type argument, using barriers which vanish on the boundary of certain cones or by an iterative type argument using Reifenberg flatness of $\partial\Omega$. We omit the details. $\qquad\square$

Lemma 3.6. *Let Ω be (δ, r_0) Reifenberg flat. Let $w \in \partial\Omega, 0 < r < r_0$, and let $u > 0$ be p harmonic in $\Omega \cap B(w, 2r)$, continuous in $B(w, 2r)$, with $u \equiv 0$ on $B(w, 2r) \setminus \Omega$. There exists $\delta' = \delta'(p, n), c = c(p, n, M) \geq 1$ such that if $0 < \delta < \delta'$, and $\hat{r} = r/c$, then $|\nabla u|^{p-2}$ extends to an $A_2(B(w, \hat{r}))$ weight with constant depending only on p, n.*

Proof. We use Lemma 3.5 to prove Lemma 3.6. Let $\{Q_j(x_j, r_j)\}$ be a Whitney decomposition of $\mathbf{R}^n \setminus \bar{\Omega}$ into open cubes with center at x_j and sidelength r_j. Then

$$\cup_j \bar{Q}(x_j, r_j) = \mathbf{R}^n \setminus \bar{\Omega}$$

$$Q(x_j, r_j) \cap Q(x_i, r_i) = \emptyset \; i \neq j$$

$$10^{-4n} d(Q_j, \partial\Omega) \leq r_j \leq 10^{-2n} d(Q_j, \partial\Omega).$$

Let $\hat{r} = r/c^2$, where c is so large that

$$c^{-1} \frac{u(x)}{d(x, \partial\Omega)} \leq |\nabla u(x)| \leq c \frac{u(x)}{d(x, \partial\Omega)} \tag{65}$$

whenever $x \in \Omega \cap B(w, c\hat{r})$. Existence of c follows from Lemma 3.4. We also assume c is so large that if $Q_j \cap B(w, 4\hat{r}) \neq \emptyset$, then $\bar{Q}_j \subset B(w, c\hat{r})$ and there exists $w_j \in \Omega \cap B(w, c\hat{r})$ with

$$|x_j - w_j| \approx d(x_j, \partial\Omega) \approx d(w_j, \partial\Omega). \tag{66}$$

Existence follows from the fact that Ω is an NTA domain.

Let
$$\lambda(x) = |\nabla u(x)|^{p-2}, \ x \in \Omega \cap B(w, 4\hat{r})$$

$$\lambda(x) = |\nabla u(w_j)|^{p-2}, \ x \in Q_j \cap B(w, 4\hat{r}).$$

From (66) we see that

$$\lambda(x) = \lambda(w_j) \approx \lambda(z) \tag{67}$$

whenever $x \in Q_j$ and $z \in B(w_j, d(w_j, \partial\Omega)/2)$.

To complete the proof of Lemma 3.6 we prove that λ satisfies the A_2 condition given in Sect. 2.2. Let $\tilde{w} \in B(w, \hat{r})$ and $0 < \tilde{r} < \hat{r}$. We consider several cases. If $\tilde{r} < d(\tilde{w}, \partial\Omega)/2$, then the A_2 condition follows from (65) and Harnack's inequality. On the other hand, if $\tilde{r} \geq d(\tilde{w}, \partial\Omega)/2$ then we choose $z \in \partial\Omega$ with $d(\tilde{w}, \partial\Omega) = |\tilde{w} - z|$ and thus

$$B(\tilde{w}, \tilde{r}) \subset B(z, 3\tilde{r}) \subset B(\tilde{w}, 8\tilde{r}).$$

First suppose $p > 2$. From Hölder's inequality, Lemmas 1.2, 1.3, and (67) we see that

$$\int_{B(\tilde{w}, \tilde{r})} \lambda dx \leq \int_{B(z, 3\tilde{r})} \lambda dx \leq c \int_{\Omega \cap B(z, c^*\tilde{r})} |\nabla u|^{p-2} dx$$

$$\leq c \left(\int_{\Omega \cap B(z, c^*\tilde{r})} |\nabla u|^p dx \right)^{(1-2/p)} \tilde{r}^{2n/p} \tag{68}$$

$$\leq cu(a_{\tilde{r}}(z))^{p-2} \tilde{r}^{n+2-p}.$$

Let $\eta = \min\{1, |p-2|^{-1}\}/20$. To estimate the integral in involving λ^{-1} observe from Lemma 3.5 and once again Harnack's inequality, that if $y \in \Omega \cap B(z, c^*\tilde{r})$, and δ' is small, then

$$cu(y) \geq u(a_{\tilde{r}}(z)) \left(\frac{d(y, \partial\Omega)}{\tilde{r}} \right)^{1+\eta}. \tag{69}$$

Therefore, using (67) and (69) we obtain

$$\int_{B(\tilde{w},\tilde{r})} \lambda^{-1} dx \le c \tilde{r}^{(1+\eta)(p-2)} u(a_{\tilde{r}}(z))^{2-p}$$

$$\times \int_{\Omega \cap B(z,c^*\tilde{r})} d(y,\partial\Omega)^{-\eta(p-2)} dy. \tag{70}$$

To estimate the integral involving the distance function in (70) set

$$I(z,s) = \int_{\Omega \cap B(z,s)} d(y,\partial\Omega)^{-\eta(p-2)} dy$$

whenever $z \in \partial\Omega \cap B(w,r)$, $0 < s < r$. Let

$$E_k = \Omega \cap B(z,s) \cap \{y : d(y,\partial\Omega) \le \delta^k s\}$$

for $k = 1, 2, \ldots$ We claim that

$$\int_{E_k} dy \le c_+^k \delta^k s^n \text{ for } k = 1, 2, \ldots \tag{71}$$

where $c_+ = c_+(p,n)$. Indeed, from δ Reifenberg flatness it is easily seen that this statement holds for E_1. Moreover, E_1 can be covered by at most c/δ^{n-1} balls of radius $100\delta s$ with centers in $\partial\Omega \cap B(z,s)$. We can then repeat the argument in each ball to get that (71) holds for E_2. Continuing in this way we get (71) for all positive integers k. Using (71) and writing $I(z,s)$ as a sum over $E_k \setminus E_{k+1}, k = 1, 2, \ldots$ we get

$$I(z,s) \le cs^{n-\eta(p-2)} + \delta^{\eta(p-2)-n} \sum_{k=1}^{\infty} (c_+^k \delta^k s)^{n-\eta(p-2)} < \tilde{c} s^{n-\eta(p-2)},$$

where $\tilde{c} = \tilde{c}(p,n)$, provided δ' is small enough. Using this estimate with $s = \tilde{r}$, we can continue our calculation in (70) and conclude that

$$\int_{B(\tilde{w},\tilde{r})} \lambda^{-1} dx \le c \tilde{r}^{n+p-2} u(a_{\tilde{r}}(z))^{(2-p)} \tag{72}$$

To complete the proof of Lemma 3.6 in the case $p > 2$, we simply combine (68) and (72). Note that the case $p = 2$ is trivial and in case $p < 2$ the argument above can be repeated with $p - 2$ replaced by $2 - p < p$. Hence Lemma 3.6 and Step 3 are complete.   □

4.5 Proof of Step 4 and Theorem 3.1

To justify the claims in Step 4, first choose $c' = c'(p, n)$ so large that if $r^* = r/c'$, then

$$\max_{\Omega \cap B(w, 4r^*)} h \leq c h(a_{r^*}(w)) \tag{73}$$

for some $c = c(p, n)$ whenever $h > 0$ is p harmonic in $\Omega \cap B(w, 2r)$. This choice is possible as we see from Lemma 1.5 and Harnack's inequality in Lemma 1.2. We also suppose $c' > 1000c_1$, where c_1 is the constant in Lemma 3.4. Hence from Lemma 3.4 we have

$$c^{-1} \frac{k(y)}{d(y, \partial\Omega)} \leq |\nabla k(y)| \leq c \frac{k(y)}{d(y, \partial\Omega)}. \tag{74}$$

whenever $y \in \Omega \cap B(w, r^*/c')$ and $k > 0$ is p harmonic in $\Omega \cap B(w, 2r^*)$, continuous in $B(w, 2r^*)$, with $k \equiv 0$ on $B(w, 2r^*) \setminus \Omega$. We temporarily assume that

$$0 < u \leq v/2 \leq c \text{ on } \Omega \cap B(w, 4r^*). \tag{75}$$

and also that

$$c^{-1} \leq u(a_{r^*}(w)). \tag{76}$$

Next if $t \in [0, 1]$, let $\tilde{u}(\cdot, t)$ be the p harmonic function in $\Omega \cap B(w, 2r^*)$ with continuous boundary values,

$$\tilde{u}(\cdot, t) = (1 - t)u + tv \tag{77}$$

on $\partial[\Omega \cap B(w, 2r^*)]$. Extend $\tilde{u}(\cdot, t), t \in (0, 1)$, to be Hölder continuous in $B(w, 2r^*)$ by putting $\tilde{u}(\cdot, t) \equiv 0$ on $B(w, 2r^*) \setminus \Omega$. Let $r' = r^*/c'$ and observe that (74) holds whenever $k = u(\cdot, t), t \in [0, 1]$, on $\Omega \cap B(w, r')$. Thus from Lemma 1.3, we see that $\tilde{u}(\cdot, t)$ is infinitely differentiable in $B(w, r')$ and so

$$\nabla \left(\cdot |\nabla \tilde{u}(\cdot, t)|^{p-2} \nabla \tilde{u}(\cdot, t) \right) = 0 \tag{78}$$

in $\Omega \cap B(w, r')$. Set

$$U(x) = U(x, t, \tau) = \frac{\tilde{u}(x, t) - \tilde{u}(x, \tau)}{t - \tau}.$$

and note from (75), (76), for fixed $t, \tau \in [0, 1], t \neq \tau$, that

$$0 \leq v/2 \leq U(x) = v - u \leq v \leq c(p, n) \tag{79}$$

on $\partial(\Omega \cap B(w, 2r^*))$, so by Lemma 1.1 we have

$$0 \leq U \leq c \text{ in } \Omega \cap B(w, r'), U \equiv 0 \text{ on } \partial\Omega \cap B(w, r'). \tag{80}$$

From (80) we see for fixed $x \in \Omega \cap B(w, 2r^*)$ that $t \to \tilde{u}(x, t)$, is Lipschitz with norm $\leq c$. Thus $\tilde{u}_\tau(x, \cdot)$ exists almost everywhere in [0,1]. Let (x_ν) be a dense sequence of $\Omega \cap B(w, 2r^*)$ and let W be the set of all $t \in [0, 1]$ for which $\tilde{u}_t(x_m, \cdot)$ exists, in the sense of difference quotients, whenever $x_m \in (x_\nu)$. We note that $H^1([0, 1] \setminus W) = 0$ where H^1 is one-dimensional Hausdorff measure. Next we note from (79) that for $t \in (0, 1]$

$$\tilde{u}(\cdot, t)/2 \leq U(\cdot, t, \tau) \leq t^{-1} \tilde{u}(\cdot, t) \tag{81}$$

on $\partial \Omega \cap B(w, 2r^*)$, so by Lemma 1.1, this inequality also holds in $\Omega \cap B(w, 2r^*)$. To find a divergence form PDE that U satisfies let $\xi = (\xi_1, \ldots, \xi_n)$, $w = (w_1, \ldots, w_n) \in \mathbf{R}^n \setminus \{0\}$, and $1 \leq i \leq n$. Then

$$|\xi|^{p-2} \xi_i - |w|^{p-2} w_i$$

$$= \int_0^1 \frac{d}{d\lambda} \{ |\lambda \xi + (1 - \lambda) w|^{p-2} [\lambda \xi_i + (1 - \lambda) w_i] \} d\lambda$$

$$= \sum_{j=1}^n (\xi - w)_j \left(\int_0^1 a_{ij} [\lambda \xi + (1 - \lambda) w] d\lambda \right),$$

where for $1 \leq i, j \leq n$, $\eta \in \mathbf{R}^n \setminus \{0\}$,

$$a_{ij}(\eta) = |\eta|^{p-4} [(p - 2) \eta_i \eta_j + \delta_{ij} |\eta|^2]. \tag{82}$$

In this display δ_{ij}, once again, denotes the Kronecker delta. Using (78), (80)–(82) we find for fixed t, τ that if

$$A_{ij}(x) = A_{ij}(x, t, \tau)$$

$$= \int_0^1 a_{ij} [\lambda \nabla \tilde{u}(x, t) + (1 - \lambda) \nabla \tilde{u}(x, \tau)] \, d\lambda,$$

then, for $x \in \Omega \cap B(w, r')$, $t, \tau \in [0, 1]$,

$$\tilde{L} U(x) = \sum_{i,j=1}^n \frac{\partial}{\partial x_i} [A_{ij}(x) U_{x_j}] = 0. \tag{83}$$

Moreover, if $x \in \Omega \cap B(w, r')$, then

$$c^{-1} |\xi|^2 ||\nabla \tilde{u}(x, t)| + |\nabla \tilde{u}(x, \tau)||^{p-2}$$

$$\leq \sum_{i,j=1}^n A_{ij}(x) \xi_i \xi_j \text{ whenever } \xi \in \mathbf{R}^n \setminus \{0\}. \tag{84}$$

Also,

$$\sum_{i,j=1}^{n} |A_{ij}(x)| \le c||\nabla \tilde{u}(x,t)| + |\nabla \tilde{u}(x,\tau)||^{p-2}, \tag{85}$$

where c depends only on p, n. From Lemma 1.3, (82)–(85), and (74) for $\tilde{u}(\cdot,t), \tilde{u}(\cdot,\tau)$, we see that U is a solution on $\Omega \cap B(w,r')$ to a locally uniformly elliptic divergence form PDE with C^∞ coefficients. Since $\tilde{u}(x,t) - \tilde{u}(x,\tau) = (t-\tau)(v(x)-u(x))$ on $\partial(\Omega \cap B(w,2r))$ it follows from the maximum principle for p harmonic functions that $\tilde{u}(x,\tau) \to \tilde{u}(x,t)$ uniformly in the closure of $\Omega \cap B(x,2r^*)$ as $\tau \to t$. Also from Lemma 1.2 we deduce that $\nabla \tilde{u}(\cdot,\tau) \to \nabla \tilde{u}(\cdot,t)$ on compact subsets of $\Omega \cap B(w,2r^*)$. Using these facts and (74) we see for $1 \le i,j \le n$,

$$A_{i,j}(x,t,\tau) \to \tilde{b}_{ij}(x) \tag{86}$$

as $\tau \to t$ uniformly on compact subsets of $\Omega \cap B(w,r')$, where $\tilde{b}_{ij}$ are defined as in (9) relative to $\tilde{u}(\cdot,t)$. Finally we note from linear elliptic PDE theory that $U(\cdot,t,\tau)$ is locally in $W^{1,2}$ and locally Hölder continuous on $\Omega \cap B(w,r')$ with norms independent of τ. From (81) we see that $U(\cdot,t,\tau), t \in (0,1]$, has a Hölder continuous extension to $B(w,r')$ obtained by putting $U \equiv 0$ in $B(w,r') \setminus \Omega$. Also Hölder constants can be chosen independent of τ for τ near t. Using these facts we see for fixed $t \in (0,1]$, that there is a sequence $U(\cdot,t,\tau_k) \to f(\cdot,t)$ on $\Omega \cap B(w,r')$ as $\tau_k \to t$. Put $f \equiv 0$ on $B(w,r') \setminus \Omega$. Then from (80)–(86), Lemma 1.2, and Schauder estimates, we conclude that f has the following properties:

(a) $\tilde{L}f = 0$ in $\Omega \cap B(w,r')$ where $\tilde{L}$ is as in (48),
(b) f is continuous in $B(w,r')$ with $f \equiv 0$ on $B(w,r') \setminus \Omega$,
(c) $f(x_m,t) = \tilde{u}_t(x_m,t)$ when $x_m \in (x_\nu)$, $t \in W$,   (87)
(d) $u/2 \le \tilde{u}(\cdot,t)/2 \le f(\cdot,t) \le c$ in $\Omega \cap B(w,r')$,
(e) $f \in C^\infty[\Omega \cap B(w,r')]$.

From (87) (c) we get

$$\ln\left(\frac{v(x_m)}{u(x_m)}\right) = \ln\left(\frac{\tilde{u}(x_m,1)}{\tilde{u}(x_m,0)}\right) = \int_0^1 \frac{f(x_m,t)}{\tilde{u}(x_m,t)} \, dt \tag{88}$$

whenever $x_m \in (x_\nu)$. Since this sequence is dense in $\Omega \cap B(w,r')$, we conclude from (87), (88) that Claims (48), (89) are true. From (88) and Lemmas 3.4, 3.6 we see that $\tilde{u}_\tau(\cdot,\tau), \tilde{u}(\cdot,\tau)$ are solutions to a degenerate divergence form elliptic PDE whose degeneracy is given in terms of an A_2 weight. Thus Lemma 1.8 can be used with r replaced by $\tilde{r} = \min(r',\hat{r})$, $v_1 = f(\cdot,t), v_2 = \tilde{u}(\cdot,t)$, where f is as in (87), (88). Let $r'' = \tilde{r}/c$ where c is the constant in Lemma 1.8. From (88)(d), (76), and Harnack's inequality we get

$$c^{-1} \leq \frac{f(a_{r''}(w),t)}{\tilde{u}(a_{r''}(w),t)} \leq c \tag{89}$$

where c depends only on p, n. Using this fact and Lemma 1.8, we find that

$$\frac{f(y,t)}{\tilde{u}(y,t)} \leq c.$$

whenever $y \in \Omega \cap B(w, r'')$ and thereupon

$$\left| \frac{f(z,t)}{\tilde{u}(z,t)} - \frac{f(y,t)}{\tilde{u}(y,t)} \right| \leq c' \left(\frac{|z-y|}{r} \right)^{\alpha} \tag{90}$$

whenever $y, z \in \Omega \cap B(w, r'')$. Hence,

$$\left| \log \left(\frac{v(y)}{u(y)} \right) - \log \left(\frac{v(z)}{u(z)} \right) \right|$$
$$\leq \int_0^1 \left| \frac{f(z,t)}{\tilde{u}(z,t)} - \frac{f(y,t)}{\tilde{u}(y,t)} \right| dt \leq c' \left(\frac{|z-y|}{r} \right)^{\alpha}. \tag{91}$$

From (91) we conclude that Theorem 3.1 is valid under assumptions (75), (76). To remove (75), (76), suppose u, v are as in Theorem 1.3. We assume as we may that

$$1 = u(a_{r^*}(w)) = v(a_{r^*}(w)) \tag{92}$$

since otherwise we divide u, v by their values at this point and use invariance of the p Laplacian under scaling. Let $\tilde{u}, \tilde{v}$ be p harmonic functions in $\Omega \cap B(w, 2r^*)$, continuous in $\bar{B}(w, 2r^*)$ with

$$\tilde{u} = \min(u,v), \tilde{v} = 2 \max(u,v) \text{ on } \partial[B(w, 2r^*) \setminus \Omega].$$

Let $r^{**} = r^*/c$. From Lemmas 1.1, 1.5, and (92) we see that

$$\tilde{u} \leq min(u,v) \leq \max(u,v) \leq \tilde{v}/2 \leq c \text{ in } \Omega \cap B(w, 2r^{**}).$$

Also, using (92), Harnack's inequality, and Lemmas 1.4, 1.5 we see that (76) is valid with r^* replaced by r^{**}. Hence (75), (76), are valid with r^*, replaced by r^{**} so Theorem 3.1 can be applied to get

$$\max[v/u, u/v] \leq \tilde{v}/\tilde{u} \leq c \tag{93}$$

in $\Omega \cap B(w, r'')$. It follows for suitably chosen c, that u, cv satisfy (75), (76) in $\Omega \cap B(w, r'')$. Consequently Theorem 3.1 is valid without assumptions (75), (76).   □

4.6   More on Boundary Harnack Inequalities

In [62] we generalize Theorem 3.1 to weak solutions u of $\nabla \cdot A(x, \nabla u(x)) = 0$, where $A = (A_1, ..., A_n) : \mathbf{R}^n \times \mathbf{R}^n \to \mathbf{R}^n$. Also $A = A(x, \eta)$ is continuous in $\mathbf{R}^n \times (\mathbf{R}^n \setminus \{0\})$ and $A(x, \eta)$, for fixed $x \in \mathbf{R}^n$, is continuously differentiable in η_k, for every $k \in \{1, ..., n\}$, whenever $\eta \in \mathbf{R}^n \setminus \{0\}$. Moreover,

(i)

$$\alpha^{-1} |\eta|^{p-2} |\xi|^2 \le \sum_{i,j=1}^{n} \frac{\partial A_i}{\partial \eta_j}(x, \eta) \xi_i \xi_j,$$

(ii)

$$\left| \frac{\partial A_i}{\partial \eta_j}(x, \eta) \right| \le \alpha |\eta|^{p-2}, 1 \le i, j \le n,$$

(iii)

$$|A(x, \eta) - A(y, \eta)| \le \beta |x - y|^\gamma |\eta|^{p-1},$$

(iv)

$$A(x, \eta) = |\eta|^{p-1} A(x, \eta/|\eta|).$$

Under these assumptions, we prove

Theorem 3.7. *Let $\Omega \subset \mathbf{R}^n$ be a (δ, r_0)-Reifenberg flat domain. Suppose that u, v are positive A-harmonic functions in $\Omega \cap B(w, 4r)$, that u, v are continuous in $B(w, 4r)$ and $u = 0 = v$ on $B(w, 4r) \setminus \Omega$. For fixed p, α, β, γ, there exist $\tilde{\delta}, \sigma > 0$ and $c_1 \ge 1$, all depending only on $p, n, \alpha, \beta, \gamma$ such that*

$$\left| \log \frac{u(y_1)}{v(y_1)} - \log \frac{u(y_2)}{v(y_2)} \right| \le c_1 \left(\frac{|y_1 - y_2|}{r} \right)^\sigma$$

whenever $y_1, y_2 \in \Omega \cap B(w, r/c_1)$. Here $w \in \partial\Omega, 0 < r < r_0$, and $0 < \delta < \tilde{\delta}$.

For completeness we now give the definition of a Lipschitz domain.

Definition C. *$\Omega \subset \mathbf{R}^n$ is said to be a bounded Lipschitz domain provided there exists a finite set of balls $\{B(x_i, r_i)\}$, with $x_i \in \partial\Omega$ and $r_i > 0$, such that $\{B(x_i, r_i)\}$ constitutes a covering of an open neighbourhood of $\partial\Omega$ and such that, for each i,*

$$\Omega \cap B(x_i, 4r_i) = \{y = (y', y_n) \in \mathbf{R}^n : y_n > \phi_i(y')\} \cap B(x_i, 4r_i),$$

$$\partial\Omega \cap B(x_i, 4r_i) = \{y = (y', y_n) \in \mathbf{R}^n : y_n = \phi_i(y')\} \cap B(x_i, 4r_i),$$

in an appropriate coordinate system and for a Lipschitz function ϕ_i. The Lipschitz constants of Ω are defined to be $M = \max_i \|\|\nabla \phi_i\|\|_\infty$ and $r_0 = \min r_i$.

In [52] we prove

Theorem 3.8. *Let $\Omega \subset \mathbf{R}^n$ be a bounded Lipschitz domain with constant M. Given $p, 1 < p < \infty, w \in \partial\Omega, 0 < r < r_0$, suppose that u and v are positive p harmonic functions in $\Omega \cap B(w, 2r)$. Assume also that u and v are continuous in $B(w, 2r)$ and $u = 0 = v$ on $B(w, 2r) \setminus \Omega$. Under these assumptions there exist $c_1, 1 \le c_1 < \infty$, and $\alpha, \alpha \in (0, 1)$, both depending only on p, n, and M, such that if $y_1, y_2 \in \Omega \cap B(w, r/c_2)$ then*

$$\left| \log \frac{u(y_1)}{v(y_1)} - \log \frac{u(y_2)}{v(y_2)} \right| \le c_1 \left(\frac{|y_1 - y_2|}{r} \right)^\alpha.$$

Remark. To prove Theorem 3.7 we argue as in Steps 1–4. The techniques and arguments are similar to the proof of Theorem 3.1.

In the proof of Theorem 3.8 we follow a somewhat similar game plan although in this case the techniques are necessarily more sophisticated. For example there is no readily available comparison function (x_n in Theorems 3.1 or 3.7) from which we can extrapolate the fundamental inequality in Step 2. Thus we study p harmonic capacitary functions in starlike Lipschitz ring domains and show these functions satisfy the fundamental inequality near boundary points where they vanish. Also, in the Lipschitz case simple examples show $|\nabla u|^{p-2}$ need not extend to an A_2 weight locally, so that Lemma 1.8 cannot be applied. Instead for given $w, r \in \partial\Omega$ we show that $|\nabla u|^{p-2}$ satisfies a Carleson measure condition in a certain starlike Lipschitz subdomain $\tilde{\Omega} \subset \Omega$. This fact and a theorem in [42] (see also [35]) can then be used to conclude that a boundary Harnack inequality (as in Lemma 1.8) holds for solutions to (8), (9) in the Lipschitz case.

Finally we note that Theorems 3.7 and 3.8 are weaker than the corresponding results for $p = 2$. Thus for example does Theorem 3.8 hold in an NTA domain? Can the assumptions on A in Theorem 3.7 be weakened?

4.7   The Martin Boundary Problem

The Martin boundary for harmonic functions was first introduced by Martin [67]. Over the years, it has been of considerable interest to researchers in potential theory. Unfortunately Martin did not receive many accolades for his contribution, as according to math. sci. net, the above is the only paper he ever wrote. In order to define the p Martin boundary of a NTA domain, we need to first define a minimal positive p harmonic function.

Definition D. *Fix $p, 1 < p < \infty$. Then $\hat{u}$ is said to be a minimal positive p harmonic function in the NTA domain Ω relative to $w \in \partial\Omega$, provided $\hat{u} > 0$ is p harmonic in Ω and $\hat{u}$ has continuous boundary value 0 on $\partial\Omega \setminus \{w\}$. $\hat{u}$ is said to be unique up to constant multiples if $\hat{v} = \lambda\hat{u}$, for some constant λ,* *whenever $\hat{v}$ is a minimal positive p harmonic function relative to $w \in \partial\Omega$. The p Martin boundary of Ω is equivalence classes of all minimal positive p harmonic functions defined relative to boundary points of Ω. Two minimal positive p harmonic functions are in the same equivalence class if and only if they are constant multiples of each other. We say that the p Martin boundary of Ω can be identified with $\partial\Omega$ provided each $w \in \partial\Omega$ corresponds to a unique (up to constant multiples) minimal positive p harmonic function.*

Theorem 3.9. *Let Ω, δ, p, r_0, be as in Theorem 3.1. There exists $\delta_+ = \delta_+(p, n)$ such that if $0 < \delta < \delta_+$, then the p Martin boundary of Ω can be identified with $\partial\Omega$.*

Remark. Theorem 3.9 for $p = 2$, i.e., harmonic functions, in an NTA domain G is an easy consequence of the boundary Harnack inequality for harmonic functions in NTA domains. Indeed, if $w \in \partial G$ and if u, v are minimal harmonic functions corresponding to w, one first uses the boundary Harnack inequality for harmonic functions to show that $\gamma = \inf_G u/v > 0$. Next one applies this result to $u - \gamma v, v$ in order to conclude that $u = \gamma v$. Note however that this argument depends heavily on linearity of the Laplacian and thus the argument fails for the p Laplacian when $p \neq 2$. We also note that if $r_0 = \infty$ in Theorem 3.1, i.e., Ω is an unbounded Reifenberg flat NTA domain, and u, v are minimal positive p harmonic functions relative to ∞, then we can let $r \to \infty$ in the conclusion of Theorem 3.1, to get $u = \lambda v, \lambda = $ constant. To make this idea work when $w \in \partial\Omega \setminus \{\infty\}$, we need to prove an analogue of Theorem 3.1 for positive p harmonic functions in $\Omega \setminus B(w, r')$, ($r'$ small) vanishing on $\partial\Omega \setminus B(w, r')$. We could get this analogue, by arguing as in the proof of Theorem 3.1, except we do not know a priori, that our functions have the fundamental nondegeneracy property (45) in an appropriate domain. We overcome this deficiency, using arguments from the proof of Theorem 3.1, as well as an induction—bootstrap type argument. We start by showing that if one such p harmonic function has the fundamental nondegeneracy property then all such functions have this property.

Lemma 3.10. *Let Ω be (δ, r_0) Reifenberg flat. Let $\hat{u}, \hat{v} > 0$ be p harmonic in $\Omega \setminus B(w, r')$, continuous in $\mathbf{R}^n \setminus B(w, r')$, with $\hat{u} \equiv \hat{v} \equiv 0$ on $\mathbf{R}^n \setminus [\Omega \cup B(w, r')]$. Suppose for some $r_1, r' < r_1 < r_0$, and $A \geq 1$, that*

$$A^{-1} \frac{\hat{u}(x)}{d(x, \partial\Omega)} \leq |\nabla\hat{u}(x)| \leq A \frac{\hat{u}(x)}{d(x, \partial\Omega)}.$$

whenever $x \in \Omega \cap [B(w, r_1) \setminus B(w, r')]$. There exists $\alpha > 0, \lambda, c \geq 1$, depending on p, n, A, such that if $0 < \delta < \hat{\delta}$ ($\hat{\delta}$ as in Theorem 3.1),

$$\lambda^{-1} \frac{\hat{v}(x)}{d(x, \partial\Omega)} \leq |\nabla\hat{v}(x)| \leq \lambda \frac{\hat{v}(x)}{d(x, \partial\Omega)}.$$

for $x \in \Omega \cap [B(w, r_1/c) \setminus B(w, cr')]$. Moreover,

$$\left| \log\left(\frac{\hat{u}(z)}{\hat{v}(z)}\right) - \log\left(\frac{\hat{u}(y)}{\hat{v}(y)}\right) \right| \leq c \left(\frac{r'}{\min(r_1, |z-w|, |z-y|)} \right)^\alpha$$

whenever $z, y \in \Omega \setminus B(w, cr')$.

Proof. We assume that $r'/r_1 \ll 1$, since otherwise there is nothing to prove. Let $\tilde{r} = \hat{c}r'$. If $\hat{c} = \hat{c}(p, n)$ is large enough, we may assume

$$\hat{u} \leq \hat{v}/2 \leq \hat{c}u \text{ in } \Omega \setminus B(w, \tilde{r}), \tag{94}$$

as we see from Theorem 3.1, Harnack's inequality, and the maximum principle. As in Step 4 of Sect. 1, let $u(\cdot, t), t \in [0, 1]$, be p harmonic in $\Omega \setminus B(w, \tilde{r})$, with continuous boundary values,

$$u(\cdot, t) = (1-t)\hat{u} + t\hat{v} \text{ on } \partial[\Omega \setminus B(w, \tilde{r})]. \tag{95}$$

Extend $u(\cdot, t), t \in [0, 1]$, to be continuous on $\mathbf{R}^n \setminus [\Omega \cup B(w, \tilde{r})]$ by setting $u(\cdot, t) \equiv 0$ on this set. Next we note from Lemma 3.3 that there exists $\epsilon_0 = \epsilon_0(p, n, M, A)$ such that if $\tilde{r} \leq s_1 < \rho_1/4 \leq r_1/16$, $\tau \in (0, 1]$, and

$$(1 - \epsilon_0)\tilde{L} \leq u(\cdot, \tau)/\hat{u} \leq (1 + \epsilon_0)\tilde{L}, \tag{96}$$

in $\Omega \cap [B(w, 2\rho_1) \setminus B(w, s_1)]$, for some $\tilde{L}$, then

$$\hat{\lambda}^{-1} \frac{u(x, \tau)}{d(x, \partial\Omega)} \leq |\nabla u(x, \tau)| \leq \hat{\lambda} \frac{u(x, \tau)}{d(x, \partial\Omega)} \tag{97}$$

in $\Omega \cap [B(w, \rho_1) \setminus B(w, 2s_1)]$ where $\hat{\lambda} = \hat{\lambda}(p, n, A)$. Observe from (94), (95), that if $\tau_1, \tau_2 \in [0, 1]$,

$$c^{-1}u(\cdot, \tau_1) \leq U(\cdot, \tau_1, \tau_2) = \frac{u(\cdot, \tau_2) - u(\cdot, \tau_1)}{\tau_2 - \tau_1}$$
$$= v - u \leq c\,u(\cdot, \tau_1) \tag{98}$$

on $\partial[\Omega \setminus B(w, \tilde{r})]$, so from the maximum principle this inequality also holds in $\Omega \setminus B(w, \tilde{r})$. Thus for ϵ_0 as in (96), there exists $\epsilon_0', 0 < \epsilon_0' \leq \epsilon_0$, with the same dependence as ϵ_0, such that if $|\tau_2 - \tau_1| \leq \epsilon_0'$, then

$$1 - \epsilon_0/2 \leq \frac{u(\cdot, \tau_2)}{u(\cdot, \tau_1)} \leq 1 + \epsilon_0/2 \text{ in } \Omega \setminus B(w, \tilde{r}). \tag{99}$$

Divide [0,1] into closed intervals, disjoint except for endpoints, of length $\epsilon_0'/2$ except possibly for the interval containing 1 which is of length $\leq \epsilon_0'/2$. Let $\xi_1 = 0 < \xi_2 < ... < \xi_m = 1$ be the endpoints of these intervals. Thus [0,1] is divided into $\{[\xi_k, \xi_{k+1}]\}_1^m$. Next suppose for some $l, 1 \leq l \leq m-1$, that (97) is valid whenever $\tau \in [\xi_l, \xi_{l+1}]$ and $x \in \Omega \cap [B(w, \rho_1) \setminus B(w, 2s_1)]$. Under this assumption we claim for some $\hat{c}_1, \hat{c}_2, \alpha$, depending only on p, n, A that

$$\left| \log \frac{u(z, \xi_{l+1})}{u(z, \xi_l)} - \log \frac{u(y, \xi_{l+1})}{u(y, \xi_l)} \right| \leq \hat{c}_1 \left(\frac{s_1}{\min\{|z - w|, |w - y|\}} \right)^\alpha \tag{100}$$

whenever $z, y \in \Omega \cap [B(w, \rho_1/\hat{c}_2) \setminus B(w, \hat{c}_2 s_1)]$.

Indeed we can retrace the argument in Theorem 3.1 to get for $z, y \in \Omega \cap [B(w, \rho_1/c) \setminus B(w, cs_1)]$, that there exists f as in (87) with

$$\left| \log \frac{u(z, \xi_{l+1})}{u(z, \xi_l)} - \log \frac{u(y, \xi_{l+1})}{u(y, \xi_l)} \right|$$

$$\tag{101}$$

$$\leq \int_{\xi_l}^{\xi_{l+1}} \left| \frac{f(z, \tau)}{u(z, \tau)} - \frac{f(y, \tau)}{u(y, \tau)} \right| d\tau \leq c \left(\frac{s_1}{\min\{|z - w|, |y - w|\}} \right)^\alpha.$$

The last inequality in (101) follows from a slightly more general version of Lemma 1.8.

We now proceed by induction. Observe from (99) as well as $u(\cdot, \xi_1) = \hat{u}$ that (96) holds whenever $\tau \in [\xi_1, \xi_2]$. Thus (97) and consequently (100) are true for $l = 1$ with $s_1 = \tilde{r}, \rho_1 = r_1/4$. Let $s_2 = \hat{c}_2 s_1, \rho_2 = \rho_1/\hat{c}_2$. By induction, suppose for some $2 \leq k < m$,

$$\left| \log \frac{u(z, \xi_k)}{\hat{u}(z)} - \log \frac{u(y, \xi_k)}{\hat{u}(y)} \right| \leq (k-1)\hat{c}_1 \left(\frac{s_k}{\min\{|z - w|, |y - w|\}} \right)^\alpha \tag{102}$$

whenever $z, y \in \Omega \cap [B(w, \rho_k) \setminus B(w, s_k)]$, where $\alpha, \hat{c}_1$ are the constants in (100). Choose $s_k' \geq 2s_k$, so that

$$\left| \frac{u(z, \xi_k)}{\hat{u}(z)} - \frac{u(y, \xi_k)}{\hat{u}(y)} \right| \leq \eta \frac{u(z, \xi_k)}{\hat{u}(z)}$$

whenever $z, y \in \Omega \cap [B(w, \rho_k) \setminus B(w, s_k')]$. Fix z as above and choose $\eta > 0$ so small that

$$(1 - \epsilon_0) \frac{u(z, \xi_k)}{\hat{u}(z)} \leq \frac{u(y, \tau)}{\hat{u}(y)} \leq (1 + \epsilon_0) \frac{u(z, \xi_k)}{\hat{u}(z)} \tag{103}$$

whenever $y \in \Omega \cap [B(w, \rho_k) \setminus B(w, s_k')]$ and $\tau \in [\xi_k, \xi_{k+1}]$. To see the size of η observe for $\tau \in [\xi_k, \xi_{k+1}]$ that

$$\frac{u(y, \tau)}{\hat{u}(y)} = \frac{u(y, \tau)}{u(y, \xi_k)} \cdot \frac{u(y, \xi_k)}{\hat{u}(y)}$$

$$\leq (1 + \epsilon_0/2)(1 + \eta) \frac{u(z, \xi_k)}{\hat{u}(z)}.$$

Thus if $\eta = \epsilon_0/4$ (ϵ_0 small), then the right hand inequality in (103) is valid. A similar argument gives the left hand inequality in (103) when $\eta = \epsilon_0/4$. Also since $k \leq 2/\epsilon_0'$, and ϵ_0', α depend only on p, n, M, A, we deduce from (102) that one can take $s_k' = \hat{c}_3 s_k$ for $\hat{c}_3 = \hat{c}(p, n, M, A)$ large enough. From (103) we find that (96) holds with $\tilde{L} = \frac{u(z,\xi_k)}{\hat{u}(z)}$ in $\Omega \cap [B(w, \rho_k) \setminus B(w, s_k')]$. From (97) we now get that (100) is valid for $l = k$ in $\Omega \cap [B(w, \frac{\rho_k}{2\hat{c}_2}) \setminus B(w, \hat{c}_2 s_k')]$. Let $s_{k+1} = \hat{c}_3 \hat{c}_2 s_k$ and $\rho_{k+1} = \frac{\rho_k}{2\hat{c}_2}$. Using (100) and the induction hypothesis we have

$$\left| \log \frac{u(z, \xi_{k+1})}{\hat{u}(z)} - \log \frac{u(y, \xi_{k+1})}{\hat{u}(y)} \right| \leq \left| \log \frac{u(z, \xi_{k+1})}{u(z, \xi_k)} - \log \frac{u(y, \xi_{k+1})}{u(y, \xi_k)} \right|$$

$$+ \left| \log \frac{u(z, \xi_k)}{\hat{u}(z)} - \log \frac{u(y, \xi_k)}{\hat{u}(y)} \right| \leq k \hat{c}_1 \left(\frac{s_{k+1}}{\min\{|z - w|, |w - y|\}} \right)^\alpha \tag{104}$$

whenever $z, y \in \Omega \cap [B(w, \rho_{k+1}) \setminus B(w, s_{k+1})]$. Thus by induction we get (102) with $k = m$. Since $u(\cdot, \xi_m) = \hat{v}$ and $s_m \leq cr'$, $\rho_m \geq r_1/c$, for some large $c = c(p, n, A)$, we can now argue as in (103) to first get (96) with $u(\cdot, \tau)$ replaced by $\hat{v}$ and then (97) for $\hat{v}$. We conclude that Lemma 3.10 is valid for $z, y \in \Omega \cap [B(w, r_1/c) \setminus B(w, cr')]$ provided c is large enough. Using the maximum principle it follows that the last display in Lemma 3.49 is also valid for $z, y \in \Omega \setminus B(w, r_1/c)$. $\square$

4.8   Proof of Theorem 3.9

Note that if $\hat{u}$ is a minimal p harmonic function satisfying (45) in $\Omega \cap B(w, r_1)$, then one can let $r' \to 0$ in Lemma 3.10 to get Theorem 3.9. Thus to complete the proof of Theorem 3.9 it suffices to show the existence of a minimal positive p harmonic function $\hat{u}$ relative to $w \in \partial\Omega$ and $0 < r_1 < r_0$ for which the fundamental nondegeneracy property in (45) holds in $\Omega \cap B(w, r_1)$. To this end we introduce,

Definition E. *We call $\tilde{\Omega} \subset \Omega$ a non tangential approach region at $w \in \partial\Omega$ if for some $\tilde{\eta} > 0$, $d(x, \partial\Omega) \geq \tilde{\eta}|x - w|$ for all $x \in \tilde{\Omega}$.*

If we wish to emphasize $w, \tilde{\eta}$ in Definition E we write $\tilde{\Omega}(w, \tilde{\eta})$. Now let $\hat{u}$ be a minimal positive p harmonic function in Ω relative to $w \in \partial\Omega$ and $0 < \delta < \delta^*$. Then we can apply Lemma 3.3 to conclude for each $\hat{x} \in \partial\Omega \setminus \{w\}$ that for some $\tilde{c} = \tilde{c}(p, n) \geq 1$,

$$\tilde{c}^{-1} \frac{\hat{u}(x)}{d(x, \partial\Omega)} \leq |\nabla \hat{u}(x)| \leq \tilde{c} \frac{\hat{u}(x)}{d(x, \partial\Omega)} \tag{105}$$

whenever $x \in \partial\Omega \cap B(\hat{x}, |\hat{x} - w|/\tilde{c}) \cap B(w, r_0)$. Using this fact we see that if $0 < \delta_+ < \delta^*$ in Theorem 3.9 then there exists $\tilde{\eta}$ depending only on p, n, such that

$$\hat{u} \text{ satisfies } (105) \text{ in } [\Omega \setminus \tilde{\Omega}(w, \tilde{\eta})] \cap B(w, r_0). \tag{106}$$

From (106) we see that if (105) holds in $\Omega(w, \tilde{\eta}) \cap B(w, r_1)$, then $\hat{u}$ can be used in Lemma 3.10 for each small $r' > 0$ so Theorem 3.9 is true. To prove the above statement, we first extend $\hat{u}$ continuously to $\mathbf{R}^n \setminus \{w\}$ by letting $\hat{u} = 0$ in $\mathbf{R}^n \setminus \Omega$. Let $0 < r < r_0/n$ and $\sigma = 100n\delta$. Using translation, rotation invariance of the p Laplacian and Reifenberg flatness in Definition B, we assume as we may that $w = 0$ and for given $\sigma > 0$ (sufficiently small) that

$$B(0, nr) \cap \{y : y_n \geq \sigma r\} \subset \Omega$$
$$\tag{107}$$
$$B(0, nr) \cap \{y : y_n \leq -\sigma r\} \subset \mathbf{R}^n \setminus \Omega$$

Extend $\hat{u}$ to a continuous function in $\mathbf{R}^n \setminus \{0\}$ by putting $\hat{u} \equiv 0$ on $\mathbf{R}^n \setminus (\Omega \cup \{0\})$. Let

$$Q = \{y : |y_i| < r, 1 \leq i \leq n - 1\} \cap \{y : \sigma r < y_n < r\} \setminus B(0, \sqrt{\sigma} r)$$

and let v_1 be the p harmonic function in Q with the following continuous boundary values,

$$v_1(y) = \hat{u}(y) \quad \text{if } y \in \partial Q \cap \{y : 2\sigma r \leq y_n\},$$
$$v_1(y) = \frac{(y_n - \sigma r)}{\sigma r}\hat{u}(y) \quad \text{if } y \in \partial Q \cap \{y : \sigma r \leq y_n < 2\sigma r\}.$$

Comparing boundary values and using the maximum principle for p harmonic functions, we deduce

$$v_1 \leq \hat{u} \text{ in } Q. \tag{108}$$

Let $\sigma(\epsilon) = \exp(-1/\epsilon)$. To complete the proof of Theorem 3.9 we will make use of the following lemmas.

Lemma 3.11. *Let $0 < \epsilon \leq \hat{\epsilon}$, let $\sigma = \sigma(\epsilon)$ be as above and let $\tilde{\eta}$ be as in (106). If $\hat{\epsilon}$ is small enough, then there exists $\hat{\theta} = \hat{\theta}(p, n), 0 < \hat{\theta} \leq 1/4$, such that if $\hat{\rho} = \sigma^{1/2 - \hat{\theta}}r$, then*

$$1 \leq \hat{u}(y)/v_1(y) \leq 1 + \epsilon$$

whenever $y \in \tilde{\Omega}(0, \tilde{\eta}/16) \cap [B(0, \hat{\rho}) \setminus B(0, 4\sqrt{\sigma}r)]$.

Lemma 3.12. *Let $v_1, \epsilon, \hat{\epsilon}, \hat{\theta}, r, \sigma$, be as in Lemma 3.11 and let $\tilde{\eta}$ be as in (106). If $\hat{\epsilon}$ is small enough, then there exist $\theta = \theta(p, n), 0 < \theta \leq \hat{\theta}/10$, and $c = c(p, n) > 1$ such that if $\rho = \sigma^{1/2 - 4\theta}r, b = \sigma^{-\theta}$, then*

$$c^{-1}\frac{v_1(x)}{d(x,\partial\Omega)} \leq |\nabla v_1(x)| \leq c\frac{v_1(x)}{d(x,\partial\Omega)}$$

whenever $x \in \tilde{\Omega}(0,\tilde{\eta}/4) \cap [B(0,b\rho) \setminus B(0,\rho/b)]$ and $0 < \epsilon \leq \hat{\epsilon}$.

Before proving Lemmas 3.11 and 3.12 we indicate how the proof of Theorem 3.9 follows from these lemmas. Indeed, using Lemmas 3.3, 3.11, 3.12 we see for $\hat{\epsilon}$ sufficiently small and fixed, $0 < \epsilon \leq \hat{\epsilon}$, that there exists $\tilde{c} > 1$, depending only on p, n, such that

$$\tilde{c}^{-1}\frac{\hat{u}(x)}{d(x,\partial\Omega)} \leq |\nabla\hat{u}(x)| \leq \tilde{c}\frac{\hat{u}(x)}{d(x,\partial\Omega)} \tag{109}$$

in $\tilde{\Omega}(0,\tilde{\eta}/2) \cap [B(0,b^{1/2}\rho) \setminus B(0,\rho/b^{1/2})]$. With $\epsilon > 0$ now fixed it follows from (109), (106), and arbitrariness of $\rho < r_0/c$ that $\hat{u}$ can be used in Lemma 3.10. As mentioned earlier, Theorem 3.9 follows from Lemma 3.10.

Proof of Lemma 3.11. From (108) we observe that it suffices to prove the righthand inequality in Lemma 3.11. We note that if $y \in \partial Q$ and $\hat{u}(y) \neq v_1(y)$, then y lies within $4\sigma r$ of a point in $\partial\Omega$. Also, $\max_{\partial B(0,t)} u$ is nonincreasing as a function of $t > 0$ as we see from the maximum principle for p harmonic functions. Using these notes and Lemmas 1.2–1.5 we see that

$$\hat{u} \leq v_1 + c\sigma^{\alpha/2} u(\sqrt{\sigma}e_n) \tag{110}$$

on ∂Q. By Lemma 1.1 this inequality also holds in Q. Using Lemmas 1.2–1.5 we also find that there exist $\beta = \beta(p,n) \geq 1$ and $c = c(p,n) > 1$ such that

$$\max\{\psi(z),\psi(y)\} \leq c\,(d(z,\partial Q)/d(y,\partial Q))^{\beta} \min\{\psi(z),\psi(y)\} \tag{111}$$

whenever $z \in Q, y \in Q \cap B(z,4d(z,\partial Q))$ and $\psi = \hat{u}$ or v_1. Also from Lemmas 1.2–1.5 applied to v_1 we deduce

$$v_1(2\sqrt{\sigma}\,re_n) \geq c^{-1}\hat{u}(\sqrt{\sigma}\,re_n). \tag{112}$$

Let $\hat{\rho}, \hat{\theta}$ be as in Lemma 3.11. From (110)–(112) we see that if $y \in \tilde{\Omega}(0,\tilde{\eta}/16) \cap [B(0,\hat{\rho}) \setminus B(0,4\sqrt{\sigma}\,r)]$, then

$$\hat{u}(y) \leq v_1(y) + c\sigma^{\alpha/2}u(\sqrt{\sigma}e_n) \leq (1 + c^2\sigma^{\alpha/2-\hat{\theta}\beta})v_1(y) \leq (1+\epsilon)v_1(y) \tag{113}$$

provided $\hat{\epsilon}$ is small enough and $\hat{\theta}\beta = \alpha/4$. Thus Lemma 3.11 is true. □

Proof of Lemma 3.12. Using Lemmas 1.2–1.5 we note that there exist $\gamma = \gamma(p,n) > 0, 0 < \gamma \leq 1/2$, and $c = c(p,n) > 1$ such that

$$\hat{u}(x) \leq c(s/t)^{\gamma}\hat{u}(se_n) \tag{114}$$

provided $x \in \mathbf{R}^n \setminus B(0,t), t \geq s$, and $se_n \in \Omega$ with $d(se_n, \partial\Omega) \geq c^{-1}s$. Using (114) with $t = r, s = \sqrt{\sigma}r$, we find that

$$v_1 \leq c\sigma^{\gamma/2}\,\hat{u}(\sqrt{\sigma}re_n) \text{ on } \partial Q \setminus \bar{B}(0, \sqrt{\sigma}\,r), \tag{115}$$

where c depends only on p, n. Let $\tilde{v}$ be the p harmonic function in Q with continuous boundary values $\tilde{v} = 0$ on $\partial Q \setminus \bar{B}(0, \sqrt{\sigma}r)$ and $\tilde{v} = v_1$ on $\partial Q \cap \partial B(0, \sqrt{\sigma}r)$. From Lemma 1.1 and (115) it follows that

$$0 \leq \tilde{v} \leq v_1 \leq \tilde{v} + c\sigma^{\gamma/2}u(\sqrt{\sigma}re_n) \text{ in } Q. \tag{116}$$

From Lemmas 1.2–1.5 we observe that

$$\tilde{v}(2\sqrt{\sigma}re_n) \geq c^{-1}v_1(\sqrt{\sigma}re_n) = c^{-1}u(\sqrt{\sigma}re_n). \tag{117}$$

Using (116), (117), and (111) applied to $\psi = \tilde{v}$ we obtain for $\rho = \sigma^{1/2-4\theta}r, \theta$ small, $b = \sigma^{-\theta}$, and $\hat{b} = 8b^2$, that

$$\tilde{v} \leq v_1 \leq (1 + c\sigma^{\gamma/2-6\theta\beta})\tilde{v} \leq (1+\epsilon)\tilde{v} \tag{118}$$

on $\tilde{\Omega}(0, \tilde{\eta}/8) \cap [B(0, \hat{b}\rho) \setminus B(0, \rho/\hat{b})]$, provided $\hat{\epsilon}$ is small enough and $\theta = \min\{\gamma/(24\beta), \hat{\theta}/10\}$.

Next let v be the p harmonic function in

$$Q' = \{y : |y_i| < r, 1 \leq i \leq n-1\} \cap \{y : \sigma r < y_n < r\} \setminus \bar{B}(2\sqrt{\sigma}\,re_n, \sqrt{\sigma}\,r)$$

with continuous boundary values $v = 0$ on $\partial Q' \setminus \bar{B}(2\sqrt{\sigma}e_n, \sqrt{\sigma}\,r)$ while $v = 1$ on $\partial B(2\sqrt{\sigma}\,re_n, \sqrt{\sigma}\,r)$. One can show that

$$v(x) \leq c\langle 2\sqrt{\sigma}\,re_n - x, \nabla v(x)\rangle \tag{119}$$

when $x \in Q'$ where $c = c(p, n)$. Clearly this inequality implies that there exists $c = c(p, n, \eta) \geq 1$, for given $\eta, 0 < \eta \leq 1/2$, such that

$$c^{-1}\frac{v(x)}{d(x, \partial Q')} \leq |\nabla v(x)| \leq c\frac{v(x)}{d(x, \partial Q')} \tag{120}$$

in $\tilde{Q}'(0, \eta) \setminus B(0, 10\sqrt{\sigma}\,r)$ where $\tilde{Q}'(0, \eta)$ is the non-tangential approach region defined relative to $0, \eta, Q'$. Using Theorem 1.3 and (120) for suitable $\eta = \eta(p, n)$ we conclude that (120) actually holds in $Q' \setminus B(0, 10\sqrt{\sigma}r)$. We now use Lemma 3.10 applied to $v, \tilde{v}$ with Ω, r', replaced by $Q', 10\sqrt{\sigma}r$, in order to get, for some $a = a(p, n) > 0$ and $c = c(p, n) > 1$, that

$$c^{-1}\frac{\tilde{v}(x)}{d(x, \partial\Omega)} \leq |\nabla\tilde{v}(x)| \leq c\frac{\tilde{v}(x)}{d(x, \partial\Omega)} \tag{121}$$

in $[B(0, r/c) \setminus B(0, c) r)]$. Finally, note that if $0 \leq \epsilon \leq \hat{\epsilon}$ and if $\hat{\epsilon}$ is sufficiently small, then $r/c > b^2\rho > \rho/b^2 > c\sqrt{\sigma}\, r$. Hence, if $\hat{\epsilon}$ is small enough then we can use (121), (118), and Lemma 3.3 to conclude that Lemma 3.12 is valid. The proof of Theorem 3.9 is now complete.   □

4.9   Further Remarks

We note that Theorem 3.9 has been generalized in [62] to weak solutions of $\nabla \cdot A(x, \nabla u) = 0$ where A is as in Theorem 3.7. Also in [52] we show that the conclusion of Theorem 3.9 holds when Ω is convex or the complement of a convex domain. Finally the conclusion of Theorem 3.9 is valid when $\Omega \subset \mathbf{R}^2$ is a Lipschitz domain. This problem for Lipschitz domains remains open when $n \geq 3$. However the same argument as in Theorem 3.9 yields that a p Martin function in a Lipschitz domain is unique (up to constant multiples) at each boundary point where a tangent plane exists.

5   Uniqueness and Regularity in Free Boundary: Inverse Type Problems

We begin this section by outlining the proof of

Theorem 4.1. *Let E be a compact convex set, $a > 0$, and p fixed, $1 < p < \infty$. If $H^{n-p}(E) > 0$, then there is a unique solution to the following free boundary problem: Find a bounded domain D with $E \subset D$ and u, p harmonic in $D \setminus E$, satisfying*

(a) u has continuous boundary values 1 on E and 0 on ∂D,
(b) u is p harmonic in $D \setminus E$,
(c) $\mu = a^{p-1} H^{n-1}|_{\partial D}$ where μ is the measure associated with u as in (15),
(d) For some positive c, r_0, and all $x \in \partial D$

$$\mu(B(x, r)) \leq c r^{n-1}, 0 < r \leq r_0.$$

5.1   History of Theorem 4.1

My interest in free boundary problems of the above type started in $\approx$1989 when Andrew Vogel (my former Ph.D. student) was a graduate student at the University of Kentucky. He went to a conference where the following problem was proposed:

5.1.1   The Ball Problem

Let g be the Green's function for a domain D with smooth boundary and pole at $0 \in D$. If $|\nabla g| = a = \text{constant}$ on ∂D, show that D is a ball with center at 0. We came up with the following proof: From the above assumption and properties of g one has

$$\omega(\partial D) = 1 = aH^{n-1}(\partial D)$$

where ω is harmonic measure on ∂D relative to 0. Thus

$$H^{n-1}(\partial D) = 1/a. \tag{122}$$

Choose $B(0, R) \subset D$ so that $y \in \partial B(0, R) \cap \partial D$. Let G be the Green's function for $B(0, R)$ with pole at 0. If $b = |\nabla G|(y)$, then since $|\nabla G| = b$ on $\partial B(0, R)$ we have, as in (122),

$$H^{n-1}(B(0, R)) = 1/b. \tag{123}$$

On the other hand from $G \leq g$ one gets $b \leq a$ or $1/a \leq 1/b$. Combining this inequality with (122), (123), we deduce

$$H^{n-1}(\partial D) \leq H^{n-1}(\partial B(0, R))$$

which in view of the isoperimetric inequality or the fact that H^{n-1} decreases under a projection implies that $B(0, R) = D$. Later we found out that numerous other authors, including H. Shahgholian, had also obtained that D is a ball, under various smoothness assumptions on $g, \partial D$.

In fact Henrot and Shahgholian in [34], generalized this problem, by proving

Theorem 4.2. *Let E be a compact convex set, $a > 0$, and p fixed, $1 < p < \infty$. If $H^{n-p}(E) > 0$, then there is a unique solution to the following free boundary problem: Find a bounded domain D with $E \subset D$ and u, p harmonic in $D \setminus E$, satisfying*

(α) *u has continuous boundary values 1 on E and 0 on ∂D,*

(β) *u is p harmonic in $D \setminus E$,*

(γ) *$|\nabla u|(x) \to a$ as $x \to \partial D$.*

Moreover D is convex and ∂D is C^∞.

To prove uniqueness in Theorem 4.2, given existence, one can argue as in the ball problem using the nearest point projection onto a convex set. In fact if u', D' are solutions to the above problem for a given a, p, one can show there exists a Henrot–Shahgholian domain $D \subset D'$ with $\partial D \cap \partial D' \neq \emptyset$. Using the nearest point projection onto a convex set as in the ball problem, it follows that $D = D'$. We note that if E is a ball in Theorem 4.2, then necessarily D is a ball, since radial solutions exist. Also, Henrot–Shahgholian domains will play the same role in the proof of Theorem 4.1, as balls did in the ball problem.

Andy and I considered other generalizations of the ball problem. We first rephrased the problem as in (c) of Theorem 4.1 by

$$\omega = aH^{n-1}|_{\partial D} \tag{124}$$

where ω is harmonic measure with respect to $0, \partial D$. However some examples of Keldysh and Larrentiev in two dimensions showed that this assumption was not enough to guarantee that D was a ball, so we also assumed that

$$|\nabla g| \leq M < \infty \text{ near } \partial D.$$

Using the Riesz representation formula for subharmonic functions one can show that boundedness of $|\nabla g|$ near ∂D, is equivalent to

$$\omega(B(x, r)) \leq cr^{n-1} \text{ for all } x \in \partial D \text{ and } 0 < r \leq r_0 \tag{125}$$

(i.e., condition (d) in Theorem 4.1). Given (124), (125), and the added assumption that

$$H^{n-1} \text{ almost every point of } \partial D \text{ lies in the reduced boundary} \atop \text{of } D \text{ in the sense of geometric measure theory,} \tag{126}$$

we were able to show that

$$\limsup_{x \to \partial D} |\nabla g| \leq a \tag{127}$$

which then allowed us to repeat the argument in the smooth case and get that D is a ball. Our paper appeared in [56]. In this paper we listed a number of symmetry problems including whether hypothesis (126) was needed for the ball theorem as well as analogues for the p Laplacian. During this period we also wrote [55]. This paper used a technique in [72] to construct a bounded domain D in $\mathbf{R}^n, n \geq 3$ for which ∂D is homeomorphic to a sphere and (124), (126) hold, but $D \neq$ ball (so necessarily (125) is false). We improved this result in [58] where we showed that there exists quasi-spheres $\neq$ ball for which (124), (126), held.

Finally in [57] we were able to show that our generalization of the ball theorem remained valid without (126). Thus we obtained an endpoint result for harmonic functions. The key new idea in removing (126) was the following square function—Carleson measure estimate:

$$\int_{D\cap B(x,r)} g \sum_{i,j=1}^{n} (g_{y_iy_j})^2 \, dy \leq cr^{n-1}, 0 < r \leq r_1, \qquad (128)$$

for some positive c, r_1, and all $x \in \partial D$. Let $d(E, F)$ denote the distance between the sets E, F. Then (128) allowed us to conclude for given $\epsilon > 0$ (in a qualitative H^{n-1} sense) that there were 'lots' of tangent balls $B(x, d(x, \partial D))$ where ∇g had oscillation $\leq \epsilon$ in $B(x, (1-\epsilon)d(x, \partial D))$. Moreover using this fact and subharmonicity of $|\nabla g|$ we could also conclude that if $\Lambda = \limsup_{y \to D} |\nabla g|$, then in a certain percentage of these balls,

$$\Lambda - \epsilon \leq |\nabla g|(x) \leq \Lambda + \epsilon \qquad (129)$$

Using (124), (129), and an asymptotic argument, we then concluded that $\Lambda \leq a$, which as earlier implies $D = B(0, R)$.

5.2   Proof of Theorem 4.1

To prove Theorem 4.1 for fixed $p, 1 < p < \infty$, we first showed that (d) in Theorem 4.1 implies,

$$|\nabla u| \leq M < \infty \text{ near } \partial D. \qquad (130)$$

To prove (130) for x near ∂D we used the inequality

$$u(x) \leq c \int_0^{c'd(x,\partial D)} \left(\frac{\mu[B(w,t)]}{t^{n-p}} \right)^{1/(p-1)} \frac{dt}{t}, \qquad (131)$$

(d) of Theorem 4.1, and the interior estimate,

$$|\nabla u(x)| \leq \hat{c}u(x)/d(x).$$

Equation (131) is proved in [46]. Armed with (130) we then proved a square function estimate similar to (128) with g replaced by u. As in the $p = 2$ case, the square function estimate enabled us to conclude that ∇u had small oscillation on 'lots' of tangent balls in the H^{n-1} measure sense. In retrospect the subharmonicity of $|\nabla g|$ in the ball problem allowed us to use the Poisson integral formula on certain level sets to get a boundary type integral for which estimates could be made in terms of g, ω. To make this method work in the proof of Theorem 4.1, it was first necessary to find a suitable divergence form partial differential equation for which u is a solution and $|\nabla u|^2$ is a subsolution. For a long time we did not think there was any such PDE and that the lack of such a PDE resulted in some rather deep questions involving absolute continuity of μ with respect to H^{n-1} measure. Finally we discovered the PDE in (8), (9). It is easily checked that $|\nabla u|^2$ is a subsolution to (8), (9). Using this discovery, we were able to follow the general outline of the proof in the harmonic case and after some delicate asymptotics eventually get first (127) with g replaced by u and then Theorem 4.1.   □

5.3   Further Uniqueness Results

We note that the derivation of (8), (9) depends heavily on the fact that the p Laplacian is homogeneous in ∇u so does not work for general PDE of p Laplace type. Our earlier investigations before (8), (9) led to [60] where we consider symmetry—uniqueness problems similar to those in Theorem 4.1 for non homogeneous PDE of p Laplace type. In this case $u, |\nabla u|^2$, are not a solution, subsolution, respectively of the same divergence form PDE. Thus we were forced to tackle some rather difficult questions involving absolute continuity of elliptic measure with respect to H^{n-1} measure on $\partial\Omega$. To outline our efforts, for these PDE's we could still prove a square function estimate for u similar to the one in (128). This estimate together with the stronger assumption that

$$c^{-1} r^{n-1} \leq \mu(B(x,r)) \leq c r^{n-1}, x \in \partial\Omega, 0 < r \leq r_0 \qquad (132)$$

(where μ is the measure related to a solution u by way of an integral identity similar to (15)), enabled us to conclude that $\partial\Omega$ is uniformly rectifiable in the sense of [18]. At one time we hoped that uniform rectifiability of $\partial\Omega$ would imply absolute continuity of a certain elliptic measure with respect to H^{n-1} measure on $\partial\Omega$. Eventually however we found an illuminating example in [10]. The example showed that harmonic measure for Laplace's equation need not be absolutely continuous with respect to H^1 measure in a uniformly rectifiable domain. This example appeared to provide a negative end for our efforts.

Later we observed that in order to obtain the desired analogue of (129) it suffices to make absolute continuity type estimates for the above elliptic measure on the boundary of a certain subdomain $\Omega_1 \subset \Omega$, with $\partial\Omega_1$ uniformly rectifiable. Here Ω_1 is obtained by adding to Ω certain balls on which $|\nabla u|$ is 'small'. With this intuition we finally were able to use a rather involved stopping time argument in order to first establish the absolute continuity of our elliptic measure with respect to $H^{n-1}|_{\partial\Omega_1}$ and second get an analogue of (127).

From 2004–2006, Björn Bennewitz was my Ph.D. student. In his thesis he generalized the nonuniqueness results of Andy and I to p harmonic functions when $\Omega \subset \mathbf{R}^2$. More specifically for fixed $p, 1 < p < \infty$, he constructed an $\Omega \neq$ a Henrot–Shahgholian domain with $\partial\Omega$ a quasi circle, $E \subset \Omega$, and for which the corresponding p harmonic u satisfied all the conditions of Theorem 4.1 except (d). His construction makes important use (as in Sect. 2) of the fact that in $\mathbf{R}^2$, $v = \log|\nabla u|$ is a subsolution to (8), (9), when $p \geq 2$ and a supersolution to (8), (9), when $1 < p \leq 2$. Reference [7] is based on his thesis.

5.4   Boundary Regularity of p Harmonic Functions

Next we discuss boundary regularity and corresponding inverse problems for positive p harmonic functions vanishing on a portion of a Lipschitz domain. We first introduce some more or less standard notation for Lipschitz domains. Let $\Omega \subset \mathbf{R}^n$ be a bounded Lipschitz domain with $w \in \partial\Omega$. If $0 < \rho < r_0$ let $\Delta(w, \rho) = \partial\Omega \cap B(w, \rho)$ and for given $b, 0 < b < 1$, $0 < r < r_0$, $x \in \Delta(w, r)$, define the nontangential approach region $\Gamma(x)$ relative to w, r, b by $\Gamma(x) = \Gamma_b(x) = \{y \in \Omega : d(y, \partial\Omega) > b|x - y|\} \cap B(w, 4r)$.

Given a measurable function k on $\cup_{x \in \Delta(w, 2r)} \Gamma(x)$ we define the non tangential maximal function $N(k) : \Delta(w, 2r) \to \mathbf{R}$ for k as

$$N(k)(x) = \sup_{y \in \Gamma(x)} |k|(y) \text{ whenever } x \in \Delta(w, 2r).$$

Let σ be H^{n-1} measure on $\partial\Omega$ and let $L^q(\Delta(w, 2r))$, $1 \leq q \leq \infty$, be the space of functions which are qth power integrable, with respect to σ, on $\Delta(w, 2r)$. Furthermore, given $f : \Delta(w, 2r) \to \mathbf{R}$, we say that f is of bounded mean oscillation on $\Delta(w, r)$, $f \in BMO(\Delta(w, r))$, if there exists A, $0 < A < \infty$, such that

$$\int_{\Delta(x,s)} |f - f_\Delta|^2 d\sigma \leq A^2 \sigma(\Delta(x, s)) \tag{133}$$

whenever $x \in \Delta(w, r)$ and $0 < s \leq r$. Here f_Δ denotes the average of f on $\Delta = \Delta(x, s)$ with respect to σ. The least A for which (133) holds is denoted by $\|f\|_{BMO(\Delta(w,r))}$. Finally we say that f is of vanishing mean oscillation on $\Delta(w, r)$, $f \in VMO(\Delta(w, r))$, provided for each $\epsilon > 0$ there is a $\delta > 0$ such that (133) holds with A replaced by ϵ whenever $0 < s < \min(\delta, r)$ and $x \in \Delta(w, r)$.

Theorem 4.3. *Let $\Omega \subset \mathbf{R}^n$ be a bounded Lipschitz domain with Lipschitz constant M. Given $p, 1 < p < \infty, w \in \partial\Omega, 0 < r < r_0$, suppose that u is a positive p harmonic function in $\Omega \cap B(w, 4r)$ and u is continuous in $B(w, 4r)$ with $u \equiv 0$ on $B(w, 4r) \setminus \Omega$. Then*

$$\lim_{y \in \Gamma(x), y \to x} \nabla u(y) = \nabla u(x)$$

For σ almost every $x \in \Delta(w, 4r)$. Furthermore there exist $q = q(p, n, M) > p$ and a constant $c = c(p, n, M) \geq 1$, such that

(i) $N(|\nabla u|) \in L^q(\Delta(w, 2r))$

(ii) $\displaystyle \int_{\Delta(w,2r)} |\nabla u|^q d\sigma \leq c r^{(n-1)(\frac{p-1-q}{p-1})} \left(\int_{\Delta(w,2r)} |\nabla u|^{p-1} d\sigma \right)^{q/(p-1)}$

(iii) $\log |\nabla u| \in BMO(\Delta(w, r))$, $\| \log |\nabla u| \|_{BMO(\Delta(w,r))} \leq c$.

Next we state

Theorem 4.4. *Let* Ω, M, p, w, r *and* u *be as in the statement of Theorem 4.3. If, in addition,* $\partial \Omega$ *is* C^1 *regular then*

$$\log |\nabla u| \in VMO(\Delta(w, r)).$$

Theorem 4.5. *Let* Ω, M, p, w, r *and* u *be as in the statement of Theorem 4.3. If* $\log |\nabla u| \in VMO(\Delta(w, r))$, *then the outer unit normal to* $\Delta(w, r)$ *is in* $VMO(\Delta(w, r/2))$.

To put these results into historical perspective, we note that for harmonic functions, i.e., $p = 2$, Theorem 4.3 was proved in [17]. Theorem 4.4 for harmonic functions was proved by [37]. This theorem for harmonic functions was generalized to vanishing chord arc domains in [43]. Also a version of Theorem 4.5 in vanishing chord arc domains was proved in [44] An improved version of this theorem in chord arc domains with small constants was proved by these authors in [45].

Currently we are in the process of generalizing Theorems 4.4, 4.5 to the more general setting of chord arc domains.

5.5   Proof of Theorem 4.3

To prove Theorem 4.3 we shall need several lemmas.

Lemma 4.6. *Let* $\Omega \subset \mathbf{R}^n$ *be a bounded Lipschitz domain with constant* M. *Given* $p, 1 < p < \infty, w \in \partial \Omega, 0 < r < r_0$, *suppose* $u > 0$ *is* p *harmonic in* $\Omega \cap B(w, 4r)$ *and continuous in* $B(w, 4r)$ *with* $u = 0$ *on* $B(w, 4r) \setminus \Omega$. *Let* μ *be the measure corresponding to* u *as in (15). There exists* $c = c(p, n, M)$ *such that*

$$\bar{r}^{p-n} \mu(\Delta(w, \bar{r})) \approx u(a_{\bar{r}}(w))^{p-1} \text{ whenever } 0 < \bar{r} \leq r/c.$$

Proof. Lemma 4.6 was essentially proved in [20].   □

Step (ii) in Theorem 4.3 for smooth domains (i.e., the reverse Hölder inequality) follows from Lemma 4.6 and a Rellich inequality for the p Laplacian (as in the case $p = 2$). For example suppose

$$\Omega \cap B(w, 4r) = \{(x', x_n) : x_n > \psi(x')\} \cap B(w, 4r) \text{ where } \psi \in C_0^\infty(\mathbf{R}^{n-1}).$$
(134)

Then from results in [64] and Schauder type theory it follows that u is smooth near $\partial\Omega$ so we can apply the divergence theorem to $|\nabla u|^p \phi e_n$. Here $\phi \in C_0^\infty(B(w, 2r))$ with $\phi \equiv 1$ on $B(w, r)$ and $|\nabla\phi| \leq cr^{-1}$. Using p harmonicity of u - Lipschitzness of $\partial\Omega$, we get

$$r^{1-n} \int_{\partial\Omega \cap B(w,r)} |\nabla u|^p \, d\sigma \leq cr^{-n} \int_{\Omega \cap B(w,2r)} |\nabla u|^p dx.$$
(135)

Also, using Lemmas 1.4 and 4.6, and the fact that $d\mu = |\nabla u|^{p-1} d\sigma$ we find that

$$r^{-n} \int_{\Omega \cap B(w,2r)} |\nabla u|^p \, d\sigma \leq cr^{-p} u(a_r(w))^p$$

$$\leq c^2 \left(r^{1-n} \int_{\partial\Omega \cap B(w,r)} |\nabla u|^{p-1} \, d\sigma \right)^{p/(p-1)}$$
(136)

where constants depend only on p, n, M. Combining (135), (136) we get (ii) in Theorem 4.3 with q replaced by p. However this reverse Hölder inequality has a self improving property so actually implies the higher integrability result in (ii), as follows from a theorem originally due to Gehring. Approximating u by certain p harmonic functions in smooth domains, applying (ii), and taking weak limits it follows that $d\mu = kd\sigma$ where μ is the measure corresponding to u as in (15). Moreover for some $q' > p/(p-1)$

$$\int_{\Delta(w,\bar{r})} k^{q'} \, d\sigma \leq c\bar{r}^{-(n-1)(\frac{p-1-q'}{p-1})} \left(\int_{\Delta(w,\bar{r})} kd\sigma \right)^{1/q'}, \quad 0 < \bar{r} \leq r/c.$$
(137)

Note that we still have to prove $k = |\nabla u|^{p-1}$ on $\partial\Omega$ (σ a. e.).

Lemma 4.7. *Let Ω, M, p, w, r and u be as in Theorem 4.3. Then there exists a starlike Lipschitz domain $\tilde{\Omega} \subset \Omega \cap B(w, 2r)$, with center at a point $\tilde{w} \in \Omega \cap B(w, r)$, $d(\tilde{w}, \partial\Omega) \geq c^{-1}r$, such that*

(a) $c\sigma(\partial\tilde{\Omega} \cap \Delta(w, r)) \geq r^{n-1}$.
(b) $c^{-1}r^{-1}u(\tilde{w}) \leq |\nabla u(x)| \leq cr^{-1}u(\tilde{w})$, for $x \in \tilde{\Omega}$.

Proof. Using Lemma 4.6 and (137) one can show there exists $\tilde{\Omega}$ as in Lemma 4.7 for which (a), (b) hold with $|\nabla u(x)|$ replaced by $u(x)/d(x, \partial\Omega)$. To finish off the proof of Lemma 4.7 we need to prove the fundamental inequality for u. In fact, assuming (134), the following stronger version is available in Ω : There exists $c = c(p, n, M) \geq 1$, and $z \in B(w, r/c)$ such that

$$|\nabla u(z)| \approx u_{x_n}(z), \approx u(z)/d(z, \partial\Omega). \tag{138}$$

(138) is a consequence of a boundary Harnack inequality that was stated in Theorem 3.8. An improved version of Lemma 4.7 will be used in the proof of Theorem 4.5. $\qquad\square$

Next from (138), we see as in (6)–(9) that $u, u_{x_i}, 1 \leq i \leq n$, both satisfy

$$(\alpha)\, L\zeta = \sum_{i,j=1}^{n} \frac{\partial}{\partial x_i}\left(b_{ij}\frac{\partial\zeta}{\partial x_j}\right) \tag{139}$$

$$(\beta)\, b_{ij}(x) = |\nabla u|^{p-4}[(p-2)u_{x_i}u_{x_j} + \delta_{ij}|\nabla u|^2](x)$$

and for $\xi \in \mathbf{R}^n$,

$$c^{-1}|\nabla u|^{p-2}|\xi|^2 \leq \sum_{i,j=1}^{n} b_{ij}(x)\xi_i\xi_j \leq c|\nabla u|^{p-2}|\xi|^2. \tag{140}$$

From (138)–(140) we see once again that $u, u_{x_i}, 1 \leq i \leq n$, *are locally solutions to a uniformly elliptic PDE in divergence form.*
Lemma 4.8. Let $\Omega, \tilde{\Omega}, M, p, w, r, u$, be as in Lemma 4.7. Define, for $y \in \tilde{\Omega}$, the measure

$$d\tilde{\gamma}(y) = d(y, \partial\tilde{\Omega}) \max_{B(y, \frac{1}{2}d(y, \partial\tilde{\Omega}))} \left\{ \sum_{i,j=1}^{n} |\nabla b_{ij}(x)|^2 \right\} dy.$$

If $z \in \partial\tilde{\Omega}$ and $0 < s < r$, then

$$\tilde{\gamma}(\tilde{\Omega} \cap B(z, s)) \leq cs^{n-1}(u(\tilde{w})/r)^{2p-4}.$$

Proof. We get Lemma 4.8 from Lemma 4.7 and integration by parts. $\qquad\square$

$\tilde{\gamma}$ *in Lemma 4.8 is said to be a Carleson measure on* $\tilde{\Omega}$. *To continue the proof of Theorem 4.3, let* $\tilde{\omega}(\tilde{w}, \cdot)$ *be elliptic measure defined with respect to* $L, \tilde{\Omega},$ *and* $\tilde{w}$ *as above.*
Lemma 4.9. Let $u, \tilde{\Omega}, \tilde{w}$, be as in Lemma 4.7 and L as in (139), (140). Then there exist $c \geq 1$ and $\theta, 0 < \theta \leq 1$, such that

$$\frac{\tilde{\omega}(\tilde{w}, E)}{\tilde{\omega}(\tilde{w}, \partial\tilde{\Omega} \cap B(z, s))} \leq c \left(\frac{\sigma(\tilde{w}, E)}{\sigma(\tilde{w}, \partial\tilde{\Omega} \cap B(z, s))}\right)^{\theta}$$

for $z \in \partial\tilde{\Omega}, 0 < s < r$, and $E \subset \partial\tilde{\Omega} \cap B(z, s)$ a Borel set. We say that $\tilde{\omega}$ is an A^{∞} weight with respect to σ, on $\partial\tilde{\Omega}$. Lemma 4.9 is a direct consequence of Lemma 4.8 and a theorem in [42]. □

Next we prove by a contradiction argument that ∇u has non tangential limits for σ almost every $y \in \Delta(w, 4r)$. To begin suppose there exists a set $F \subset \Delta(w, 4r), \sigma(F) > 0$, such that if $y \in F$ then the limit of $\nabla u(z)$, as $z \to y$ with $z \in \Gamma(y)$, does not exist. Let $y \in F$ be a point of density for F with respect to σ. Then $t^{1-n}\sigma(\Delta(y, t) \setminus F) \to 0$ as $t \to 0$, so if $t > 0$ is small enough, then $c\sigma(\partial\tilde{\Omega} \cap \Delta(y, t) \cap F) \geq t^{n-1}$ where $\tilde{\Omega} \subset \Omega$ is the starlike Lipschitz domain defined in Lemma 4.7 with $w, \tilde{w}, r$ replaced by $y, \tilde{y}, t$. From (139), (140), we see that $u_{x_i}, 1 \leq i \leq n$, is a weak solution to $L\zeta = 0$ in $\tilde{\Omega}$.

We now apply a theorem in [14] to deduce that $u_{x_k}, 1 \leq k \leq n$, has nontangential limits in $\tilde{\Omega}$, almost everywhere with respect to $\tilde{\omega}(\cdot, \tilde{w})$. Recall from Lemma 4.9 that $\tilde{\omega}$ and σ are mutually absolutely continuous. Thus these limits also exist almost everywhere with respect to σ. Since nontangential limits in $\tilde{\Omega}$ agree with those in Ω, for σ almost every point in F, we have reached a contradiction. Thus ∇u has nontangential limits almost everywhere in Ω. Step (i) follows from (137), Lemmas 4.6 and 1.3. Finally we use nontangential limits of ∇u, the fact that for small $t, \{u = t\}$ is Lipschitz, the implicit function theorem, as well as (i) of Theorem 4.3, to take limits as $t \to 0$ in order to conclude that $k = |\nabla u|$ in (137). The proof of Theorem 4.3 is now complete. □

5.6   Proof of Theorem 4.4

To prove Theorem 4.4 it suffices by way of an argument of Sarason (see [38]) to show that there exist $0 < \epsilon_0$ and $\tilde{r} = \tilde{r}(\epsilon)$, for $\epsilon \in (0, \epsilon_0)$, such that whenever $y \in \Delta(w, r)$ and $0 < s < \tilde{r}(\epsilon)$ we have

$$\fint_{\Delta(y,s)} |\nabla u|^p d\sigma \leq (1 + \epsilon) \left(\fint_{\Delta(y,s)} |\nabla u|^{p-1} d\sigma\right)^{p/(p-1)}. \tag{141}$$

Here $\fint_E f d\sigma$ denotes the average of f on E with respect to σ. The proof of (141) is by contradiction. Otherwise there exist two sequences $\{y_m\}_1^{\infty}, \{s_m\}_1^{\infty}$ satisfying $y_m \in \Delta(w, r)$ and $s_m \to 0$ as $m \to \infty$ such that (141) is false with y, s replaced by y_m, s_m for $m = 1, 2, \ldots$. Let $A = e^{1/\epsilon}$ and put $y'_m = y_m + As_m n(y_m)$, where $n(y_m)$ is the inner unit normal to

Ω at y_m. Since $\partial\Omega$ is C^1 we see for $\epsilon > 0$ small and $m = m(\epsilon)$ large that if $\Omega(y'_m)$ is constructed by drawing all line segments from points in $B(y'_m, As_m/4)$ to points in $\Delta(y_m, As_m)$, then $\Omega(y'_m)$ is starlike Lipschitz with respect to y'_m. Let $D_m = \Omega(y'_m) \setminus \bar{B}(y'_m, As_m/8)$. and let u_m be the p harmonic function in D_m that is continuous in $\mathbf{R}^n$ with $u_m \equiv 1$ on $B(y'_m, As_m/8)$ and $u \equiv 0$ on $\mathbf{R}^n \setminus \Omega(y'_m)$. From the boundary Harnack inequality in Theorem 3.8 with w, r, u, v replaced by $y_m, As_m/100, u, u_m$ we deduce that if $w_1, w_2 \in \Omega \cap B(y_m, 2s_m)$ then

$$\left| \log\left(\frac{u_m(w_1)}{u(w_1)} \right) - \log\left(\frac{u_m(w_2)}{u(w_2)} \right) \right| \le cA^{-\alpha}. \tag{142}$$

Letting $w_1, w_2 \to z_1, z_2 \in \Delta(y_m, 2s_m)$ in (142) and using Theorem 4.3, we get, σ almost everywhere, that

$$\left| \log\left(\frac{|\nabla u_m(z_1)|}{|\nabla u(z_1)|} \right) - \log\left(\frac{|\nabla u_m(z_2)|}{|\nabla u(z_2)|} \right) \right| \le cA^{-\alpha}$$

or equivalently that

$$(1 - \tilde{c}A^{-\alpha}) \frac{|\nabla u_m(z_1)|}{|\nabla u_m(z_2)|} \le \frac{|\nabla u(z_1)|}{|\nabla u(z_2)|} \le (1 + \tilde{c}A^{-\alpha}) \frac{|\nabla u_m(z_1)|}{|\nabla u_m(z_2)|}. \tag{143}$$

Using (143) and the fact that (141) is false we obtain

$$\frac{\displaystyle\fint_{\Delta(y_m, s_m)} |\nabla u_m|^p d\sigma}{\left(\displaystyle\fint_{\Delta(y_m, s_m)} |\nabla u_m|^{p-1} d\sigma \right)^{p/(p-1)}} \tag{144}$$

$$\ge (1 - cA^{-\alpha}) \frac{\displaystyle\fint_{\Delta(y_m, s_m)} |\nabla u|^p d\sigma}{\left(\displaystyle\fint_{\Delta(y_m, s_m)} |\nabla u|^{p-1} d\sigma \right)^{p/(p-1)}} \ge (1 - cA^{-\alpha})(1 + \epsilon) \, .$$

Let T_m be a conformal affine mapping of $\mathbf{R}^n$ which maps the origin and e_n onto y_m and y'_m respectively and which maps $W = \{x \in \mathbf{R}^n : x_n = 0\}$ onto the tangent plane to $\partial\Omega$ at y_m. Let D'_m, u'_m be such that $T_m(D'_m) = D_m$ and $u_m(T_m x) = u'_m(x)$ whenever $x \in D'_m$. Since the p Laplace equation is invariant under translations, rotations, and dilations, we see that u'_m is p harmonic in D'_m. Letting $m \to \infty$ one can show that u'_m converges uniformly on $\mathbf{R}^n$ to u' where u' is continuous on $\mathbf{R}^n$ and p harmonic in $D' = \Omega' \setminus B(e_n, 1/8)$ with $u \equiv 1$ on $B(e_n, 1/8)$ and $u \equiv 0$ on $\mathbf{R}^n \setminus \Omega'$. Here Ω' is obtained by drawing all line segments connecting points in $B(0,1) \cap W$ to points in $B(e_n, 1/4)$. Changing variables in (144) and using Rellich type inequalities one gets

$$(1 - cA^{-\alpha})(1 + \epsilon) \leq \limsup_{m \to \infty} \frac{\fint_{\partial D'_m \cap B(0,1/A)} |\nabla u'_m|^p d\sigma}{\left(\fint_{\partial D'_m \cap B(0,1/A)} |\nabla u'_m|^{p-1} d\sigma \right)^{p/(p-1)}}$$

$$\leq \frac{\fint_{W \cap B(0,1/A)} |\nabla u'|^p d\sigma}{\left(\fint_{W \cap B(0,1/A)} |\nabla u'|^{p-1} d\sigma \right)^{p/(p-1)}}$$

(145)

Finally from interior estimates for p harmonic functions and Schwarz reflection one finds for $z \in B(0, 1/A)$ that

$$(1 - cA^{-\theta})|\nabla u'(0)| \leq |\nabla u'(z)| \leq (1 + cA^{-\theta})|\nabla u'(0)|$$

which in view of (145) yields

$$(1 + cA^{-\theta}) \geq \frac{\fint_{W \cap B(0,1/A)} |\nabla u'|^p dx'}{\left(\fint_{W \cap B(0,1/A)} |\nabla u'|^{p-1} dx' \right)^{p/(p-1)}} \geq (1 - cA^{-\alpha})(1 + \epsilon).$$

Clearly this inequality cannot hold for ϵ small since $A = e^{1/\epsilon}$. The proof of Theorem 4.4 is now complete.

5.7 Proof of Theorem 4.5

In this section we prove Theorem 4.5. We shall need the following refined version of Lemma 4.7.

Lemma 4.10. Given Ω, w, p, n, M, u as in Theorem 4.3. If $\log |\nabla u| \in VMO(\Delta(w, r))$, then for each $\epsilon > 0$ there exists, $0 < \tilde{r} = \tilde{r}(\epsilon) < r$ and $c = c(p, n, M), 1 \leq c < \infty$, such that if $0 < r' \leq \tilde{r}$ then the following :

There is a starlike Lipschitz domain $\tilde{\Omega} \subset \Omega \cap B(w, r')$, with center at a point $\hat{w} \in \Omega \cap B(w, r')$, $d(\hat{w}, \partial\Omega) \geq r'/c$, and Lipschitz constant $\leq c$, satisfying

(a)
$$\frac{\sigma(\partial\tilde{\Omega} \cap \Delta(w, r'))}{\sigma(\Delta(w, r'))} \geq 1 - \epsilon.$$

(b)
$$(1 - \epsilon)b^{p-1} \leq \frac{\mu(\Delta(y, s))}{\sigma(\Delta(y, s))} \leq (1 + \epsilon)b^{p-1}$$

where $0 < s \leq r', y \in \partial\tilde{\Omega} \cap \Delta(w, r')$, and $\log b$ is the average of $\log |\nabla u|$ on $\Delta(w, 4r')$. Moreover,

(c)
$$c^{-1}\frac{u(\hat{w})}{r'} \leq |\nabla u(x)| \leq c\frac{u(\hat{w})}{r'} \text{ for all } x \in \tilde{\Omega}.$$

Proof. Lemma 4.10 is proved in [51] as Lemma 4.1.   □

To begin the proof of Theorem 4.5 let n denote the outer unit normal to $\partial\Omega$ and put

$$\eta = \lim_{\tilde{r} \to 0} \sup_{\tilde{w} \in \Delta(w, r/2)} \|n\|_{BMO(\Delta(\tilde{w}, \tilde{r}))}. \tag{146}$$

To prove Theorem 4.5 it is enough to prove that $\eta = 0$. To do this we argue by contradiction and assume that (146) holds for some $\eta > 0$. This assumption implies that there exist a sequence of points $\{w_j\}$, $w_j \in \Delta(w, r/2)$, and a sequence of scales $\{r_j\}$, $r_j \to 0$, such that $\|n\|_{BMO(\Delta(w_j, r_j))} \to \eta$ as $j \to \infty$. To get a contradiction we use a blow-up argument. In particular, let u be as in the statement of Theorem 4.5 and extend u to $B(w, 4r)$ by putting $u = 0$ in $B(w, 4r) \setminus \Omega$. For $\{w_j\}$, $\{r_j\}$ as above we define $\Omega_j = \{r_j^{-1}(x - w_j) : x \in \Omega\}$ and

$$u_j(z) = \lambda_j u(w_j + r_j z) \text{ whenever } z \in \Omega_j.$$

Using properties of Lipschitz domains, one can show that subsequences of $\{\Omega_j\}, \{\partial\Omega_j\}$ converge to $\Omega_\infty, \partial\Omega_\infty$, in the Hausdorff distance sense, where Ω_∞ is an unbounded Lipschitz domain with Lipschitz constant bounded by M. Moreover, from (15), Lemmas 1.2–1.5, and 4.10 we deduce for an appropriate choice of (λ_j), that a subsequence of (u_j) converges uniformly on compact subsets of $\mathbf{R}^n$ to u_∞, a positive p harmonic function in Ω_∞ vanishing continuously on $\partial\Omega_j$. If $d\mu_j = |\nabla u_j|^{p-1}d\sigma|_{\partial\Omega_j}$, it also follows that a subsequence of (μ_j) converges weakly as Radon measures to μ_∞ where

$$\int_{\mathbf{R}^n} |\nabla u_\infty|^{p-2}\langle\nabla u_\infty, \nabla\phi\rangle dx = -\int_{\partial\Omega_\infty} \phi d\mu_\infty \tag{147}$$

whenever $\phi \in C_0^\infty(\mathbf{R}^{n-1})$ Moreover, using Lemma 4.10, one can show that μ_∞ and u_∞, satisfy,

$$(a)\, \mu_\infty = \sigma \text{ on } \partial\Omega_\infty,$$

$$(b)\, c^{-1} \le |\nabla u_\infty(z)| \le 1 \text{ whenever } z \in \Omega_\infty. \tag{148}$$

Finally one shows that (147), (148) imply

$$\Omega_\infty \text{ is a halfspace} \tag{149}$$

which in turn implies that $\eta = 0$, a contradiction to (146). If M is sufficiently small, then (149) follows directly from a theorem in [1]. For large M we needed to use our generalization in [53] of the work in [11].

We discuss this work further in the next section.

5.8 Regularity in a Lipschitz Free Boundary Problem

We begin our discussion of two phase free boundary problems for p harmonic functions with some notation. Let $D \subset \mathbf{R}^n$ be a bounded domain and suppose that u is continuous on D. Put

$$D^+(u) = \{x \in D : u(x) > 0\}$$

$$F(u) = \partial D^+(u) \cap D$$

$$D^-(u) = D \setminus [D^+(u) \cup F(u)].$$

$F(u)$ is called the free boundary of u in D. Let $G > 0$ be an increasing function on $[0, \infty)$ and suppose for some $N > 0$ that $s^{-N}G(s)$ is decreasing when $s \in (0, \infty)$. Let $u^+ = \max(u, 0)$ and $u^- = -\min(u, 0)$.

Definition F. *We say that u satisfies weakly the two sided boundary condition $|\nabla u^+| = G(|\nabla u^-|)$ on $F(u)$ provided the following holds. Assume that $w \in F(u)$ and there is a ball $B(\hat{w}, \hat{\rho}) = \{y : |y - \hat{w}| < \hat{\rho}\} \subset D^+(u) \cup D^-(u)$ with $w \in \partial B(\hat{w}, \hat{\rho})$. If $\nu = (\hat{w} - w)/|\hat{w} - w|$ and $B(\hat{w}, \hat{\rho}) \subset D^+(u)$, then*

$$(*)\ \ u(x) = \alpha\langle x - w, \nu\rangle^+ - \beta\langle x - w, \nu\rangle^- + o(|x - w|),$$

as $x \to w$ non-tangentially while if $B(\hat{w}, \hat{\rho}) \subset D^-(u)$, then $(*)$ holds with $x - w$ replaced by $w - x$, where, $\alpha, \beta \in [0, \infty]$ and $\alpha = G(\beta)$.

We note that if If $F(u), u$ are sufficiently smooth in D, then at $\hat{w}$,

$$\alpha = |\nabla u^+| = G(\beta) = G(|\nabla u^-|).$$

In [51] we prove,

Theorem 4.11. Let u be continuous in D, p harmonic in $D \setminus F(u), 1 < p < \infty, p \neq 2$, and a weak solution to $|\nabla u^+| = G(|\nabla u^-|)$ on $F(u)$. Suppose $B(0,2) \subset D, 0 \in F(u)$, and $F(u)$ coincides in B $(0, 2)$ with the graph of a Lipschitz function. Then $F(u) \cap B(0,1)$ is $C^{1,\sigma}$ where σ depends only on p, n, N and the Lipschitz constant for the graph function.

5.9   History of Theorem 4.11

For $p = 2$, i.e., harmonic functions, Caffarelli developed a theory for general two-phase free boundary problems in [11–13]. In [11] Lipschitz free boundaries were shown to be $C^{1,\sigma}$-smooth for some $\sigma \in (0,1)$ and in [13] it was shown that free boundaries which are well approximated by Lipschitz graphs are in fact Lipschitz. Finally, in [12] the existence part of the theory was developed. We also note that the work in [11] was generalized in [70] to solutions of fully nonlinear PDEs of the form $F(\nabla^2 u) = 0$, where F is homogeneous. Further analogues of [11] were obtained for a class of nonisotropic operators and for fully nonlinear PDE's of the form $F(\nabla^2 u, \nabla u) = 0$, where F is homogeneous in both arguments, in [24, 25]. Extension of the results in [11] were made to non-divergence form linear PDE with variable coefficients in [16], and generalized in [26] to fully nonlinear PDE's of the form $F(\nabla^2 u, x) = 0$. Finally generalizations of the work in [11] (also [13]) to linear divergence form PDE's with variable coefficients were obtained in [27, 28].

5.10   Proof of Theorem 4.11

To outline the proof of Theorem 4.11 we need another definition.

Definition G. *We say that a real valued function h is monotone on an open set O in the direction of $\tau \in \mathbf{R}^n$, provided $h(x - \tau) \leq h(x)$ whenever $x \in O$. If $x, y \in \mathbf{R}^n$, let $\theta(x, y)$ be the angle between x and y. Given $\theta_0, 0 < \theta_0 < \pi, \epsilon_0 > o$, and ν with $|\nu| = 1$, put*

$$\Gamma(\nu, \theta_0, \epsilon_0) = \{y \in \mathbf{R}^n : |\theta(y, \nu)| < \theta_0, 0 < |y| < \epsilon_0\}.$$

We note from elementary geometry that if h is monotone on O with respect to the directions in $\Gamma(\nu, \theta_0, \epsilon_0)$, then

$$(2) \qquad \sup_{B(x-\tau, |\tau|\sin(\theta_0/2))} h \leq h(x)$$

whenever $x \in O$ and $\tau \in \Gamma(\nu, \theta_0/2, \epsilon_0/2)$. To establish Theorem 4.11 we first show the existence of a cone of monotonicity. To this end, we assume as we may, that

$$\Omega \cap B(0,2) = \{(x', x_n) : x_n > \psi(x')\} \cap B(w, 4r), \ \psi \text{ Lipschitz on } \mathbf{R}^{n-1}.$$
$$(150)$$

If M is the Lipschitz norm of ψ, then as in (138) we see that Theorem 3.8 and (150) imply there exists $c = c(p, n, M) \geq 1$, such that whenever $z \in B(0, r_1), r_1 = 1/c,$

$$|\nabla u(z)| \approx u_{x_n}(z), \approx u(z)/d(z, \partial\Omega). \tag{151}$$

Clearly (151) implies the existence of $\theta_0 \in (0, \pi/2], \epsilon_0 > 0$, and $c > 1$, depending only on p, n, M, such that

$$\begin{array}{c} u \text{ is monotone in } B(0, r_1) \text{ with respect to} \\ \text{the directions in the cone } \Gamma(e_n, \theta_0, \epsilon_0). \end{array} \tag{152}$$

5.11   Enlargement of the Cone of Monotonicity in the Interior

Let $\tau \in \Gamma(e_n, \theta_0/2, \epsilon_0)$ for θ_0, ϵ_0, as in (152), put $\epsilon = |\tau| \sin(\theta_0/2)$ and set

$$v_\epsilon(x) = v_{\epsilon,\tau}(x) = \sup_{y \in B(x,\epsilon)} u(y - \tau)$$

We note from (152) that $v_\epsilon(x) \leq u(x)$, when $x \in B(0, r_1)$. Next we show that if $\rho, \mu > 0$ are small enough, depending only on p, n, and the Lipschitz constant for $\partial\Omega \cap B(0, 1)$, and $\nu = \nabla u(\frac{r_1 e_n}{8}))/|\nabla u(\frac{r_1 e_n}{8})|$, then

$$v_{(1+\mu\lambda)\epsilon}(x) \leq (1 - \mu\lambda\epsilon)u(x) \text{ whenever } x \in B(\tfrac{r_1 e_n}{8}, \rho r_1). \tag{153}$$

where $\lambda = \cos(\theta_0/2 + \theta(\nu, \tau))$, and $0 < |\tau| \leq \epsilon_0 \rho r_1$. The proof of (153) is essentially the same as in [11], thanks to basic interior estimates for p harmonic functions, Theorem 3.8, and (152).

5.12   Enlargement of the Cone of Monotonicity at the Free Boundary

In this part of the proof we show there exists $\bar{\mu} > 0$, depending only on p, n, M, such that if τ, ϵ are as defined above (153), then

$$v_{(1+\bar{\mu}\lambda)\epsilon}(x) \leq u(x) \text{ whenever } x \in B(0, r_1/100), \qquad (154)$$

It is shown in [11] that (154) implies the existence of $\omega, |\omega| \doteq 1, \bar{\theta} \in (0, \pi/2]$, $c_-, c_+ > 1$, such that

$$u \text{ is monotone in } \Gamma(\omega, \bar{\theta}, \epsilon_0/c_+) \qquad (155)$$

where $\pi/2 - \bar{\theta} = c_-^{-1}(\pi/2 - \theta_0)$, $\Gamma(e_n, \theta_0, \epsilon_0) \subset \Gamma(\omega, \bar{\theta}, \epsilon_0/c_+)$, and all constants depend only on p, n, M.

·Using (155) as well as invariance of the p Laplace equation under scalings and translations, we can replace $u(x)$ by $u(x_0 + \eta x)/\eta$ and then repeat our argument in (153), (154), in order to eventually conclude the $C^{1,\sigma}$-smoothness of $F(u) \cap B(0, 1/2)$. Hence to prove Theorem 4.11 we have to prove (154). To do this, given an $n \times n$ symmetric matrix M let

$$P(M) = \inf_{A \in A_p} \sum_{i,j=1}^{n} a_{ij} M_{ij}.$$

where A_p denotes the set of all symmetric $n \times n$ matrices $A = \{a_{ij}\}$ which satisfy

$$\min\{p - 1, 1\} |\xi|^2 \leq \sum_{i,j=1}^{n} a_{ij} \xi_i \xi_j \leq \max\{p - 1, 1\} |\xi|^2$$

whenever $\xi \in \mathbf{R}^n \setminus \{0\}$. Next we state

Lemma 4.12. *Let $\phi > 0$ be in $C^2(D), \|\nabla\phi\|_{L^\infty(D)} \leq 1/2$, p fixed, $1 < p < \infty$, and suppose that*

$$\phi(x)P(\nabla^2\phi(x)) \geq 50pn |\nabla\phi(x)|^2$$

whenever $x \in D$. Let u be continuous in an open set O containing the closure of $\bigcup_{x \in D} B(x, \phi(x))$ and define

$$v(x) \leftarrow \max_{\bar{B}(x,\phi(x))} u$$

whenever $x \in D$. If u is p-harmonic in $O \setminus \{u = 0\}$, then v is continuous and a p-subsolution in $\{v \neq 0\} \cap G$ whenever G is an open set with $\bar{G} \subset D$.

Theorem 4.13. *Let $\Omega \subset \mathbf{R}^n$ be a Lipschitz graph domain with Lipschitz constant M. Given $p, 1 < p < \infty, w \in \partial\Omega, r > 0$, suppose that $\hat{u}$ and $\hat{v}$ are non-negative p-harmonic functions in $\Omega \cap B(w, 2r)$ with $\hat{v} \leq \hat{u}$. Assume also that $\hat{u}, \hat{v}$, are continuous in $B(w, 2r)$ with $\hat{u} \equiv 0 \equiv \hat{v}$ on $B(w, 2r) \setminus \Omega$. There exists $c \geq 1$, depending only on p, n, M, such that if $y, z \in \Omega \cap B(w, r/c)$, then*

$$\frac{\hat{u}(y) - \hat{v}(y)}{\hat{v}(y)} \leq c \frac{\hat{u}(z) - \hat{v}(z)}{\hat{v}(z)}.$$

Remark. Lemma 4.12 is not much more difficult than the corresponding lemma in [11], thanks to translation, dilation and rotational invariance of the p Laplacian. Theorem 4.13 uses the full toolbox developed in [49–52] and [62]. For $p = 2$ Theorem 4.13 is equivalent to a boundary Harnack inequality for harmonic functions while for $p \neq 2$, it is stronger than the boundary Harnack inequality in Theorem 3.8. That is, Theorem 4.13 implies not only boundedness of $\hat{u}/\hat{v}$, but also Hölder continuity of the ratio in $\Omega \cap B(w, r/c^*)$, for some $c^* - c^*(p, n, M) \geq 1$. Theorem 4.13 is our main contribution to the proof of Theorem 4.11. To prove (154), using Lemma 4.12 and Theorem 4.13, let

$$\tilde{v}_t(x) := \sup_{y \in B(x, \epsilon \phi_{\mu \lambda t}(x))} u(y - \tau) \text{ for } t \in [0, 1]$$

and $x \in B(0, r_1) \setminus B(r_1 e_n/8, \rho r_1)$. Here $\{\phi_t\}$, is a family of C^2 functions, each satisfying the hypotheses of Lemma 4.12 in $B(0, r_1) \setminus B(r_1 e_n/8, \rho r_1)$. Moreover,

(a) $\phi_t \equiv 1$ on $B(0, r_1) \setminus B(0, r_1/2)$
(b) $1 \leq \phi_t \leq 1 + t\gamma$ in $B(0, r_1) \setminus B(r_1 e_n/8, \rho r_1)$
(c) $\phi_t \geq 1 + ht\gamma$ in $B(0, r_1/100)$
(d) $|\nabla \phi_t| \leq \gamma t$.

In (c), $h > 0$ depends on ρ, p, n while $\gamma > 0$ is a parameter to be chosen sufficiently small. From (c) one sees that (154) holds if

$$\tilde{v}_t \leq u \text{ in } B(0, r_1) \setminus B(r_1 e_n/8, \rho r_1) \text{ whenever } t \in [0, 1]. \tag{156}$$

From Step 1 and (b) above, this inequality holds when $t = 0$. One can now use a method of continuity type argument to show that if (156) is false then there exist $t \in [0, 1]$ for which $\tilde{v}_t \leq u$, $w \in F(u) \cap \{\tilde{v}_t = 0\}$, and $\hat{w}, \hat{\rho} > 0$ with

$$B(\hat{w}, \hat{\rho}) \subset \{\tilde{v}_t > 0\} \subset D^+(u) \text{ and } w \in \partial B(\hat{w}, \hat{\rho}).$$

One then uses in this tangent ball, the asymptotic free boundary condition for u, similar asymptotics for $\tilde{v}_t$, (153), Theorem 4.13, and a Hopf maximum principle type argument to get a contradiction. Thus (156) and so (154) are valid. This completes our outline of the proof of Theorem 4.11. □

5.13 An Application of Theorem 4.11

Recall that the proof of Theorem 4.5 was by contradiction. Indeed assuming that

$$\eta = \lim_{\tilde{r} \to 0} \sup_{\tilde{w} \in \Delta(w, r/2)} \|\nu\|_{BMO(\Delta(\tilde{w}, \tilde{r}))} \neq 0 \tag{157}$$

we obtained u_∞, a positive p-harmonic function in a Lipschitz graph domain, Ω_∞, which is Hölder continuous in $\mathbf{R}^n$ with $u_\infty \equiv 0$ on $\mathbf{R}^n \setminus \Omega_\infty$. We also had

$$\int_{\mathbf{R}^n} |\nabla u_\infty|^{p-2} \langle \nabla u_\infty, \nabla \psi \rangle dx = -\int_{\partial \Omega_\infty} \psi d\sigma_\infty \qquad (158)$$

whenever $\psi \in C_0^\infty(\mathbf{R}^n)$ and

$$c^{-1} \le |\nabla u_\infty(z)| \le 1 \text{ whenever } z \in \Omega_\infty. \qquad (159)$$

where σ_∞ is surface area on $\partial \Omega_\infty$. To get a contradiction to (157) we needed to show that (158), (159) imply Ω_∞ is a halfspace. This conclusion follows easily from applying the argument in Theorem 4.11 to $u_\infty(Rx)/R$ and letting $R \to \infty$, once it is shown that u_∞ is a weak solution to the free boundary problem in Theorem 4.11 with $G(s) = 1 + s$ for $s \in (0, \infty)$. That is, u is a weak solution to the 'one phase free boundary problem'

$$|\nabla u^+| \equiv 1 \text{ and } |\nabla u^-| \equiv 0 \text{ on } F(u). \qquad (160)$$

To prove (160) assume $w \in F(u_\infty)$ and that there exists a ball $B(\hat{w}, \hat{\rho})$, $\hat{w} \in \mathbf{R}^n \setminus \partial \Omega_\infty$ and $\hat{\rho} > 0$, such that $w \in \partial B(\hat{w}, \hat{\rho})$. Let P be the plane through w with normal $\nu = (\hat{w} - w)/|\hat{w} - w|$. We claim that P is a tangent plane to Ω_∞ at w in the usual sense. That is given $\epsilon > 0$ there exists $\hat{r}(\epsilon) > 0$ such that

$$\Psi(P \cap B(w, r), \partial \Omega_\infty \cap B(w, r)) \le \epsilon r \qquad (161)$$

whenever $0 < r \le \hat{r}(\epsilon)$. Once (161) is proved we can show that

(i)   If $B(\hat{w}, \hat{\rho}) \subset \Omega_\infty$ then $u_\infty^+(x) = \langle x - w, \nu \rangle + o(|x - w|)$ in Ω_∞

(ii)   If $B(\hat{w}, \hat{\rho}) \subset \mathbf{R}^n \setminus \Omega_\infty$ then $u_\infty^+(x) = \langle w - x, \nu \rangle + o(|x - w|)$ in Ω_∞.

$$\qquad (162)$$

To prove (162) (given (161)) we assume that $w = 0, \nu = e_n$, and $\hat{\rho} = 1$. This assumption is permissible since linear functions and the p-Laplacian are invariant under rotations, translations, and dilations. Then $\hat{w} = e_n$ and either $B(e_n, 1) \subset \Omega_\infty$ or $B(e_n, 1) \subset \mathbf{R}^n \setminus \bar{\Omega}_\infty$. We assume that $B(e_n, 1) \subset \Omega_\infty$, since the other possibility, $B(e_n, 1) \subset \mathbf{R}^n \setminus \bar{\Omega}_\infty$, is handled similarly. Let $\{r_j\}$ be a sequence of positive numbers tending to 0 and let $\hat{u}_j(z) = u_\infty(r_j z)/r_j$ whenever $z \in \mathbf{R}^n$. Let $\hat{\Omega}_j = \{z : r_j z \in \Omega_\infty\}$ be the corresponding blow-up regions. Then $\hat{u}_j$ is p-harmonic in $\hat{\Omega}_j$ and Hölder continuous in $\mathbf{R}^n$ with $\hat{u}_j \equiv 0$ on $\mathbf{R}^n \setminus \hat{\Omega}_j$. Moreover, (159) is valid for each j with u_∞ replaced by $\hat{u}_j$. Using these facts, assumption (161), and Lemmas 1.2–1.5 we see that a subsequence of $\{\hat{u}_j\}$, denoted $\{u'_j\}$, converges uniformly on compact subsets of $\mathbf{R}^n$, as $j \to \infty$, to a Hölder continuous function u'_∞. Moreover, u'_∞ is a nonnegative p-harmonic function in $H = \{x : x_n > 0\}$ with $u'_\infty \equiv 0$ on $\mathbf{R}^n \setminus H$. Let $\{\Omega'_j\}$ be the subsequence of $\{\hat{\Omega}_j\}$ corresponding to $\{u'_j\}$. From (161) we see that $\Omega'_j \cap B(0, R)$ converges to $H \cap B(0, R)$ whenever $R > 0$, in the sense of Hausdorff distance as $j \to \infty$. Finally we note that $\nabla u'_j \to \nabla u'_\infty$

uniformly on compact subsets of H and hence

$$c^{-1} \le |\nabla u'_\infty| \le 1 \tag{163}$$

where c is the constant in (159). Next we apply the boundary Harnack inequality in Theorem 4.1 with

$$\Omega = H, \hat{u}(x) = u'_\infty(x), \text{ and } \hat{v}(x) = x_n.$$

Letting $r \to \infty$ in Theorem 4.1, it follows that

$$u'_\infty(x) = lx_n \tag{164}$$

for some nonnegative l. From (163) and the above discussion we conclude that

$$c^{-1} \le l \le 1. \tag{165}$$

Next using (158) we see that if σ'_j is surface area on Ω'_j, σ surface area on H, and $\phi \ge 0 \in C_0^\infty(\mathbf{R}^n)$, then

$$\int_{\partial\{u'_j>0\}} \phi d\sigma'_j = - \int_{\mathbf{R}^n} |\nabla u'_j|^{p-2} \langle \nabla u'_j, \nabla\phi \rangle dx$$

$$\to - \int_{\mathbf{R}^n} |\nabla u'_\infty|^{p-2} \langle \nabla u'_\infty, \nabla\phi \rangle dx = l^{p-1} \int_{\{x_n=0\}} \phi d\sigma \tag{166}$$

as $j \to \infty$. Moreover, using the divergence theorem we find that

$$\int_{\partial\{u'_j>0\}} \phi d\sigma'_j \ge - \int_{\{u'_j>0\}} \nabla \cdot (\phi e_n) dx \to - \int_{\{u'_\infty>0\}} \nabla \cdot (\phi e_n) dx = \int_{\{x_n=0\}} \phi d\sigma \tag{167}$$

as $j \to \infty$. Combining (166), (167) we obtain first that $l \ge 1$ and thereupon from (165) that $l = 1$. Thus any blowup sequence of u_∞, relative to zero, tends to x_n^+ uniformly on compact subsets of $\mathbf{R}^n$, and the corresponding gradients tend uniformly to e_n on compact subsets of H. This conclusion is easily seen to imply (162). Hence (161) implies (162).

5.14 Proof of (161)

The proof of (161) is again by contradiction. We continue under the assumption that $w = 0, \nu = \hat{w} = e_n$, and $\hat\rho = 1$. First suppose that

$$B(e_n, 1) \subset \Omega_\infty. \tag{168}$$

If (161) is false, then there exists a sequence $\{s_m\}$ of positive numbers and $\delta > 0$ with $\lim\limits_{m \to \infty} s_m = 0$ and the property that

$$\Omega_\infty \cap \partial B(0, s_m) \cap \{x : x_n \le -\delta s_m\} \ne \emptyset \tag{169}$$

for each m. To get a contradiction we show that (169) leads to

$$\limsup_{t \to 0} t^{-1} u_\infty(te_n) = \infty \tag{170}$$

which in view of the mean value theorem from elementary calculus, contradicts (159). For this purpose let f be the p-harmonic function in $B(e_n, 1) \setminus \bar{B}(e_n, 1/2)$ with continuous boundary values,

$$f \equiv 0 \text{ on } \partial B(e_n, 1) \text{ and } f \equiv \min_{\bar{B}(e_n, 1/2)} u_\infty \text{ on } \partial B(e_n, 1/2).$$

Recall that n f can be written explicitly in the form,

$$f(x) = \begin{cases} A|x - e_n|^{(p-n)/(p-1)} + B & \text{when } p \ne n, \\ -A \log |x - e_n| + B & \text{when } p = n, \end{cases}$$

where A, B are constants. Doing this we see that

$$\lim_{t \to 0} t^{-1} f(te_n) > 0. \tag{171}$$

From the maximum principle for p-harmonic functions we also have

$$u_\infty \ge f \text{ in } B(e_n, 1) \setminus \bar{B}(e_n, 1/2). \tag{172}$$

Next we show that if $0 < s < 1/4$, and $u_\infty \ge kf$ in $\bar{B}(0, s) \cap B(e_n, 1)$, for some $k \ge 1$, then there exists $\xi = \xi(p, n, M, \delta) > 0$ and $s', 0 < s' < s/2$, such that

$$u_\infty \ge (1 + \xi)kf \text{ in } \bar{B}(0, s') \cap B(e_n, 1). \tag{173}$$

Clearly (171)–(173) and an iterative argument yield (170). To prove (173) we observe from a direct calculation that

$$|\nabla f(x)| \approx f(x)/(1 - |x - e_n|) \text{ when } x \in B(e_n, 1) \setminus \bar{B}(e_n, 1/2), \tag{174}$$

where proportionality constants depend only on p, n. Also, we observe from (169) and Lipschitzness of $\partial \Omega_\infty$ that if m_0 is large enough, then there exists a sequence of points $\{t_l\}_{m_0}^\infty$ in $\Omega_\infty \cap \{x : x_n = 0\}$ and $\eta = \eta(p, n, M, \delta) > 0$ such that for $l \ge m_0$,

$$\eta s_l \leq |t_l| \leq \eta^{-1} s_l \text{ and } d(t_l, \partial \Omega_\infty) \geq \eta |t_l|. \tag{175}$$

Choose $t_m \in \{t_l\}_{m_0}^\infty$ such that $\eta^{-1}|t_m| \leq s/100$. If $\rho = d(t_m, \partial \Omega_\infty)$, then from (175), and Lemmas 1.2–1.5 for u_∞ we deduce for some $C = C(p, n, M, \delta) \geq 1$ that

$$C u_\infty(t_m) \geq \max_{\bar{B}(0, 4|t_m|)} u_\infty. \tag{176}$$

From (176), the assumption that $kf \leq u_\infty$, Lemmas 1.2–1.5 for kf, and the fact that t_m lies in the tangent plane to $B(e_n, 1)$ through 0, we see there exists $\lambda = \lambda(p, n, M, \delta), 0 < \lambda \leq 10^{-2}$, and $m_1 \geq m_0$ such that if $m \geq m_1$ and $t'_m = t_m + 3\lambda \rho e_n$, then

$$B(t'_m, 2\rho\lambda) \subset B(e_n, 1) \text{ and } (1 + \lambda)kf \leq u_\infty \text{ on } \bar{B}(t'_m, \rho\lambda). \tag{177}$$

Let $\tilde{f}$ be the p-harmonic function in $G = B(0, 4|t_m|) \cap B(e_n, 1) \setminus \bar{B}(t'_m, \rho\lambda)$ with continuous boundary values $\tilde{f} = kf$ on $\partial[B(e_n, 1) \cap B(0, 4|t_m|)]$ while $\tilde{f} = (1 + \lambda)kf$ on $\partial B(t'_m, \rho\lambda)$. Using (174), Theorem 1.13, and Harnack's inequality for $\tilde{f} - kf, kf$ as in the proof of (154), we deduce the existence of $\tau > 0, \bar{c} \geq 1$, with

$$(1 + \tau\lambda)kf \leq \tilde{f} \tag{178}$$

in $B(e_n, 1) \cap \bar{B}(0, |t_m|/\bar{c})$ where $\tau = \tau(p, n, M, \delta), 0 < \tau < 1$, and $\bar{c} = \bar{c}(p, n, M) \geq 1$. Moreover, using the maximum principle for p-harmonic functions we see from (177) that

$$\tilde{f} \leq u_\infty \text{ in } G. \tag{179}$$

Combining (178), (179), we get (173) with $\xi = \tau\lambda$ and $s' = |t_m|/\bar{c}$. As mentioned earlier, (173) leads to a contradiction. Hence (161) is true when (168) holds. If $B(e_n, 1) \subset \mathbf{R}^n \setminus \bar{\Omega}_\infty$ we proceed similarly. That is, if (161) is false, then there exists a sequence $\{s_m\}$ of positive numbers and $\delta > 0$ with $\lim_{m \to \infty} s_m = 0$ and the property that

$$\mathbf{R}^n \setminus \bar{\Omega}_\infty \cap \partial B(0, s_m) \cap \{x : x_n \leq -\delta s_m\} \neq \emptyset \tag{180}$$

for each m. To get a contradiction one shows that (180) leads to

$$\liminf_{t \to 0} t^{-1} \max_{B(0,t)} u_\infty = 0 \tag{181}$$

which in view of Lipschitzness of Ω_∞ and the mean value theorem from elementary calculus, again contradicts (159). We omit the details. $\square$

5.15   Closing Remarks

To state our likely results in [54] we need a definition.

Definition H. *Given $\epsilon > 0$ we say that u is ϵ-monotone in $O \subset \mathbf{R}^n$, with respect to the directions in the cone $\tilde{\Gamma}(\nu, \theta_0) = \{y \in \mathbf{R}^n : |y| = 1 \text{ and } \theta(\nu, y) < \theta_0\}$, if*

$$\sup_{B(x, \epsilon' \sin \theta_0)} u(y - \epsilon'\nu) \leq u(x)$$

whenever $\epsilon' \geq \epsilon$ and $x \in O$ with $B(x - \epsilon'\nu, \epsilon' \sin \theta_o) \subset O$. Moreover, u is said to be monotone or fully monotone in $O \subset \mathbf{R}^n$, with respect to the directions in the cone $\Gamma(\nu, \theta_0)$, provided the above inequality holds whenever $\epsilon' > 0$.

Plausible Theorem 4.14. *Let $D, u, D^+(u), D^-(u), F(u), G$ be as in Theorem 4.11. If $\bar{\theta} \in (\pi/4, \pi/2)$, then there is a $\bar{\epsilon} = \bar{\epsilon}(\bar{\theta}, p, n, N)$ such that if u is ϵ monotone on $B(0, 2)$ with respect to the directions in the cone $\Gamma(e_n, \bar{\theta})$, for some $\epsilon \in (0, \bar{\epsilon})$, then u is monotone in $B(0, 1/2)$ with respect to the directions in the cone $\Gamma(e_n, \bar{\theta}_1)$, where $\bar{\theta}_1$ has the same dependence as $\bar{\epsilon}$. Equivalently, $F(u) \cap B(0, 1/2)$ is the graph of a Lipschitz function with Lipschitz norm depending on $\bar{\theta}, p, n, N$.*

We note that we can use Theorems 4.14 and 4.11 to conclude that ϵ monotonicity of u implies $F(u)$ is $C^{1,\sigma}$ provided ϵ is small enough.

Plausible Theorem 4.15. *Let $D, F(u), D^+(u), D^-(u), G$ be as in Plausible Theorem 4.14 except for the following changes:*

(a) Assume only that u^+ is ϵ monotone in $\Gamma(e_n, \bar{\theta})$,
(b) Assume $0 < \delta \leq |\nabla u| \leq \delta^{-1}$ on $D^+(u) \cap B(0, 2)$,
(c) G is also Lipschitz continuous.

There exists $\hat{\epsilon} > 0$ and $\hat{\theta} \in (\pi/4, \pi/2)$, both depending on p, n, δ, N, such that if $\hat{\theta} < \bar{\theta} \leq \pi/2$, and $0 < \epsilon \leq \hat{\epsilon}$, then u^+ is monotone in $B(0, 2)$ with respect to the directions in the cone $\Gamma(e_n, \theta_1)$ for some $\theta_1 > 0$, depending on $p, n, \delta, N, \hat{\epsilon}, \hat{\theta}$.

As a corollary to Plausible Theorem 4.15 we also plan to show that

Plausible Corollary 4.16. *Replace the ϵ monotonicity assumption in Plausible Theorem 4.15 by*

$$\Psi(F(u) \cap B(0, 2), \Lambda \cap B(0, 2)) \leq \epsilon,$$

where Λ is the graph of a Lipschitz function with Lipschitz norm $\leq \tan(\pi/2 - \hat{\theta})$. Then the same conclusion holds as in Plausible Theorem 4.15.

One can also ask if Theorem 4.11 generalizes to equations of p Laplace type, as in the $p = 2$ case. In this case, the analogue of Lemma 4.12 may be difficult since the proof of this lemma made important use of the invariance of the

p Laplace equation under rotations and translations. Also, the analogue of Theorem 4.13 could be difficult.

Acknowledgements Work partially supported by NSF DMS-0900291.

References

1. H.W. Alt, L.A. Caffarelli, A. Friedman, A free boundary problem for quasilinear elliptic equations. Ann. Scuola Norm. Sup. Pisa Cl. Sci. (4) **11**(1), 1–44 (1984)
2. A. Ancona, Principe de Harnack à la Frontière et Théorème de Fatou pour un Opéet Elliptique Dans un Domain Lipschitzien. Ann. Inst. Fourier (Grenoble) **28**(4), 169–213 (1978)
3. A. Batakis, Harmonic measure of some cantor type sets. Ann. Acad. Sci. Fenn. **21**(2), 255–270 (1996)
4. A. Batakis, A continuity property of the dimension of harmonic measure under perturbations. Ann. Inst. H. Poincaré Probab. Stat. **36**(1), 87–107 (2000)
5. A. Batakis, Continuity of the dimension of the harmonic measure of some cantor sets under perturbations. Annales de l' Institut Fourier **56**(6), 1617–1631 (2006)
6. B. Bennewitz, J. Lewis, On the dimension of p harmonic measure. Ann. Acad. Sci. Fenn. **30**, 459–505 (2005)
7. B. Bennewitz, Nonuniqueness in a free boundary problem. Rev. Mat. Iberoam. **24**(2), 567–595 (2008)
8. J. Lewis, K. Nyström, A. Vogel, p harmonic measure in space (submitted)
9. C. Bishop, L. Carleson, J. Garnett, P. Jones, Harmonic measures supported on curves. Pac. J. Math. **138**, 233–236 (1989)
10. C. Bishop, P. Jones, Harmonic measure and arclength. Ann. Math. (2) **132**(3), 511–547 (1990)
11. L. Caffarelli, A Harnack inequality approach to the regularity of free boundaries. Part I, Lipschitz free boundaries are $C^{1,\alpha}$. Revista Mathematica Iberoamericana **3**, 139–162 (1987)
12. L. Caffarelli, A Harnack inequality approach to the regularity of free boundaries. Part III. Existence theory, compactness, and dependence on X. Ann. Scuola Norm. Sup. Pisa Cl. Sci. (4) **15**(4), 583–602 (1988)
13. L. Caffarelli, Harnack inequality approach to the regularity of free boundaries. Part II. Flat free boundaries are Lipschitz. Comm. Pure Appl. Math. **42**(1), 55–78 (1989)
14. L. Caffarelli, E. Fabes, Mortola, S. Salsa, Boundary behavior of nonnegative solutions of elliptic operators in divergence form. Indiana J. Math. **30**(4), 621–640 (1981)
15. L. Carleson, On the support of harmonic measure for sets of cantor type. Ann. Acad. Sci. Fenn. **10**, 113–123 (1985)
16. M.C. Cerutti, F. Ferrari, S. Salsa, Two-phase problems for linear elliptic operators with variable coefficients: Lipschitz free boundaries are $C^{1,\gamma}$. Arch. Ration. Mech. Anal. **171**, 329–448 (2004)
17. B. Dahlberg, On estimates of harmonic measure. Arch. Ration. Mech. Anal. **65**, 275–288 (1977)
18. G. David, S. Semmes, Analysis of and on uniformly rectifiable sets. Am. Math. Soc. Surv. Mono. **38** (1993)
19. E. DiBenedetto, Degenerate Parabolic Equations, Universitext (Springer, New York, 1993)
20. A. Eremenko, J. Lewis, Uniform limits of certain A harmonic functions with applications to quasiregular mappings. Ann. Acad. Sci. Fenn. AI Math. **16**, 361–375 (1991)

21. E. Fabes, C. Kenig, R. Serapioni, The local regularity of solutions to degenerate elliptic equations. Comm. Part. Differ. Equat. **7**(1), 77–116 (1982)
22. E. Fabes, D. Jerison, C. Kenig, The Wiener test for degenerate elliptic equations. Ann. Inst. Fourier (Grenoble) **32**, 151–182 (1982)
23. E. Fabes, D. Jerison, C. Kenig, *Boundary Behavior of Solutions to Degenerate Elliptic Equations*. Conference on Harmonic Analysis in Honor of Antonio Zygmund, vol. I, II, Chicago, IL, 1981. Wadsworth Math. Ser (Wadsworth Belmont, CA, 1983), pp. 577–589
24. M. Feldman, Regularity for nonisotropic two phase problems with Lipschitz free boundaries. Differ. Integr. Equat. **10**, 227–251 (1997)
25. M. Feldman, Regularity of Lipschitz free boundaries in two-phase problems for fully non-linear elliptic equations. Indiana Univ. Math. J. **50**(3), 1171–1200 (2001)
26. F. Ferrari, Two-phase problems for a class of fully nonlinear elliptic operators, Lipschitz free boundaries are $C^{1,\gamma}$. Am. J. Math. **128**(3), 541–571 (2006)
27. F. Ferrari, S. Salsa, Regularity of the free boundary in two-phase problems for linear elliptic operators. Adv. Math. **214**, 288–322 (2007)
28. F. Ferrari, S. Salsa, Subsolutions of elliptic operators in divergence form and applications to two-phase free boundary problems. Bound. Value Probl. **2007**, 21 (2007). Article ID 57049
29. R.M. Gabriel, An extended principle of the maximum for harmonic functions in 3-dimension. J. Lond. Math. Soc. **30**, 388–401 (1955)
30. R. Gariepy, W. Ziemer, A regularity condition at the boundary for solutions of quasilinear elliptic equations. Arch. Ration. Mech. Anal. **6**, 25–39 (1977)
31. W.K. Hayman, in *Research Problems in Function Theory*, ed. by W.K. Hayman (The Athlone Press, London, 1967)
32. H. Hedenmalm, I. Kayamov, On the Makarov law of the iterated logarithm. Proc. Am. Math. Soc. **135**(7), 2235–2248 (2007)
33. J. Heinonen, T. Kilpelainen, O. Martio, *Nonlinear Potential Theory of Degenerate Elliptic Equations* (Dover, NY, 2006)
34. A. Henrot, H. Shahgholian, Existence of classical solutions to a free boundary problem for the p Laplace operator: (I) the exterior convex case. J. Reine Angew. Math. **521**, 85–97 (2000)
35. S. Hofmann, J. Lewis, The Dirichlet problem for parabolic operators with singular drift term. Mem. Am. Math. Soc. **151**(719), 1–113 (2001)
36. T. Iwaniec, J. Manfredi, Regularity of p harmonic functions in the plane. Revista Mathematica Iberoamericana **5**(1–2), 1–19 (1989)
37. D. Jerison, C. Kenig, Boundary behavior of harmonic functions in nontangentially accessible domains. Adv. Math. **46**, 80–147 (1982)
38. D. Jerison, C. Kenig, The logarithm of the Poisson kernel of a C^1 domain has vanishing mean oscillation. Trans. Am. Math. Soc. **273**, 781–794 (1982)
39. P. Jones, T. Wolff, Hausdorff dimension of harmonic measures in the plane. Acta Math. **161**, 131–144 (1988)
40. R. Kaufman, J.M. Wu, On the snowflake domain. Ark. Mat. **23**, 177–183 (1985)
41. J. Kemper, A boundary Harnack inequality for Lipschitz domains and the principle of positive singularities. Comm. Pure Appl. Math. **25**, 247–255 (1972)
42. C. Kenig, J. Pipher, The Dirichlet problem for elliptic operators with drift term. Publ. Mat. **45**(1), 199–217 (2001)
43. C. Kenig, T. Toro, Harmonic measure on locally flat domains. Duke Math. J. **87**, 501–551 (1997)
44. C. Kenig, T. Toro, Free boundary regularity for harmonic measure and Poisson kernels. Ann. Math. **150**, 369–454 (1999)
45. C. Kenig, T. Toro, Poisson kernel characterization of Reifenberg flat chord arc domains. Ann. Sci. Ecole Norm. Sup. (4) **36**(3), 323–401 (2003)
46. T. Kilpeläinen, X. Zhong, Growth of entire A – subharmonic functions. Ann. Acad. Sci. Fenn. AI Math. **28**, 181–192 (2003)

47. J. Lewis, Capacitary functions in convex rings. Arch. Ration. Mech. Anal. **66**, 201–224 (1977)
48. J. Lewis, Note on p harmonic measure. Comput. Meth. Funct. Theor. **6**(1), 109–144 (2006)
49. J. Lewis, K. Nyström, Boundary behavior for p harmonic functions in Lipschitz and starlike Lipschitz ring domains. Ann. Sci. École Norm. Sup. (4) **40**(4), 765–813 (2007)
50. J. Lewis, K. Nyström, Boundary behavior of p harmonic functions in domains beyond Lipschitz domains. Adv. Calc. Var. **1**, 1–38 (2008)
51. J. Lewis, K. Nyström, Regularity and free boundary regularity for the p Laplacian in Lipschitz and C^1 domains. Ann. Acad. Sci. Fenn. **33**, 1–26 (2008)
52. J. Lewis, K. Nyström, *Boundary Behaviour and the Martin Boundary Problem for p Harmonic Functions in Lipschitz Domains* (with Kaj Nyström), Annals of Mathematics **172**, 1907–1948 (2010)
53. J. Lewis, K. Nyström, *Regularity of Lipschitz Free Boundaries in Two Phase Problems for the p Laplace Operator*, Advances in Mathematics **225**, 2565–2597 (2010)
54. J. Lewis, K. Nyström, *Regularity of Flat Free Boundaries in Two Phase Problems for the p Laplace Operator*, to appear Annales de l'Institut Henri Poincare, Analyse non lineaire.
55. J. Lewis, A. Vogel, On pseudospheres. Revista Mathematica Iberoamericana **7**, 25–54 (1991)
56. J. Lewis, A. Vogel, *On Some Almost Everywhere Symmetry Theorems*. Nonlinear Diffusion Equations and Their Equilibrium, vol. 3 (Birkhäuser, Basel, 1992), pp. 347–374
57. J. Lewis, A. Vogel, A symmetry theorem revisited. Proc. Am. Math. Soc. **130**(2), 443–451 (2001)
58. J. Lewis, A. Vogel, On pseudospheres that are quasispheres. Revista Mathematica Iberoamericana **17**, 221–255 (2001)
59. J. Lewis, A. Vogel, Uniqueness in a free boundary problem. Comput. Meth. Funct. Theor. **31**, 1591–1614 (2006)
60. J. Lewis, A. Vogel, Symmetry theorems and uniform rectifiability. Bound. Value Probl. **2007**, 1–59 (2007)
61. J. Lewis, G. Verchota, A. Vogel, On Wolff snowflakes. Pac. J. Math. **218**(1), 139–166 (2005)
62. J. Lewis, N. Lunstrom, K. Nyström, Boundary Harnack inequalities for operators of p Laplace type in Reifenberg flat domains. Proc. Symp. Pure Math. **79**, 229–266 (2008)
63. J. Lewis, K. Nyström, P. Poggi Corradini, *p Harmonic Measure in Simply Connected Domains*, Ann. Inst. Fourier Grenoble **61**, 2, 689–715 (2011)
64. G. Lieberman, Boundary regularity for solutions of degenerate elliptic equations. Nonlinear Anal. **12**(11), 1203–1219 (1988)
65. J. Llorente, J. Manfredi, J.M. Wu, p harmonic measure is not Additive on null sets. Ann. Scuola Norm. Sup. Pisa Cl. Sci. (5) **4**(2), 357–373 (2005)
66. N. Makarov, Distortion of boundary sets under conformal mapping. Proc. Lond. Math. Soc. **51**, 369–384 (1985)
67. R.S. Martin, Minimal positive harmonic functions. Trans. Am. Math. Soc. **49**, 137–172 (1941)
68. V.G. Maz'ya, The continuity at a boundary point of the solutions of quasilinear elliptic equations (Russian). Vestnik Leningrad. Univ. **25**(13), 42–55 (1970)
69. J. Serrin, Local behavior of solutions of quasilinear elliptic equations. Acta Math. **111**, 247–302 (1964)
70. P. Wang, Regularity of free boundaries of two-phase problems for fully non-linear elliptic equations of second order. Part 1: Lipschitz free boundaries are $C^{1,\alpha}$. Comm. Pure Appl. Math. **53**, 799–810 (2000)
71. T. Wolff, Plane harmonic measures live on sets of σ finite length. Ark. Mat. **31**(1), 137–172 (1993)

72. T. Wolff, in *Counterexamples with harmonic gradients in* $\mathbf{R}^3$. Essays in honor of Elias M. Stein. Princeton Mathematical Series, vol. 42 (Princeton University Press, Princeton, 1995), pp. 321–384
73. J.M. Wu, Comparisons of kernel functions, boundary Harnack principle and relative Fatou theorem on Lipschitz domains. Ann. Inst. Fourier (Grenoble) **28**(4), 147–167 (1978)

Regularity of Supersolutions

Peter Lindqvist

1 Introduction

The regularity theory for *solutions* of certain parabolic differential equations of the type

$$\frac{\partial u}{\partial t} = \operatorname{div} \mathbf{A}(x, t, u, \nabla u) \tag{1}$$

is a well developed topic, but when it comes to (semicontinuous) *supersolutions* and *subsolutions* a lot remains to be done. Supersolutions are often auxiliary tools as in the celebrated Perron method, for example, but they are also interesting in their own right. They appear as solutions to obstacle problems and variational inequalities.

As a mnemonic rule $v_t \geq \operatorname{div} \mathbf{A}(x, t, v, \nabla v)$ for smooth supersolutions. Our supersolutions are required to be lower semicontinuous but are not assumed to be differentiable in any sense: part of the theory is to prove that they have Sobolev derivatives. If one instead studies weak supersolutions that by definition belong to a Sobolev space, then one has the task to prove that they are semicontinuous. Unfortunately, the weak supersolutions do not form a good closed class under monotone convergence. For bounded functions the definitions yield the same class of supersolutions.

The modern theory of viscosity solutions, created by Lions, Crandall, Evans, Ishii, Jensen, and others, relies on the appropriately defined *viscosity supersolutions*, which are merely lower semicontinuous functions by their definition. For second order equations, these are often the same functions as those supersolutions that are encountered in potential theory. The

P. Lindqvist (✉)
Department of Mathematical Sciences, Norwegian University of Science and Technology,
7491 Trondheim, Norway
e-mail: lqvist@math.ntnu.no

J. Lewis et al., *Regularity Estimates for Nonlinear Elliptic and Parabolic Problems*, Lecture Notes in Mathematics 2045, DOI 10.1007/978-3-642-27145-8_2,
© Springer-Verlag Berlin Heidelberg 2012

link enables one to study the regularity properties also of the viscosity supersolutions. This is the case for the so-called p-Laplace equation.

As an example of what we have in mind, consider the Laplace equation $\Delta u = 0$ and recall that a superharmonic function is a lower semicontinuous function satisfying a comparison principle with respect to the harmonic functions. An analogous definition comes from the super meanvalue property. General superharmonic functions are not differentiable in the classical sense. Nonetheless, the following holds.

Proposition 1. *Suppose that v is a superharmonic function defined in $\mathbb{R}^n$. Then the Sobolev derivative ∇v exists and*

$$\int_{B_R} |\nabla v|^q \, dx < \infty$$

whenever $0 < q < \frac{n}{n-1}$. Moreover,

$$\int_{\mathbb{R}^n} \langle \nabla v, \nabla \eta \rangle \, dx \geq 0$$

for $\eta \geq 0$, $\eta \in C_0^\infty(\mathbb{R}^n)$.

The fundamental solution $v(x) = |x|^{2-n}$ $(= -\log(|x|)$, when $n = 2)$ is a superharmonic function showing that the summability exponent q is sharp. We seize the opportunity to mention that the superharmonic functions are exactly the same as the viscosity supersolutions of the Laplace equation. In other words, *a viscosity supersolution has a gradient in Sobolev's sense.* As an example, the Newtonian potential

$$v(x) = \sum \frac{c_j}{|x - q_j|^{n-2}},$$

where the rational points q_j are numbered and the c_j's are convergence factors, is a superharmonic function, illustrating that functions in the Sobolev space can be infinite in a dense set. The proof of the proposition follows from Riesz's representation theorem, a classical result according to which we have a harmonic function plus a Newtonian potential. This was about the Laplace equation.

A similar theorem holds for the viscosity supersolutions (= the p-superharmonic functions) of the so-called p-Laplace equation

$$\nabla \cdot \left(|\nabla v|^{p-2} \nabla v \right) = 0$$

but now $0 < q < \frac{n(p-1)}{n-1}$. (Strictly speaking, we obtain a proper Sobolev space only for $p > 2 - \frac{1}{n}$, because $q < 1$ otherwise.) The principle of superposition is not valid and, in particular, Riesz's representation theorem is no longer available. The original proof in [20] was based on the obstacle problem in the calculus of variations and on the so-called weak Harnack inequality. At present, the simplest proof seems to rely upon an approximation with so-called infimal convolutions

$$v_\varepsilon(x) = \inf_y \left\{ v(y) + \frac{|x-y|^2}{2\varepsilon} \right\}, \quad \varepsilon > 0.$$

At each point $v_\varepsilon(x) \nearrow v(x)$. They are viscosity supersolutions, if the original v is. Moreover, they are (locally) Lipschitz continuous and hence differentiable a.e. Therefore the approximants satisfy expedient a priori estimates, which, to some extent, can be passed over to the original function v itself.

Another kind of results is related to the *pointwise behaviour*. The viscosity supersolutions are pointwise defined. At *each* point we have

$$v(x) = \operatorname{ess\,liminf}_{y \to x} v(y)$$

where *essential limes inferior* means that sets of measure zero are neglected in the calculation of the lower limit. In the linear case $p = 2$ the result seems to be due to Brelot, cf. [2]. So much about the p-Laplace equation for now. The theory extends to a wider class of elliptic equations of the type

$$\operatorname{div} \mathbf{A}(x, u, \nabla u) = 0.$$

For parabolic equations like

$$\frac{\partial u}{\partial t} = \sum_{i,j} \frac{\partial}{\partial x_i} \left(\left| \sum_{k,m} a_{k,m} \frac{\partial u}{\partial x_k} \frac{\partial u}{\partial x_m} \right|^{\frac{p-2}{2}} a_{i,j} \frac{\partial u}{\partial x_j} \right),$$

where the matrix $(a_{i,j})$ satisfies the ellipticity condition

$$\sum a_{i,j} \xi_i \xi_j \geq \gamma |\xi|^2,$$

the situation is rather similar, although technically much more demanding. Now the use of infimal convolutions as approximants offers considerable simplification, at least in comparison with the original proofs in [15]. We will study the **Evolutionary p-Laplace Equation**

$$\frac{\partial u}{\partial t} = \operatorname{div}(|\nabla u|^{p-2} \nabla u) \tag{2}$$

where $u = u(x, t)$, restricting ourselves to the slow diffusion case $p > 2$. The celebrated Barenblatt solution[1]

$$B_p(x, t) = \begin{cases} t^{-n/\lambda} \left[C - \frac{p-2}{p} \lambda^{1/(1-p)} \left(\frac{|x|}{t^{1/\lambda}} \right)^{\frac{p}{p-1}} \right]_+^{\frac{p-1}{p-2}} & \text{if } t > 0 \\ 0 & \text{if } t \leq 0 \end{cases} \quad (3)$$

where $\lambda = n(p-2) + p$ is the leading example of a viscosity supersolution (= p-supercaloric function). It has a compact support in the x-variable for each fixed instance t. Disturbances propagate with finite speed and an interface (moving boundary) appears. Notice that

$$\int_0^T \int_{|x|<1} |\nabla B_p(x, t)|^p \, dx \, dt = \infty$$

due to the singularity at the origin. Thus B_p fails to be a weak supersolution in a domain containing the origin.[2] Our main theorem is:

Theorem 2. *Let $p > 2$ and suppose that $v = v(x, t)$ is a viscosity supersolution in the domain Ω_T in $\mathbb{R}^n \times \mathbb{R}$. Then*

$$v \in L^q_{loc}(\Omega_T) \text{ whenever } q < p - 1 + \frac{p}{n}$$

and the Sobolev derivative

$$\nabla v = \left(\frac{\partial v}{\partial x_1}, \cdots, \frac{\partial v}{\partial x_n} \right)$$

exists and

$$\nabla v \in L^q_{loc}(\Omega_T) \text{ whenever } q < p - 1 + \frac{1}{n+1}.$$

The summability exponents are sharp. Moreover,

$$\iint_{\Omega_T} \left(-v\eta_t + \langle |\nabla v|^{p-2} \nabla v, \nabla \eta \rangle \right) dx \, dt \geq 0$$

for all $\eta \geq 0$, $\eta \in C_0^\infty(\Omega_T)$.

---

[1] "Einen wahren wissenschaftlichen Werth erkenne ich—auf dem Felde der *Mathematik*— nur in concreten mathematischen Wahrheiten, oder schärfer ausgedrückt, 'nur in mathematischen Formeln'. Diese allein sind, wie die Geschichte der Mathematik zeigt, das Unvergängliche. Die verschiedenen Theorien für die Grundlagen der Mathematik (so die von Lagrange) sind von der Zeit weggeweht, aber die Lagrangesche Resolvente ist geblieben!" Kronecker 1884.

[2] In my opinion, a definition of "supersolutions" that excludes the fundamental solution cannot be regarded as entirely satisfactory.

The Barenblatt solution shows that the exponents are sharp. Notice that the time derivative is not included in the statement. Actually, the time derivative need not be a function, as the example $v(x,t) = 0$, when $t \leq 0$, and $v(x,t) = 1$, when $t > 0$ shows. Dirac's delta appears! It is worth our while emphasize that the gradient ∇v is not present in the definition of the viscosity supersolutions (= the p-supercaloric functions).

An important feature is that the viscosity supersolutions are defined at each point, not just almost everywhere in their domain. When it comes to the pointwise behaviour, one may even exclude all future times so that only the instances $\tau < t$ are used for the calculation of $v(x,t)$, as in the next theorem. (It is also, of course, valid without restriction to the past times.)

Theorem 3. *Let $p \geq 2$. A viscosity supersolution of the Evolutionary p-Laplace Equation satisfies*

$$v(x,t) = \operatorname*{ess\,liminf}_{\substack{(y,\tau)\to(x,t) \\ \tau<t}} v(y,\tau)$$

at each interior point (x,t).

In the calculation of *essential limes inferior* sets of $(n+1)$-dimensional Lebesgue measure zero are neglected. We mention an immediate consequence, which does not seem to be easily obtained by other methods.

Corollary 4. *Two viscosity supersolutions that coincide almost everywhere are equal at each point.*

A general comment about the method employed in these notes is appropriate. We do not know about proofs for viscosity supersolutions that would totally stay within that framework. It must be emphasized that the proofs are carried out for those supersolutions that are defined as one does in Potential Theory, namely through comparison principles, and then the results are valid even for the viscosity supersolutions, just because, incidentally, they are the same functions. The identification[3] of these two classes of "supersolutions" is not a quite obvious fact. This limits the applicability of the method.

In passing, we also treat the measure data equation

$$\frac{\partial v}{\partial t} - \nabla \cdot (|\nabla v|^{p-2}\nabla v) = \mu$$

where the right-hand side is a Radon measure. It follows quite easily from Theorem 2 that each viscosity supersolution induces a measure and is a

---

[3]Unfortunately, the proof of this fundamental identification is not included in these notes, only the reference [10] is given. There is an as yet unpublished much simpler proof by P. Juutinen.

solution to the measure data equation. (The reversed problem, which starts with a given measure μ instead of a given function v, is a much investigated topic, cf. [1].)

Some other equations that are susceptible of this kind of analysis are the Porous Medium Equation[4]

$$\frac{\partial u}{\partial t} = \Delta(|u|^{m-1}u)$$

and

$$\frac{\partial(|u|^{p-2}u)}{\partial t} = \nabla \cdot (|\nabla u|^{p-2}\nabla u),$$

but it does not seem to be known which equations of the form

$$\frac{\partial u}{\partial t} = F(x, t, u, \nabla u, D^2 u)$$

enjoy the property of having their viscosity supersolutions in some local Sobolev x-space. I hope that this could be a fruitful research topic for the younger readers. I thank T. Kuusi and M. Parviainen for a careful reading of the manuscript.

2   The Stationary Equation

For reasons of exposition,[5] we begin with the stationary equation

$$\Delta_p u \equiv \operatorname{div}(|\nabla u|^{p-2}\nabla u) = 0, \tag{4}$$

which offers some simplifications not present in the time dependent situation. In principle, here we keep $p \geq 2$, although the theory often allows that $1 < p \leq 2$, at least with minor changes. Moreover, the cases $p > n$, $p = n$, and $p < n$ often require separate proofs. We sometimes skip the borderline case $p = n$.

Some general references are [8], [9], and [21].

Definition 5. *We say that $u \in W_{loc}^{1,p}(\Omega)$ is a weak solution in Ω, if*

---

[4]The Porous Medium Equation is not well suited for the viscosity theory (it is not "proper"), but the comparison principle works well. It is not ∇v but $\nabla(|v|^{m-1}v)$ that is guaranteed to exist.

[5]Chapter "Introduction to Random Tug-of-War Games and PDEs" is rather independent of the present chapter.

$$\int_{\Omega} \langle |\nabla u|^{p-2}\nabla u, \nabla \eta \rangle \, dx = 0$$

for all $\eta \in C_0^{\infty}(\Omega)$. If, in addition, u is continuous, it is called a p-harmonic function.

The weak solutions can, in accordance with the elliptic regularity theory, be made continuous after a redefinition in a set of Lebesgue measure zero. The Hölder continuity estimate

$$|u(x) - u(y)| \le L\,|x - y|^{\alpha} \tag{5}$$

holds when $x, y \in B(x_0, r)$, $B(x_0, 2r) \subset\subset \Omega$; here α depends on n and p but L also depends on the norm $\|u\|_{p,B(x_0,2r)}$. We omit the proof. The continuous weak solutions are called *p-harmonic functions.*[6] In fact, even the gradient is continuous. One has $u \in C_{loc}^{1,\alpha}(\Omega)$, where $\alpha = \alpha(n,p)$. This deep result of N. Ural'tseva will not be needed here. According to [27] positive solutions obey the Harnack inequality.

Lemma 6 (Harnack's Inequality). *If the p-harmonic function u is non-negative in the ball $B_{2r} = B(x_0, 2r)$, then*

$$\max_{\overline{B_r}} u \le C_{n,p} \min_{\overline{B_r}} u \,.$$

The p-Laplace equation is the *Euler-Lagrange equation* of a variational integral. Let us recall the *Dirichlet problem* in a bounded domain Ω. Let $f \in C(\overline{\Omega}) \cap W^{1,p}(\Omega)$ represent the boundary values. Then there exists a unique function u in $C(\Omega) \cap W^{1,p}(\Omega)$ such that $u - f \in W_0^{1,p}(\Omega)$ and

$$\int_{\Omega} |\nabla u|^p \, dx \le \int_{\Omega} |\nabla(u + \eta)|^p \, dx$$

for all $\eta \in C_0^{\infty}(\Omega)$. The minimizer is p-harmonic. If the boundary $\partial\Omega$ is regular enough, the boundary values are attained in the classical sense:

$$\lim_{x \to \xi} u(x) = f(\xi), \quad \xi \in \Omega.$$

When it comes to the super- and subsolutions, several definitions are currently being used. We need the following ones:

(1) **Weak supersolutions** (test functions under the integral sign)

---

[6]Thus the 2-harmonic functions are the familiar harmonic functions encountered in Potential Theory.

(2) **p-superharmonic functions** (defined via a comparison principle)
(3) **Viscosity supersolutions** (test functions evaluated at points of contact)

The p-superharmonic functions and the viscosity supersolutions are exactly the same functions, see [10]. They are not assumed to have any derivatives. In contrast, the weak supersolutions are by their definition required to belong to the Sobolev space $W_{loc}^{1,p}(\Omega)$ and therefore their Caccioppoli estimates are at our disposal. As we will see, locally *bounded* p-superparabolic functions (= viscosity supersolutions) are, indeed, weak supersolutions, having Sobolev derivatives as they should. To this one may add that the weak supersolutions are p-superharmonic functions, provided that the issue of semicontinuity be properly handled.

Definition 7. *We say that a function* $v : \Omega \to (-\infty, \infty]$ *is p-superharmonic in* Ω, *if*

(i) *v is finite in a dense subset*
(ii) *v is lower semicontinuous*
(iii) *In each subdomain* $D \subset\subset \Omega$ *v obeys the comparison principle:*
    if $h \in C(\overline{D})$ *is p-harmonic in D, then the implication*

$$v|_{\partial D} \geq h|_{\partial D} \quad \Rightarrow \quad v \geq h$$

   is valid.

Remarks. For $p = 2$ this is the classical definition of superharmonic functions due to F. Riesz. It is sufficient[7] to assume that $v \not\equiv \infty$ instead of (i). The fundamental solution $|x|^{(p-n)/(p-1)}$, is not a weak supersolution in $\mathbb{R}^n$, merely because it fails to belong to the right Sobolev space, but it is p-superharmonic.

The next definition is from the theory of viscosity solutions. See [5] and [17] for an overview. One defines them as being both viscosity super- and subsolutions, since it is not practical to do it in one stroke.

Definition 8. *Let* $p \geq 2$. *A function* $v : \Omega \to (-\infty, \infty]$ *is called a* viscosity supersolution, *if*

(i) *v is finite in a dense subset*
(ii) *v is lower semicontinuous*
(iii) *Whenever* $x_0 \in \Omega$ *and* $\phi \in C^2(\Omega)$ *are such that* $v(x_0) = \phi(x_0)$ *and* $v(x) > \phi(x)$ *when* $x \neq x_0$, *we have*

$$\mathrm{div}\left(|\nabla\phi(x_0)|^{p-2}\nabla\phi(x_0)\right) \leq 0.$$

---

[7]This is not quite that simple in the parabolic case.

Remarks. The differential operator is evaluated only at the point of contact. The singular case $1 < p < 2$ requires a modification,[8] if it so happens that $\nabla\phi(x_0) = 0$. Notice that each point has its own family of test functions. If there is no test function touching from below at x_0, then there is no requirement: the point passes for free. Please, notice that nothing is said about the gradient ∇v, it is $\nabla\phi(x_0)$ that appears.

As we mentioned, both definitions lead to the same class of "supersolutions". Some examples are the following functions.

$$(n - p)|x|^{-\frac{n-p}{p-1}} \quad (n \neq p), \quad V(x) = \int \frac{\varrho(y)\,dy}{|x - y|^{n-2}} \quad (p = 2, n \geq 3),$$

$$V(x) = \sum \frac{c_j}{|x - q_j|^{(n-p)/(p-1)}} \quad (2 < p < n),$$

$$V(x) = \int \frac{\varrho(y)\,dy}{|x - y|^{(n-p)/(p-1)}} \quad (2 < p < n), \quad v(x) = \min\{v_1, v_2, \cdots, v_m\}.$$

The first example is *the fundamental solution*, which fails to belong to the "natural" Sobolev space $W^{1,p}_{loc}(\mathbb{R}^n)$.[9] The second is the Newtonian potential. In the third example the c_j's are positive convergence factors and the q_j's are the rational points; the superposition of fundamental solutions is credited to Crandall and Zhang, cf. [3]. The last example says that one may take the pointwise minimum of a finite number of p-superharmonic functions, which is an essential ingredient in the celebrated Perron method.

The functions in the next lemma, the continuous weak supersolutions, form a more tractable subclass, when it comes to a priori estimates, since they are differentiable.

Lemma 9. *Let $v \in C(\Omega) \cap W^{1,p}(\Omega)$. Then the following conditions are equivalent:*

(i) $\int_\Omega |\nabla v|^p\,dx \leq \int_\Omega |\nabla(v + \eta)|^p\,dx$ when $\eta \geq 0, \eta \in C^\infty_0(\Omega)$,
(ii) $\int_\Omega \langle|\nabla v|^{p-2}\nabla v, \nabla\eta\rangle\,dx \geq 0$ when $\eta \geq 0, \eta \in C^\infty_0(\Omega)$,
(iii) v is p-superharmonic.

Proof. The equivalence of (i) and (ii) is plain. So is the necessity of (iii), stating that the comparison principle must hold. The crucial part is the sufficiency of (iii), which will be established by the help of an *obstacle problem* in the calculus of variations. The function v itself will act as an obstacle for the admissible functions in the minimization of the p-energy $\int_D |\nabla v|^p\,dx$ and it also induces the boundary values in the subdomain D. If D is a regular

---

[8]There is no requirement when $\nabla\phi$ is 0, see [10].

[9]Therefore it is not a weak supersolution, but it is a viscosity supersolution and a p-superharmonic function.

subdomain of Ω, then there exists a unique minimizer, say w_v, in the class

$$\mathcal{F}_v = \{w \in C(\overline{D}) \cap W^{1,p}(D) | \ w \geq v, \ w = v \ on \ \partial D\}.$$

The crucial part is the continuity of w_v cf. [24]. The solution of the obstacle problem automatically has the property (i), and hence also (ii). We claim that $w_v = v$ in D, from which the desired conclusion thus follows. The minimizer is a p-harmonic function in the open set $\{w_v > v\}$ where the obstacle does not hinder. On the boundary of this set $w_v = v$. Hence the comparison principle, which v is assumed to obey, can be applied. It follows that $w_v \leq v$ in the same set. To avoid a contradiction it must be the empty set. The conclusion is that $w_v = v$ in D, as desired. One can now deduce that (iii) is sufficient. □

A function, whether continuous or not, belonging to $W^{1,p}(\Omega)$ and satisfying (ii) in the previous lemma is called a *weak supersolution*. For completeness we record below that weak supersolutions are semicontinuous "by nature".

Proposition 10. *A weak supersolution* $v \in W^{1,p}(\Omega)$ *is lower semicontinuous (after redefinition in a set of measure zero). We can define*

$$v(x) = \operatorname{ess\,liminf}_{y \to x} v(y)$$

pointwise.

Proof. The case $p > n$ is clear, since then the Sobolev space contains only continuous functions (Morrey's inequality). In the range $p < n$ we claim that

$$v(x) = \operatorname{ess\,liminf}_{y \to x} v(y)$$

at a.e. $x \in \Omega$. The proof follows from this, because the right-hand side is always lower semicontinuous. We omit two demanding steps. First, it is required to establish that v is locally bounded. This is standard regularity theory. Second, for non-negative functions we use "the weak Harnack estimate"

$$\left(\frac{1}{|B_{2r}|} \int_{B_{2r}} v^q \, dx\right)^{\frac{1}{q}} \leq C \operatorname{ess\,inf}_{B_r} v,$$

when $q < n(p-1)/(n-p)$, $C = C(n,p,q)$. This comes from the celebrated Moser iteration, cf. [27].[10] Taking $q = 1$ and using the non-negative function

---

[10]Harnack's inequality can be replaced by the more elementary estimate

$$\operatorname{ess\,sup}_{B_r} (v(x_0) - v(x))_+ \leq \frac{C}{|B_{2r}|} \int_{B(x_0,2r)} (v(x_0) - v(x))_+ \, dx$$

as a starting point for the proof. It follows immediately that also

$v(x) - m(2r)$, where

$$m(r) = \operatorname*{ess\,inf}_{B_r} v,$$

we have

$$0 \le \frac{1}{|B_{2r}|} \int_{B_{2r}} v \, dx - m(2r)$$

$$= \frac{1}{|B_{2r}|} \int_{B_{2r}} (v(x) - m(2r)) \, dx \le C(m(r) - m(2r)).$$

Since $m(r)$ is monotone, $m(r) - m(2r) \to 0$ as $r \to 0$. It follows that

$$\operatorname*{ess\,liminf}_{y \to x_0} v(y) = \lim_{r \to 0} m(2r) = \lim_{r \to 0} \frac{1}{|B_{2r}|} \int_{B(x_0, 2r)} v(x) \, dx$$

at *each* point x_0. Lebesgue's differentiation theorem states that the limit of the average on the right-hand side coincides with $v(x_0)$ at almost every point x_0. □

Lemma 11 (Caccioppoli). *Let $v \in C(\Omega) \cap W^{1,p}(\Omega)$ be a p-superharmonic function. Then*

$$\int_\Omega \zeta^p |\nabla v|^p \, dx \le p^p \left(\operatorname*{osc}_{\zeta \ne 0} v \right)^p \int_\Omega |\nabla \zeta|^p \, dx$$

holds for non-negative $\zeta \in C_0^\infty(\Omega)$. If $v \ge 0$, then

$$\int_\Omega \zeta^p v^{-1-\alpha} |\nabla v|^p \, dx \le \left(\frac{p}{\alpha} \right)^p \int_\Omega v^{p-1-\alpha} |\nabla \zeta|^p \, dx$$

when $\alpha > 0$.

Proof. To prove the first estimate, fix ζ and let $L = \sup v$ taken over the set where $\zeta \ne 0$. Use the test function

$$\eta = (L - v(x)) \zeta(x)^p$$

in Lemma 9(ii) and arrange the terms.

$$\operatorname*{ess\,sup}_{B_r} (v(x_0) - v(x)) \le \frac{C}{|B_{2r}|} \int_{B(x_0, 2r)} |v(x_0) - v(x)| \, dx.$$

If x_0 is a Lebesgue point, the integral approaches zero as $r \to 0$ and it follows that

$$\operatorname*{ess\,liminf}_{x \to x_0} v(x) \ge v(x_0).$$

The opposite inequality holds for "arbitrary" functions at their Lebesgue points. (See the end of chapter "The Problems of the Obstacle in Lower Dimension and for the Fractional Laplacian.")

To prove the second estimate, first replace $v(x)$ by $v(x)+\varepsilon$, if needed, and use

$$\eta = v^{-\alpha}\zeta^p.$$

The rest is clear.   □

The special case $\alpha = p - 1$ is appealing, since the right-hand member of the inequality

$$\int_\Omega \zeta^p |\nabla \log v|^p \, dx \le \left(\frac{p}{p-1}\right)^p \int_\Omega |\nabla \zeta|^p \, dx \tag{6}$$

is independent of the non-negative function v itself.

We aim at approximating v with functions for which Lemma 11 is valid. To this end, let v be lower semicontinuous and bounded in Ω:

$$0 \le v(x) \le L.$$

Define

$$v_\varepsilon(x) = \inf_{y\in\Omega} \left\{ v(y) + \frac{|x-y|^2}{2\varepsilon} \right\}, \quad \varepsilon > 0. \tag{7}$$

Then

- $v_\varepsilon(x) \nearrow v(x)$ as $\varepsilon \to 0+$
- $v_\varepsilon(x) - |x|^2/2\varepsilon$ is locally concave in Ω
- v_ε is locally Lipschitz continuous in Ω
- The Sobolev gradient ∇v_ε exists and belongs to $L^\infty_{loc}(\Omega)$

The last assertion follows from Rademacher's theorem about Lipschitz functions, cf. [7]. Thus these "infimal convolutions" are rather regular. A most interesting property for a bounded viscosity supersolution is the following:

Proposition 12. *The approximant v_ε is a viscosity supersolution in the open subset of Ω where*

$$\mathrm{dist}(\mathrm{x}, \partial\Omega) > \sqrt{2L\varepsilon}.$$

Proof. First, notice that for x as required above, the infimum is attained at some point $y = x^\star$ comprised in Ω. The possibility that $x^\star$ escapes to the boundary of Ω is prohibited by the inequalities

$$\frac{|x-x^\star|^2}{2\varepsilon} \le \frac{|x-x^\star|^2}{2\varepsilon} + v(x^\star) = v_\varepsilon(x) \le v(x) \le L,$$

$$|x - x^\star| \le \sqrt{2L\varepsilon} < \mathrm{dist}(\mathrm{x}, \partial\Omega).$$

This explains why the domain shrinks a little. Now we give two proofs.

Viscosity proof: Fix a point x_0 so that also $x_0^\star \in \Omega$. Assume that the test function φ touches v_ε from below at x_0. Using

$$\varphi(x_0) = v_\varepsilon(x_0) = \frac{|x_0 - x_0^\star|^2}{2\varepsilon} + v(x_0^\star)$$

$$\varphi(x) \le v_\varepsilon(x) \le \frac{|x - y|^2}{2\varepsilon} + v(y)$$

we can verify that the function

$$\psi(x) = \varphi(x + x_0 - x_0^\star) - \frac{|x_0 - x_0^\star|^2}{2\varepsilon}$$

touches the original function v from below at the point $x_0^\star$. Since $x_0^\star$ is an interior point, the inequality

$$\operatorname{div}\left(|\nabla \psi(x_0^\star)|^{p-2} \nabla \psi(x_0^\star)\right) \le 0$$

holds by assumption. Because

$$\nabla \psi(x_0^\star) = \nabla \varphi(x_0), \quad D^2 \psi(x_0^\star) = D^2 \varphi(x_0),$$

we also have that

$$\operatorname{div}\left(|\nabla \varphi(x_0)|^{p-2} \nabla \varphi(x_0)\right) \le 0$$

at the original point x_0, where φ was touching v_ε. Thus v_ε fulfills the requirement in the definition.

Proof by Comparison: Denote

$$\Omega_\varepsilon = \left\{ x \in \Omega \mid \operatorname{dist}(x, \partial\Omega) > \sqrt{2L\varepsilon} \right\}.$$

We have to verify the comparison principle for v_ε. To this end, let $D \subset\subset \Omega_\varepsilon$ be a subdomain and suppose that $h \in C(\overline{D})$ is a p-harmonic function so that $v_\varepsilon(x) \ge h(x)$ on the boundary ∂D or, in other words,

$$\frac{|x - y|^2}{2\varepsilon} + v(y) \ge h(x) \text{ when } x \in \partial D, \ y \in \Omega.$$

Thus, writing $y = x + z$, we have

$$w(x) \equiv v(x + z) + \frac{|z|^2}{2\varepsilon} \ge h(x), \quad x \in \partial D$$

whenever z is a small fixed vector. But also $w = w(x)$ is a p-superharmonic function in Ω_ε. By the comparison principle $w(x) \ge h(x)$ in D. Given any point x_0 in D, we may choose $z = x_0^\star - x_0$. This yields $v_\varepsilon(x_0) \ge h(x_0)$. Since x_0 was arbitrary, we have verified that

$$v_\varepsilon(x) \ge h(x), \quad \text{when } x \in D.$$

This concludes the proof. We record the following result.

Corollary 13. *The approximant v_ε is a weak supersolution in Ω_ε, i.e.,*

$$\int_{\Omega_\varepsilon} \langle |\nabla v_\varepsilon|^{p-2} \nabla v_\varepsilon, \nabla \eta \rangle \, dx \geq 0 \tag{8}$$

when $\eta \geq 0$, $\eta \in C_0^\infty(\Omega_\varepsilon)$.

Proof. This is a combination of the Proposition and Lemma 9 provided that v_ε obeys the comparison principle. This is clear, if v is p-superharmonic. However, when the original assumption is that v is a viscosity supersolution, then one needs the piece of knowledge that v is also p-superharmonic. This is proved in [10]. □

The Caccioppoli estimate for v_ε reads

$$\int_\Omega \zeta^p |\nabla v_\varepsilon|^p \, dx \leq (pL)^p \int_\Omega |\nabla \zeta|^p \, dx,$$

when ε is so small that the support of ζ is in Ω_ε. By a compactness argument (a subsequence of) ∇v_ε is locally weakly convergent in $L^p(\Omega)$. We conclude that ∇v exists in Sobolev's sense and that

$$\nabla v_\varepsilon \quad \rightharpoonup \quad \nabla v \quad \text{weakly in } L^p_{loc}(\Omega).$$

By the weak lower semicontinuity of the integral also

$$\int_\Omega \zeta^p |\nabla v|^p \, dx \leq (pL)^p \int_\Omega |\nabla \zeta|^p \, dx.$$

We have proved the first part of the next theorem.

Theorem 14. *Suppose that v is a bounded viscosity supersolution in Ω. Then the Sobolev gradient ∇v exists and $v \in W^{1,p}_{loc}(\Omega)$. Moreover,*

$$\int_\Omega \langle |\nabla v|^{p-2} \nabla v, \nabla \eta \rangle \, dx \geq 0 \tag{9}$$

for all $\eta \geq 0, \eta \in C_0^\infty(\Omega)$.

Proof. To conclude the proof, we show that the convergence $\nabla v_\varepsilon \to \nabla v$ is strong in $L^p_{loc}(\Omega)$, so that we may pass to the limit under the integral sign in (8). To this end, fix a function $\theta \in C_0^\infty(\Omega)$, $0 \leq \theta \leq 1$ and use the test function $\eta = (v - v_\varepsilon)\theta$ in the equation for v_ε. Then

$$\int_\Omega \langle |\nabla v|^{p-2} \nabla v - |\nabla v_\varepsilon|^{p-2} \nabla v_\varepsilon, \nabla((v - v_\varepsilon)\theta) \rangle \, dx$$

$$\leq \int_\Omega \langle |\nabla v|^{p-2} \nabla v, \nabla((v - v_\varepsilon)\theta) \rangle \, dx \longrightarrow 0,$$

where the last integral approaches zero because of the weak convergence. The
first integral splits into the sum

$$\int_\Omega \theta \langle |\nabla v|^{p-2}\nabla v - |\nabla v_\varepsilon|^{p-2}\nabla v_\varepsilon, \nabla(v - v_\varepsilon)\rangle \, dx$$

$$+ \int_\Omega (v - v_\varepsilon)\langle |\nabla v|^{p-2}\nabla v - |\nabla v_\varepsilon|^{p-2}\nabla v_\varepsilon, \nabla\theta\rangle \, dx.$$

The last integral approaches zero because its absolute value is majorized by

$$\left(\int_D (v - v_\varepsilon)^p \, dx\right)^{1/p}\left[\left(\int_D |\nabla v|^p \, dx\right)^{(p-1)/p} + \left(\int_D |\nabla v_\varepsilon|^p \, dx\right)^{(p-1)/p}\right]\|\nabla\theta\|,$$

where D contains the support of θ and $\|v - v_\varepsilon\|_p$ approaches zero. Thus we
have established that

$$\int_\Omega \theta \langle |\nabla v|^{p-2}\nabla v - |\nabla v_\varepsilon|^{p-2}\nabla v_\varepsilon, \nabla(v - v_\varepsilon)\rangle \, dx$$

approaches zero. Now the strong convergence of the gradients follows from
the vector inequality

$$2^{2-p}|b - a|^p \le \langle |b|^{p-2}b - |a|^{p-2}a, b - a\rangle \tag{10}$$

valid for $p > 2$.   □

It also follows that the Caccioppoli estimates in Lemma 11 are valid for
locally bounded p-superharmonic functions. The case when v is unbounded
can be reached via the truncations

$$v_k = \min\{v(x), k\}, \quad k = 1, 2, 3, \ldots,$$

because Theorem 14 holds for these locally bounded functions. Aiming at a
local result, we may just by adding a constant assume that $v \ge 0$ in Ω. The
situation with $v = 0$ on the boundary $\partial\Omega$ offers expedient simplifications. We
shall describe an iteration procedure, under this extra assumption. See [13].

Lemma 15. *Assume that $v \ge 0$ and that $v_k \in W_0^{1,p}(\Omega)$ when $k = 1, 2, \ldots$
Then*

$$\int_\Omega |\nabla v_k|^p \, dx \le k \int_\Omega |\nabla v_1|^p \, dx$$

and, in the case $1 < p < n$

$$\int_\Omega v^\alpha \, dx \le C_\alpha \left(1 + \int_\Omega |\nabla v_1|^p \, dx\right)^{\frac{n}{n-p}}$$

whenever $\alpha < \frac{n(p-1)}{n-p}$.

Proof. Let j be a large index and use the test functions

$$\eta_k = (v_k - v_{k-1}) - (v_{k+1} - v_k), \quad k = 1, 2, \cdots, j-1$$

in the equation for v_j, i.e.,

$$\int_\Omega \langle |\nabla v_j|^{p-2} \nabla v_j, \nabla \eta_k \rangle \, dx \geq 0.$$

Indeed, $\eta_k \geq 0$. We obtain

$$A_{k+1} = \int_\Omega \langle |\nabla v_j|^{p-2} \nabla v_j, \nabla v_{k+1} - \nabla v_k \rangle \, dx$$

$$\leq \int_\Omega \langle |\nabla v_j|^{p-2} \nabla v_j, \nabla v_k - \nabla v_{k-1} \rangle \, dx = A_k.$$

Thus

$$A_{k+1} \leq A_1 = \int_\Omega |\nabla v_1|^p \, dx$$

and hence

$$A_1 + A_2 + \cdots + A_j \leq jA_1.$$

The "telescoping" sum becomes

$$\int_\Omega |\nabla v_j|^p \, dx \leq j \int_\Omega |\nabla v_1|^p \, dx.$$

This was the first claim.

If $1 < p < n$, it follows from Tshebyshev's and Sobolev's inequalities that

$$j|j \leq v \leq 2j|^{\frac{1}{p*}} \leq \left(\int_\Omega v_{2j}^{p*} \, dx \right)^{\frac{1}{p*}} \leq S \left(\int_\Omega |\nabla v_{2j}|^p \, dx \right)^{\frac{1}{p}} \leq S(2j)^{\frac{1}{p}} A_1^{\frac{1}{p}},$$

where $p* = np/(n-p)$. We arrive at the estimate

$$|j \leq v \leq 2j| \leq Cj^{-\frac{n(p-1)}{n-p}} A_1^{\frac{n}{n-p}}$$

for the measure of the level sets. To conclude the proof we write

$$\int_\Omega v^\alpha \, dx = \int_{v \leq 1} v^\alpha \, dx + \sum_{j=1}^\infty \int_{2^{j-1} < v \leq 2^j} v^\alpha \, dx.$$

Since

$$\int_{2^{j-1}<v\le 2^j} v^\alpha \, dx \le C \, 2^{j\alpha} 2^{-(j-1)\frac{n(p-1)}{n-p}} A_1^{\frac{n}{n-p}},$$

the series converges when α is as prescribed. $\square$

It remains to abandon the restriction about zero boundary values and to estimate

$$\int_\Omega |\nabla v_1|^p \, dx.$$

The **reduction to zero boundary values** is done locally in a ball $B_{2r} \subset\subset \Omega$. Suppose first that $v \in C(\overline{B_{2r}}) \cap W^{1,p}(B_{2r})$, $v \ge 0$, and define

$$w = \begin{cases} v \text{ in } \overline{B_r} \\ h \text{ in } B_{2r}\backslash\overline{B_r} \end{cases}$$

where h is the p-harmonic function in the annulus having outer boundary values zero and inner boundary values v. Now $h \le v$. The so defined w is p-superharmonic in B_{2r}, which follows by comparison. It is quite essential that the original v was defined in a domain larger than B_r! We also have

$$\int_{B_{2r}} |\nabla w|^p \, dx \le C r^{n-p} (\max_{B_{2r}} w)^p$$

after some estimation.[11]

Finally, if $v \in W^{1,p}(B_{2r})$ is semicontinuous and bounded (but not necessarily continuous), then we first modify the approximants v_ε defined as in (10) and obtain p-superharmonic functions w_ε. Since $0 \le w_\varepsilon \le v_\varepsilon \le v$, the previous estimate becomes

$$\int_{B_{2r}} |\nabla w_\varepsilon|^p \, dx \le C r^{n-p} (\max_{B_{2r}} v)^p$$

and, by the weak lower semicontinuity of the integral, we can pass to the limit as ε approaches zero. We end up with a p-superharmonic function $w \in W_0^{1,p}(B_{2r})$ such that $w = v$ in B_r and, in particular,

$$\int_{B_r} |\nabla v|^p \, dx \le \int_{B_{2r}} |\nabla w|^p \, dx \le C r^{n-p} (\max_{B_{2r}} v)^p.$$

---

[11] It is important to include the whole B_{2r}. Of course, the Caccioppoli estimate (Lemma 11) will do over any smaller ball B_ϱ, $\varrho < 2r$. To get the missing estimate, say over the boundary annulus $B_{2r} \setminus B_{3r/2}$, the test function $\eta = \zeta h$ works in Definition 5, where $\zeta = 1$ in the annulus and $= 0$ on ∂B_r. The zero boundary values of the weak solution h were essential.

This is the desired modified function. Now, repeat the procedure with every function $\min\{v(x), k\}$ in sight. We obtain

$$\int_{B_r} |\nabla v_1|^p \, dx \leq \int_{B_{2r}} |\nabla w_1|^p \, dx \leq C r^{n-p} 1^p$$

for the modification of $v_1 = \min\{v(x), 1\}$. We have achieved that the bounds in the previous lemma hold for the modified function over the domain B_{2r} and a fortiori for the original v, estimated only over the smaller ball B_r. Such a *local* estimate is all that is needed in the proof of the theorem below.

Theorem 16. *Suppose that v is a viscosity supersolution ($=$ p-super-harmonic function) in Ω. Then*

$$v \in L^q_{loc}(\Omega), \quad whenever \quad q < \frac{n(p-1)}{n-p}$$

in the case $1 < p \leq n$ and v is continuous if $p > n$. Moreover, ∇v exists in Sobolev's sense[12] and

$$\nabla v \in L^q_{loc}(\Omega), \quad whenever \quad q < \frac{n(p-1)}{n-1}$$

in the case $1 < p \leq n$. In the case $p > n$ we have $\nabla v \in L^p_{loc}(\Omega)$. Finally,

$$\int_\Omega \langle |\nabla v|^{p-2} \nabla v, \nabla \eta \rangle \, dx \geq 0 \tag{11}$$

when $\eta \geq 0$, $\eta \in C_0^\infty(\Omega)$.

Proof. In view of the local nature of the theorem we may assume that $v > 0$. According to the previous construction we can further reduce the proof to the case $v_k \in W_0^{1,p}(B_{2r})$ for each truncation at height k. The first part of the theorem is included in Lemma 15 when $1 < p < n$. We skip the borderline case $p = n$. The case $p > n$ is related to the fact that then all functions in the Sobolev space $W^{1,p}$ are continuous.

We proceed to the **estimation of the gradient**. First we keep $1 < p < n$ and write

$$\int_B |\nabla v_k|^q \, dx = \int_B v_k^{\frac{(1+\alpha)q}{p}} \left| \frac{\nabla v_k}{v_k^{(1+\alpha)/p}} \right|^q \, dx$$

$$\leq \left\{ \int_B v_k^{\frac{(1+\alpha)q}{p-q}} \, dx \right\}^{1-\frac{q}{p}} \left\{ \int_B v_k^{-1-\alpha} |\nabla v_k|^p \, dx \right\}^{\frac{q}{p}}.$$

---

[12]Strictly speaking, one needs $p > 2 - \frac{1}{n}$ so that $q \geq 1$. This can be circumvented.

Take $q < n(p-1)/(n-1)$ and fix α so that

$$\frac{(1+\alpha)q}{p-q} < \frac{n(p-1)}{n-p}.$$

Continuing, the Caccioppoli estimate yields the majorant

$$\leq \left\{ \int_B v_k^{\frac{(1+\alpha)q}{p-q}} \, dx \right\}^{1-\frac{q}{p}} C \left\{ \int_{2B} v_k^{p-1-\alpha} \, dx \right\}^{\frac{q}{p}}. \tag{12}$$

We can take $v \geq 1$. Then let $k \longrightarrow \infty$. Clearly, the resulting majorant is finite (Lemma 15). This concludes the case $1 < p < n$.

If $p > n$ we obtain that

$$\int_{B_r} |\nabla \log v_k|^p \, dx \leq C r^{n-p}$$

from (6), where C is independent of k. Hence $\log v_k$ is continuous. So is v itself. Now

$$\int_{B_r} |\nabla v_k|^p \, dx = \int_{B_r} v_k^p |\nabla \log v_k|^p \, dx \leq C \|v\|_\infty^p r^{n-p}.$$

implies the desired p-summability of the gradient.   □

It stands to reason that the lower semicontinuous solutions of (9) are p-superharmonic functions. However, this is not known under the summability assumption $\nabla v \in L^q_{loc}(\Omega)$ accompanying the differential equation, if $q < p$ and $p < n$. In fact, an example of Serrin indicates that even for solutions to linear equations strange phenomena occur, cf. [26]. False solutions appear, when the a priori summability of the gradient is too poor. About this topic there is nowadays a theory credited to Iwaniec, cf. [19].[13]

3   The Evolutionary Equation

This chapter is rather independent of the previous one. See [6], [11], [12], [28], [30], and [31] for some background. After some definitions we first treat bounded supersolutions and then the unbounded ones. As a mnemonic rule, $v_t \geq \Delta_p v$ for smooth supersolutions, $u_t \leq \Delta_p u$ for smooth subsolutions. We need the following classes of supersolutions:

(1) **Weak supersolutions** (test functions under the integral sign)
(2) p-**supercaloric functions** (defined via a comparison principle)
(3) **Viscosity supersolutions** (test functions evaluated at points of contact)

---

[13]The "pathological solutions" of Serrin are now called "very weak solutions".

The weak supersolutions do not form a good closed class under monotone convergence.

3.1   Definitions

We first define the concept of solutions, p-supercaloric functions and viscosity supersolutions. Then we state the main theorem. The section ends with an outline of the procedure for the proof.

Suppose that Ω is a bounded domain in $\mathbf{R^n}$ and consider the space-time cylinder $\Omega_T = \Omega \times (0,T)$. Its *parabolic boundary* consists of the portions $\Omega \times \{0\}$ and $\partial\Omega \times [0,T]$.

Definition 17. *In the case[14] $p \geq 2$ we say that $u \in L^p(0,T;W^{1,p}(\Omega))$ is a* weak solution *of the Evolutionary p-Laplace Equation, if*

$$\int_0^T \int_\Omega \left(-u\phi_t + \langle |\nabla u|^{p-2}\nabla u, \nabla\phi\rangle\right) dx\,dt = 0 \qquad (13)$$

for all $\phi \in C_0^1(\Omega_T)$. If the integral in (13) is ≥ 0 for all test functions $\phi \geq 0$, we say that u is a weak supersolution.

In particular, one has the requirement

$$\int_0^T \int_\Omega \left(|u|^p + |\nabla u|^p\right) dx\,dt < \infty.$$

By the regularity theory one may regard a weak solution $u = u(x,t)$ as continuous.[15] For simplicity we call the continuous weak solutions for *p-caloric functions*.[16]

The *interior Hölder estimate*[17] takes the following form for solutions according to [6]. In the subdomain $D \times (\delta, T - \delta)$

$$|u(x_1,t_1) - u(x_2,t_2)| \leq \gamma \|u\|_{L^\infty(\Omega_T)} \left(|x_1 - x_2|^\alpha + |t_1 - t_2|^{\alpha/p}\right), \qquad (14)$$

where the positive exponent α depends only on n and p, while the constant γ depends, in addition, on the distance to the subdomain.

---

[14]The singular case $1 < p < 2$ requires an extra a priori assumption, for example, $u \in L^\infty(0,T;L^2(\Omega))$ will do.

[15]The weak supersolutions are lower semicontinuous according to [18], see chapter "The Problems of the Obstacle in Lower Dimension and for the Fractional Laplacian."

[16]One may argue that this is more adequate than "p-parabolic functions", which is in use.

[17]This is weaker than the estimate in [6]. See also [29] for intrinsic scaling.

Definition 18. *We say that a function $v : \Omega_T \to (-\infty, \infty]$ is p-supercaloric in Ω_T, if*

(i) *v is finite in a dense subset*
(ii) *v is lower semicontinuous*
(iii) *In each cylindrical subdomain $D \times (t_1, t_2) \subset\subset \Omega_T$ v obeys the comparison principle:*
if $h \in C(\overline{D} \times [t_1, t_2])$ is p-caloric in $D \times (t_1, t_2)$, then $v \geq h$ on the parabolic boundary of $D \times (t_1, t_2)$ implies that $v \geq h$ in the whole subdomain.

The next definition is from the theory of viscosity solutions. One defines them as being both viscosity super- and subsolutions, since it is not practical to do it in one stroke.

Definition 19. *Let $p \geq 2$. A function $v : \Omega_T \to (-\infty, \infty]$ is called a viscosity supersolution, if*

(i) *v is finite in a dense subset*
(ii) *v is lower semicontinuous*
(iii) *Whenever $(x_0, t_0) \in \Omega_T$ and $\phi \in C^2(\Omega_T)$ are such that $v(x_0, t_0) = \phi(x_0, t_0)$ and $v(x, t) > \phi(x, t)$ when $(x, t) \neq (x_0, t_0)$, we have*

$$\phi_t(x_0, t_0) \geq \nabla \cdot \left(|\nabla \phi(x_0, t_0)|^{p-2} \nabla \phi(x_0, t_0) \right).$$

The p-supercaloric functions are exactly the same as the viscosity supersolutions. For a proof of this fundamental equivalence we refer to [10]. The leading example is the Barenblatt solution, which is a viscosity supersolution in the whole $\mathbf{R}^{n+1}$. Another example is any function of the form

$$v(x, t) = g(t),$$

where $g(t)$ is an arbitrary monotone increasing lower semicontinuous function. We also mention

$$v(x, t) + \frac{\varepsilon}{T - t}, \quad 0 < t < T,$$

$$v(x, t) = \min\{v_1(x, t), \dots, v_2(x, t)\}.$$

Finally, if $v \geq 0$ is a viscosity supersolution, so is the function obtained by redefining $v(x, t) = 0$ when $t \leq 0$. –These were some examples.

An **outline of the procedure** is the following. We aim at proving the summability results (Theorem 2) for a viscosity supersolution v.

Step 1. Assume first that v is bounded.
Step 2. Approximate v locally with infimal convolutions v_ε. These are differentiable.

Step 3.   The infimal convolutions are viscosity supersolutions and they are shown to be weak supersolutions of the equation (with test functions under the integral sign).

Step 4.   Estimates of the Caccioppoli type for v_ε are extracted from the equation.

Step 5.   The Caccioppoli estimates are passed over from v_ε to v. This concludes the proof for bounded functions.

Step 6.   The unbounded case is reached via the bounded viscosity supersolutions $v_k = \min\{v, k\}, k = 1, 2, \cdots$, for which the results in Step 5 already are available.

Step 7.   An iteration with respect to the index k is designed so that the final result does not blow up as $k \to \infty$. This works well when the parabolic boundary values (in the subdomain studied) are zero.

Step 8.   An extra construction is performed to reduce the proof to the situation of zero parabolic boundary values (so that the iterated result in Step 7 is at our disposal).

3.2   Bounded Supersolutions

We aim at proving Theorem 2. The first step is to consider *bounded* viscosity supersolutions. We want to prove that they are weak supersolutions. We first approximate them with their infimal convolutions. Then estimates mainly of the Caccioppoli type are proved for these approximants. Finally, the so obtained estimates are passed over to the original functions. Assume therefore that

$$0 \le v(x,t) \le L, \quad (x,t) \in \Omega_T = \Omega \times (0,T).$$

The approximants

$$v_\varepsilon(x,t) = \inf_{(y,\tau)\in\Omega_T} \left\{ v(y,\tau) + \frac{|x-y|^2 + |t-\tau|^2}{2\varepsilon} \right\}, \quad \varepsilon > 0,$$

have the properties

- $v_\varepsilon(x,t) \nearrow v(x,t)$ as $\varepsilon \to 0+$
- $v_\varepsilon(x,t) - \frac{|x|^2 + t^2}{2\varepsilon}$ is locally concave in Ω_T
- v_ε is locally Lipschitz continuous in Ω_T
- The Sobolev derivatives $\frac{\partial v_\varepsilon}{\partial t}$ and ∇v_ε exist and belong to $L^\infty_{loc}(\Omega_T)$

The last assertion follows from Rademacher's theorem about Lipschitz functions. Thus these "infimal convolutions" are rather regular. The existence of the time derivative is very useful. A most interesting property for a bounded viscosity supersolution is the following:

Proposition 20. *The approximant v_ε is a viscosity supersolution in the open subset of Ω_T where*

$$\text{dist}((\mathrm{x},\mathrm{t}),\partial\Omega_T) > \sqrt{2L\varepsilon}.$$

Proof. First, notice that for (x,t) as required above, the infimum is attained at some point $(y,\tau) = (x^\star,t^\star)$ comprised in Ω_T. The possibility that $(x^\star,t^\star)$ escapes to the boundary of Ω is prohibited by the inequalities

$$\frac{|x-x^\star|^2 + |t-t^\star|^2}{2\varepsilon} \leq \frac{|x-x^\star|^2 + |t-t^\star|^2}{2\varepsilon} + v(x^\star,t^\star)$$

$$= v_\varepsilon(x,t) \leq v(x,t) \leq L,$$

$$\sqrt{|x-x^\star|^2 + |t-t^\star|^2} \leq \sqrt{2L\varepsilon} < \text{dist}((\mathrm{x},\mathrm{t}),\partial\Omega_T).$$

Thus the domain shrinks a little. Again there are two proofs.

Viscosity proof: Fix a point (x_0,t_0) so that also $(x_0^\star,t_0^\star) \in \Omega_T$. Assume that the test function φ touches v_ε from below at $(x_0^\star,t_0^\star)$. Using

$$\varphi(x_0,t_0) = v_\varepsilon(x_0,t_0) = \frac{|x_0-x_0^\star|^2 + |t_0-t_0^\star|^2}{2\varepsilon} + v(x_0^\star,t_0^\star)$$

$$\varphi(x,t) \leq v_\varepsilon(x,t) \quad \leq \frac{|x-y|^2 + |t-\tau|^2}{2\varepsilon} + v(y,\tau)$$

we can verify that the function

$$\psi(x,t) = \varphi(x + x_0 - x_0^\star, t + t_0 - t_0^\star) - \frac{|x_0-x_0^\star|^2 + |t_0-t_0^\star|^2}{2\varepsilon}$$

touches the original function v from below at the point $(x_0^\star,t_0^\star)$. Since $(x_0^\star,t_0^\star)$ is an interior point, the inequality

$$\text{div}\big(|\nabla\psi(x_0^\star,t_0^\star)|^{p-2}\nabla\psi(x_0^\star,t_0^\star)\big) \leq \psi_t(x_0^\star,t_0^\star)$$

holds by assumption. Because

$$\psi_t(x_0^\star,t_0^\star) = \varphi_t(x_0,t_0), \quad \nabla\psi(x_0^\star,t_0^\star) = \nabla\varphi(x_0,t_0), \quad D^2\psi(x_0^\star,t_0^\star) = D^2\varphi(x_0,t_0)$$

we also have that

$$\text{div}\big(|\nabla\varphi(x_0,t_0)|^{p-2}\nabla\varphi(x_0,t_0)\big) \leq \varphi_t(x_0,t_0)$$

at the original point (x_0,t_0), where φ was touching v_ε. Thus v_ε fulfills the requirement in the definition.

Proof by Comparison: We have to verify the comparison principle for v_ε in a subcylinder D_{t_1,t_2} having at least the distance $\sqrt{2L\varepsilon}$ to the boundary of Ω_T. To this end, assume that $h \in C(\overline{D_{t_1,t_2}})$ is a p-caloric function such that $v_\varepsilon \geq h$ on the parabolic boundary. It follows that the inequality

$$\frac{|x-y|^2 + |t-\tau|^2}{2\varepsilon} + v(y,\tau) \geq h(x,t)$$

is available when $(y,\tau) \in \Omega_T$ and (x,t) is on the parabolic boundary of D_{t_1,t_2}. Fix an arbitrary point (x_0, t_0) in D_{t_1,t_2}. Then we can take $y = x + x_0^\star - x_0$ and $\tau = t + t_0^\star - t_0$ in the inequality above so that

$$w(x,t) \equiv v(x + x_0^\star - x_0, t + t_0^\star - t_0) + \frac{|x_0 - x_0^\star|^2 + |t_0 - t_0^\star|^2}{2\varepsilon} \geq h(x,t)$$

when (x,t) is on the parabolic boundary. But the translated function w is p-supercaloric in the subcylinder D_{t_1,t_2}. By the comparison principle $w \geq h$ in the whole subcylinder. In particular,

$$v_\varepsilon(x_0, t_0) = w(x_0, t_0) \geq h(x_0, t_0).$$

This proves the comparison principle for v_ε.   □

The "viscosity proof" did not contain any comparison principle while the "proof by comparison" required the piece of knowledge that the original v obeys the principle. To proceed, this comparison is unevitable. It is needed for v_ε. As a viscosity supersolution v_ε obeys the comparison principle in its domain according to [10]. This *parabolic* comparison principle allows comparison in space-time cylinders. We will encounter domains of a more general shape, but the following *elliptic* version of the principle turns out to be enough for our purpose. Instead of the expected parabolic boundary, the whole boundary (the "Euclidean" boundary) appears.

Proposition 21. *Given a domain $\Upsilon \subset\subset \Omega_\varepsilon$ and a p-caloric function $h \in C(\overline{\Upsilon})$, then $v_\varepsilon \geq h$ on the whole boundary $\partial\Upsilon$ implies that $v_\varepsilon \geq h$ in Υ.*

Now Υ does not have to be a space-time cylinder and $\partial\Upsilon$ is the total boundary in $\mathbb{R}^{n+1}$.

Proof. It is enough to realize that the proof is immediate when Υ is a finite union of space-time cylinders $D_j \times (a_j, b_j)$. To verify this, just start with the earliest cylinder(s) and pay due attention to the passages of t over the a_j's and b_j's. Then the general case follows by exhausting Υ with such unions. Indeed, given $\alpha > 0$ the compact set $\{h(x,t) \geq v_\varepsilon(x,t)\}$ is contained in an open finite union

$$\bigcup D_j \times (a_j, b_j)$$

comprised in Υ so that $h < v_\varepsilon + \alpha$ on the (Euclidean) boundary of the union. It follows that $h \leq v_\varepsilon + \alpha$ in the whole union. Since α was arbitrary, we conclude that $v_\varepsilon \geq h$ in Υ. $\qquad\qquad\qquad\qquad\qquad\qquad\qquad\qquad\qquad\quad$ □

The above *elliptic* comparison principle does not acknowledge the presence of the parabolic boundary. The reasoning above can easily be changed so that the latest portion of the boundary is exempted. For this improvement, suppose that $t < T^\star$ for all $(x,t) \in \Upsilon$; in this case $\partial\Upsilon$ may have a plane portion with $t = T^\star$. It is now sufficient to verify that

$$v_\varepsilon \geq h \quad \text{on} \quad \partial\Upsilon \quad \text{when} \quad t < T^\star$$

in order to conclude that $v_\varepsilon \geq h$ in Υ. To see this, just use

$$v_\varepsilon + \frac{\sigma}{T^\star - t}$$

in the place of v_ε and then let $\sigma \to 0+$. This variant of the comparison principle is convenient for the proof of the following conclusion.

Lemma 22. *The approximant v_ε is a weak supersolution in the shrunken domain, i.e.,*

$$\int_0^T \int_\Omega \left(-v_\varepsilon \frac{\partial\phi}{\partial t} + \langle |\nabla v_\varepsilon|^{p-2}\nabla v_\varepsilon, \nabla\phi\rangle \right) dx\, dt \geq 0 \tag{15}$$

whenever $\phi \in C_0^\infty(\Omega_\varepsilon \times (\varepsilon, T - \varepsilon))$, $\phi \geq 0$.

Proof. We show that in a given subdomain $D_{t_1,t_2} = D \times (t_1, t_2)$ of the "shrunken domain" our v_ε coincides with the solution of an obstacle problem. The solutions of the obstacle problem are *per se* weak supersolutions. Hence, so is v_ε. Consider the class of all functions

$$\begin{cases} w \in C(\overline{D_{t_1,t_2}}) \cap L^p(t_1, t_2; W^{1,p}(D)), \\ w \geq v_\varepsilon \text{ in } D_{t_1,t_2}, \text{ and} \\ w = v_\varepsilon \text{ on the parabolic boundary of } D_{t_1,t_2}. \end{cases}$$

The function v_ε itself acts as an obstacle and induces the boundary values. There exists a (unique) weak supersolution w_ε in this class satisfying the variational inequality

$$\int_{t_1}^{t_2} \int_D \left((\psi - w_\varepsilon)\frac{\partial\psi}{\partial t} + \langle |\nabla w_\varepsilon|^{p-2}\nabla w_\varepsilon, \nabla(\psi - w_\varepsilon)\rangle \right) dx\, dt$$

$$\geq \frac{1}{2}\int_D (\psi(x,t_2) - w_\varepsilon(x,t_2))^2\, dx$$

for all smooth ψ in the aforementioned class. Moreover, w_ε is p-caloric in the open set $A_\varepsilon = \{w_\varepsilon > v_\varepsilon\}$, where the obstacle does not hinder. We refer to [4]. On the boundary ∂A_ε we know that $w_\varepsilon = v_\varepsilon$ except possibly when $t = t_2$. By the elliptic comparison principle we have $v_\varepsilon \geq w_\varepsilon$ in A_ε. On the other hand $w_\varepsilon \geq v_\varepsilon$. Hence $w_\varepsilon = v_\varepsilon$.

To finish the proof, let $\varphi \in C_0^\infty(D_{t_1,t_2})$, $\varphi \geq 0$, and choose $\psi = v_\varepsilon + \varphi = w_\varepsilon + \varphi$ above. An easy manipulation yields (15). $\square$

Recall that $0 \leq v \leq L$. Then also $0 \leq v_\varepsilon \leq L$. An estimate for ∇v_ε is provided in the well-known lemma below.

Lemma 23 (Caccioppoli). *The inequality*

$$\int_0^T \int_\Omega \zeta^p |\nabla v_\varepsilon|^p \, dx \, dt \leq CL^p \int_0^T \int_\Omega |\nabla \zeta|^p \, dx \, dt + CL^2 \int_0^T \int_\Omega \left| \frac{\partial \zeta^p}{\partial t} \right| dx \, dt$$

holds whenever $\zeta \in C_0^\infty(\Omega_\varepsilon \times (\varepsilon, T - \varepsilon))$, $\zeta \geq 0$.

Proof. Use the test function

$$\phi(x,t) = (L - v_\varepsilon(x,t)) \zeta^p(x,t). \quad \square$$

The Caccioppoli estimate above leads to the conclusion that, keeping $0 \leq v \leq L$, the Sobolev gradient $\nabla v \in L_{loc}^p$ exists and

$$\nabla v_\varepsilon \rightharpoonup \nabla v \quad \text{weakly in} \quad L_{loc}^p,$$

at least for a subsequence. For v the Caccioppoli estimate

$$\int_0^T \int_\Omega \zeta^p |\nabla v|^p \, dx \, dt \leq CL^p \int_0^T \int_\Omega |\nabla \zeta|^p \, dx \, dt + CL^2 \int_0^T \int_\Omega \left| \frac{\partial \zeta^p}{\partial t} \right| dx \, dt,$$

is immediate, because of the lower semicontinuity of the integral under weak convergence. However the corresponding passage to the limit under the integral sign of

$$\int_0^T \int_\Omega \left(-v_\varepsilon \frac{\partial \phi}{\partial t} + \langle |\nabla v_\varepsilon|^{p-2} \nabla v_\varepsilon, \nabla \phi \rangle \right) dx \, dt \geq 0$$

requires a justification, as $\varepsilon \to 0$. The elementary vector inequality

$$\left| |b|^{p-2} b - |a|^{p-2} a \right| \leq (p-1)|b-a| \, (|b| + |a|)^{p-2}, \tag{16}$$

$p > 2$, and Hölder's inequality show that it is sufficient to establish that

$$\nabla v_\varepsilon \longrightarrow \nabla v \quad \text{strongly in} \quad L_{loc}^{p-1},$$

to accomplish the passage. Notice the exponent $p-1$ in place of p. This strong convergence is given in the next theorem, where the sequence is renamed to v_k. The proof is from [22].

Theorem 24. *Suppose that $v_1, v_2, v_3, \ldots$ is a sequence of Lipschitz continuous weak supersolutions, such that*

$$0 \leq v_k \leq L \quad in \quad \Omega_T = \Omega \times (0,t), \quad v_k \to v \quad in \quad L^p(\Omega_T).$$

Then

$$\nabla v_1, \nabla v_2, \nabla v_3, \ldots$$

is a Cauchy sequence in $L_{loc}^{p-1}(\Omega_T)$.

Proof. The central idea is that the measure of the set where $|v_j - v_k| > \delta$ is small. Given $\delta > 0$, we have, in fact,

$$\text{mes}\{|v_j - v_k| > \delta\} \leq \delta^{-p}\|v_j - v_k\|_p^p \tag{17}$$

according to Tshebyshef's inequality. Fix a test function $\theta \in C_0^\infty(\Omega_T)$, $0 \leq \theta \leq 1$. From the Caccioppoli estimate we can extract a bound of the form

$$\iint_{\{\theta \neq 0\}} |\nabla v_k|^p \, dx \, dt \leq A^p, \quad k = 1, 2, \ldots,$$

since the support is a compact subset. Fix the indices k and j and use the test function

$$\varphi = (\delta - w_{jk})\theta$$

where

$$w_{jk} = \begin{cases} \delta, & \text{if } v_j - v_k > \delta \\ v_j - v_k, & \text{if } |v_j - v_k| \leq \delta \\ -\delta, & \text{if } v_j - v_k < -\delta \end{cases}$$

in the equation

$$\int_0^T \int_\Omega \left(-v_j \frac{\partial \phi}{\partial t} + \langle |\nabla v_j|^{p-2} \nabla v_j, \nabla \phi \rangle \right) dx \, dt \geq 0.$$

Since $|w_{jk}| \leq \delta$, we have $\varphi \geq 0$. In the equation for v_k use

$$\varphi = (\delta + w_{jk})\theta.$$

Add the two equations and arrange the terms:

$$\iint_{|v_j - v_k| \leq \delta} \theta \langle |\nabla v_j|^{p-2} \nabla v_j - |\nabla v_k|^{p-2} \nabla v_k, \nabla v_j - \nabla v_k \rangle \, dx \, dt$$

$$\leq \delta \int_0^T \int_\Omega \langle |\nabla v_j|^{p-2} \nabla v_j + |\nabla v_k|^{p-2} \nabla v_k, \nabla \theta \rangle \, dx \, dt$$

$$- \int_0^T \int_\Omega w_{jk} \langle |\nabla v_j|^{p-2} \nabla v_j - |\nabla v_k|^{p-2} \nabla v_k, \nabla \theta \rangle \, dx \, dt$$

$$+ \int_0^T \int_\Omega (v_j - v_k) \frac{\partial}{\partial t}(\theta w_{jk}) \, dx \, dt - \delta \int_0^T \int_\Omega (v_j + v_k) \frac{\partial \theta}{\partial t} \, dx \, dt$$

$$= I - II + III - IV.$$

The left-hand side is familiar from inequality (22). As we will see, the right-hand side is of the magnitude $O(\delta)$. We begin with term III, which contains time derivatives that ought to be avoided. Integration by parts yields

$$III = \int_0^T \int_\Omega \theta \frac{\partial \theta}{\partial t} \left(\frac{w_{jk}^2}{2} \right) dx \, dt + \int_0^T \int_\Omega (v_j - v_k) w_{jk} \frac{\partial \theta}{\partial t} \, dx \, dt$$

$$= -\frac{1}{2} \int_0^T \int_\Omega w_{jk}^2 \frac{\partial \theta}{\partial t} \, dx \, dt + \int_0^T \int_\Omega (v_j - v_k) w_{jk} \frac{\partial \theta}{\partial t} \, dx \, dt.$$

We obtain the estimate

$$|III| \leq \frac{1}{2}\delta^2 \|\theta_t\|_1 + 2L\delta \|\theta_t\|_1 \leq \delta C_3.$$

For the last term we immediately have

$$|IV| \leq 2\delta L \|\theta_t\|_1 = \delta C_4.$$

The two first terms are easy,

$$|I| \leq \delta C_1, \quad |II| \leq \delta C_2.$$

Summing up,

$$|I| + |II| + |III| + |IV| \leq C\delta.$$

Using the vector inequality (10) to estimate the left hand side, we arrive at

$$\iint_{|v_j - v_k| \leq \delta} \theta |\nabla v_j - \nabla v_k|^p \, dx \, dt \ \leq 2^{p-2} \delta C,$$

$$\iint_{|v_j - v_k| \leq \delta} \theta |\nabla v_j - \nabla v_k|^{p-1} \, dx \, dt = O(\delta^{1 - \frac{1}{p}}).$$

We also have in virtue of (17)

$$\iint_{|v_j - v_k| > \delta} \theta |\nabla v_j - \nabla v_k|^{p-1} \, dx \, dt$$

$$\leq \delta^{-1} \|v_j - v_k\|_p \left(\|\nabla v_j\|_p + \|\nabla v_k\|_p \right)^{p-1}$$

$$\leq (2A)^{p-1} \delta^{-1} \|v_j - v_k\|_p.$$

Finally, combining the estimates over the sets $|v_j - v_k| \leq \delta$ and $|v_j - v_k| > \delta$, we have an integral over the whole domain:

$$\int_0^T \int_\Omega \theta |\nabla v_j - \nabla v_k|^{p-1} \, dx \, dt \le O(\delta^{1-\frac{1}{p}}) + C_5 \delta^{-1} \|v_j - v_k\|_p.$$

Since the left-hand side is independent of δ, we can make it as small as we please, by first fixing δ small enough and then taking the indices large enough. $\square$

We have arrived at the following result for bounded supersolutions.

Theorem 25. *Let v be a bounded viscosity supersolution of $v_t = \Delta_p v$ in Ω_T, $p \ge 2$. Then*

$$\nabla v = \left(\frac{\partial v}{\partial x_1}, \ldots, \frac{\partial v}{\partial x_n} \right)$$

exists in Sobolev's sense, $\nabla v \in L^p_{loc}(\Omega_T)$, and

$$\int_0^T \int_\Omega \left(-v \frac{\partial \phi}{\partial t} + \langle |\nabla v|^{p-2} \nabla v, \nabla \phi \rangle \right) dx \, dt \ge 0$$

for all non-negative compactly supported test functions φ.

Notice that so far the exponent is p, as it should for v bounded.

Remark. It was established that the bounded viscosity supersolutions are also weak supersolutions. In the opposite direction, according to [18] every weak supersolution is lower semicontinuous upon a redefinition in a set of $(n+1)$-dimensional measure 0. Moreover, the representative obtained as

$$\underset{(y,\tau)\to(x,t)}{\text{ess liminf }} v(y, \tau)$$

will do. A proof is given in chapter "The Problems of the Obstacle in Lower Dimension and for the Fractional Laplacian."

We need a few auxiliary results.

Lemma 26 (Sobolev's inequality). *If $u \in L^p(0, t; W_0^{1,p}(\Omega))$, then*

$$\int_0^T \int_\Omega |u|^{p(1+\frac{2}{n})} \, dx \, dt \le C \int_0^T \int_\Omega |\nabla u|^p \, dx \, dt \left\{ \underset{0<t<T}{\text{ess sup}} \int_\Omega |u(x,t)|^2 \, dx \right\}^{\frac{p}{n}}.$$

Proof. See for example [6, Chap. 1, Proposition 3.1].

If the test function ϕ is zero on the lateral boundary $\partial \Omega \times [t_1, t_2]$, then the differential inequality for the weak supersolution takes the form

$$\int_{t_1}^{t_2} \int_\Omega \left(-v \frac{\partial \phi}{\partial t} + \langle |\nabla v|^{p-2} \nabla v, \nabla \phi \rangle \right) dx \, dt$$

$$+ \int_\Omega v(x, t_2) \phi(x, t_2) \, dx \ge \int_\Omega v(x, t_1) \phi(x, t_1) \, dx.$$

Thus, if v is zero on the lateral boundary, we may take $\phi = v$ above. We obtain

$$\frac{1}{2}\int_\Omega v(x,t_1)^2\,dx \le \int_{t_1}^{t_2}\int_\Omega |\nabla v|^p\,dx\,dt + \frac{1}{2}\int_\Omega v(x,t_2)^2\,dx, \qquad (18)$$

which estimates the *past* in terms of the *future* and an "energy term".

3.3   Unbounded Supersolutions

We proceed to study an unbounded viscosity supersolution ($=p$-supercaloric function) v. Let us briefly describe the method taken from [14]. The starting point is to apply Theorem 24 on the functions $v_k = \min\{v, k\}$ so that estimates depending on $k = 1, 2, \cdots$ are obtained. Then an iterative procedure is used to gradually increase the summability exponent of v. First, we achieve that v^α for some small exponent $\alpha < p-2$. That result is iterated, again using the v_k's till we come close to the exponent $\alpha = p - 1 - 0$. Then the passage over $p-1$ requires a special, although simple, device. At the end we will reach the desired summability for the function v itself. From this it is not difficult to obtain the corresponding result also for the gradient ∇v. Again the v_k's are employed.

The considerations are in a bounded subdomain, which we again call $\Omega_T = \Omega \times (0, T)$, for simplicity. The situation is easier when the function is zero on the parabolic boundary:

$$v(x,0) = 0 \text{ when } x \in \Omega, \quad v = 0 \text{ on } \partial\Omega \times [0,T].$$

We may assume that $v \ge 0$. The functions

$$v_k(x,t) = \min\{v(x,t), k\}, \qquad k = 0, 1, 2, \ldots$$

cut off at the height k are bounded, whence the previous results apply for them. Fix a large index j. We may use the test functions

$$\phi_k = (v_k - v_{k-1}) - (v_{k+1} - v_k), \quad k = 1, 2, \ldots, j$$

in the equation

$$\int_0^\tau\int_\Omega \langle |\nabla v_j|^{p-2}\nabla v_j, \nabla\phi_j\rangle\,dx\,dt + \int_0^\tau\int_\Omega \phi_k\frac{\partial v_j}{\partial t}\,dx\,dt \ge 0,$$

where $0 < \tau \le T$. Indeed, $\phi_k \ge 0$. The "forbidden" time derivative can be avoided through an appropriate regularization. In principle v_j is first replaced by its convolution with a mollifier and later the limit is to be taken. We postpone this complication in order to keep the exposition more transparent. The insertion of the test function yields

$$\int_0^T \int_\Omega \left(\langle |\nabla v_j|^{p-2} \nabla v_j, \nabla (v_{k+1} - v_k) \rangle + (v_{k+1} - v_k) \frac{\partial v_j}{\partial t} \right) dx\, dt$$

$$\leq \int_0^T \int_\Omega \left(\langle |\nabla v_j|^{p-2} \nabla v_j, \nabla (v_k - v_{k-1}) \rangle + (v_k - v_{k-1}) \frac{\partial v_j}{\partial t} \right) dx\, dt,$$

succinctly written as

$$a_{k+1}(\tau) \leq a_k(\tau).$$

It follows that

$$\sum_{k=1}^{j} a_k(\tau) \leq j a_1(\tau)$$

and, since the sum is "telescoping", we have the result below.

Lemma 27. *If each* $v_k \in L^p(0, T; W_0^{1,p}(\Omega))$ *and* $v_k(x, 0) = 0$ *when* $x \in \Omega$, *then*

$$\int_0^T \int_\Omega |\nabla v_j|^p \, dx\, dt + \frac{1}{2} \int_\Omega v_j^2(x, \tau)\, dx$$

$$\leq j \int_0^T \int_\Omega |\nabla v_1|^p \, dx\, dt + j \int_\Omega v_j(x, \tau)\, dx$$

holds for a.e. τ *in the range* $0 < \tau \leq T$.

Before continuing, we justify the use of the time derivative in the previous reasoning.

Regularisation of the equation. We use the convolution

$$(f \star \rho_\varepsilon)(x, t) = \int_{-\infty}^{\infty} f(x, t - s) \rho_\varepsilon(s)\, ds, \tag{19}$$

where ρ_ε is, for instance, Friedrich's mollifier defined as

$$\rho_\varepsilon(t) = \begin{cases} \frac{C}{\varepsilon} e^{-\varepsilon^2/(\varepsilon^2 - t^2)}, & |t| < \varepsilon, \\ 0, & |t| \geq \varepsilon. \end{cases}$$

If the function v_j is extended as 0 when $t \leq 0$, $x \in \Omega$, the new function is p-supercaloric in $\Omega \times (-\infty, T)$, indeed. To see this, one has only to verify the comparison principle.

We have, when $\tau \leq T - \varepsilon$

$$\int_{-\infty}^{\tau} \int_\Omega \left(\langle (|\nabla v_j|^{p-2} \nabla v_j) \star \rho_\varepsilon, \nabla \varphi \rangle + \varphi \frac{\partial}{\partial t} (v_j \star \rho_\varepsilon) \right) dx\, dt \geq 0$$

for all test functions $\varphi \geq 0$ vanishing on the lateral boundary. Replace v_k in the previous proof by

$$\tilde{v}_k = \min\{v_j \star \rho_\varepsilon, k\}$$

and choose

$$\varphi_k = (\tilde{v}_k - \tilde{v}_{k-1}) - (\tilde{v}_{k+1} - \tilde{v}_k).$$

Since the convolution with respect to the time variable does not affect the zero boundary values on the lateral boundary, we conclude that $\tilde{v}_k$ vanishes on the parabolic boundary of $\Omega \times (-\delta/2, T - \delta/2)$, when $\varepsilon < \delta/2$ and δ can be taken as small as we wish. (The functions $v_k \star \rho_\varepsilon$ instead of the employed $(v \star \rho_\varepsilon)_k$ do not work well in this proof.) The same calculations as before yield

$$\tilde{a}_{k+1}(\tau) \leq \tilde{a}_k(\tau) \quad \text{and} \quad \sum_{k=1}^{j} \tilde{a}_k(\tau) \leq j\, \tilde{a}_1(\tau),$$

where

$$\tilde{a}_k(\tau) = \int_{-\delta/2}^{\tau} \int_{\Omega} \langle (|\nabla v_j|^{p-2}\nabla v_j) \star \rho_\varepsilon, \nabla(\tilde{v}_k - \tilde{v}_{k-1})\rangle\, dx\, dt$$

$$+ \int_{-\delta/2}^{\tau} \int_{\Omega} (\tilde{v}_k - \tilde{v}_{k-1})\frac{\partial}{\partial t}(v_j \star \rho_\varepsilon)\, dx\, dt.$$

Summing up, we obtain

$$\sum_{k=1}^{j} \tilde{a}_k(\tau) = \int_{-\delta/2}^{\tau} \int_{\Omega} \langle (|\nabla v_j|^{p-2}\nabla v_j) \star \rho_\varepsilon, \nabla \tilde{v}_j\rangle\, dx\, dt$$

$$+ \int_{-\delta/2}^{\tau} \int_{\Omega} \tilde{v}_j \frac{\partial}{\partial t}(v_j \star \rho_\varepsilon)\, dx\, dt,$$

where the last integral can be written as

$$\frac{1}{2} \int_{\Omega} (v_j \star \rho_\varepsilon)^2(x, \tau)\, dx.$$

Also for $\tilde{a}_1(\tau)$ we get an expression free of time derivatives. Therefore we can safely first let $\varepsilon \to 0$ and then $\delta \to 0$. This leads to the lemma.

Let us return to the lemma. Provided that we already have a majorant for the term

$$\int_0^T \int_{\Omega} |\nabla v_1|^p\, dx\, dt,$$

we see that
$$\int_0^T \int_\Omega |\nabla v_j|^p \, dx \, dt = O(j^2).$$

Yet, the right magnitude is $O(j)$.

Lemma 28. *Suppose that* $v_j \in L^p(0,T; W_0^{1,p}(\Omega))$ *and*

$$\int_0^T \int_\Omega |\nabla v_j|^p \, dx \, dt \leq K j^2, \quad j = 1, 2, 3, \dots$$

Then $v \in L_{loc}^q(\Omega_T)$, *whenever* $q < p - 2$. *(Here* $p > 2$.*)*

Proof. (Recall that the desired bound is $p - 1 + \frac{p}{n}$ and not the above $p - 2$.)
The assumption and Sobolev's inequality will give us a bound on the measure
of the level sets
$$E_j = \{(x,t) | \, j \leq v(x,t) \leq 2j\}$$

so that the integral can be controlled. To this end, denote

$$\kappa = 1 + \frac{2}{n}.$$

We have

$$j^{\kappa p}|E_j| \leq \iint_{E_j} v_{2j}^{\kappa p} \, dx \, dt \leq \int_0^T \int_\Omega v_{2j}^{\kappa p} \, dx \, dt$$

$$\leq C \int_0^T \int_\Omega |\nabla v_{2j}|^p \, dx \, dt \cdot \left(\operatorname*{ess\,sup}_{0<t<T} \int_\Omega v_{2j}^2 \, dx \right)^{\frac{p}{n}} \leq C K j^2 \left(4|\Omega|j^2 \right)^{\frac{p}{n}}.$$

It follows that
$$|E_j| \leq \text{Constant } j^{2-p}.$$

We use this to estimate the L^q-norm using a dyadic division of the domain.
Thus

$$\int_0^T \int_\Omega v^q \, dx \, dt \leq T|\Omega| + \sum_{j=1}^\infty \iint_{E_{2^{j-1}}} v^q \, dx \, dt$$

$$\leq T|\Omega| + \sum_{j=1}^\infty 2^{jq}|E_{2^{j-1}}|$$

$$\leq T|\Omega| + C \sum_{j=1}^\infty 2^{j(q+2-p)},$$

which is a convergent majorant when $q < p - 2$. $\qquad\square$

Remark. If the majorant Kj^2 in the assumption is replaced by a better Kj^γ, then the procedure yields that $|E_j| \approx j^{\gamma-p}$ resulting in $q < p - \gamma$.

The previous lemma guarantees that v^ε is summable for some small positive power ε, since[18] $p > 2$. To improve the exponent, we start from Lemma 27 and write the estimate in the form

$$\int_0^{t_1} \int_\Omega |\nabla v_j|^p \, dx \, dt \leq j \int_0^T \int_\Omega |\nabla v_1|^p \, dx \, dt + j \int_\Omega v_j(x, \tau) \, dx,$$

where $0 < t_1 \leq \tau \leq T$. Integrate with respect to τ over the integral $[t_1, T]$:

$$(T - t_1) \int_0^{t_1} \int_\Omega |\nabla v_j|^p \, dx \, dt \leq j(T - t_1)K + j \int_{t_1}^T \int_\Omega v_j(x, t) \, dx \, dt$$

$$\leq j(T - t_1)K + j^{2-\varepsilon} \int_{t_1}^T \int_\Omega v^\varepsilon(x, t) \, dx \, dt.$$

Thus we have reached the estimate

$$\int_0^{t_1} \int_\Omega |\nabla v_j|^p \, dx \, dt \leq j^{2-\varepsilon} K_1, \qquad (t_1 < T). \tag{20}$$

This is an improvement from j^2 to $j^{2-\varepsilon}$, but we have to obey the restriction that $\varepsilon \leq 1$, because the term $j(T - t_1)K$ was absorbed. Estimating again the measures $|E_j|$, but starting with the bound $j^\gamma K$, $\gamma = 2 - \varepsilon$ in place of $j^2 K$, yields

$$|E_j| \approx j^{2-p-\varepsilon} \qquad 0 < \varepsilon \leq 1.$$

The result is that

$$\int_0^{t_1} \int_\Omega v^q \, dx \, dt < \infty \qquad \text{when} \qquad 0 < q < p - \gamma = p - 2 + \varepsilon.$$

Iterating, we have the scheme

$$q_0 = \varepsilon \qquad\qquad\qquad\qquad T$$
$$q_1 = p - 2 + \varepsilon \qquad\qquad\qquad t_1$$
$$q_2 = 2(p - 2) + \varepsilon \qquad\qquad t_2$$

We can continue till we reach

$$q_k = k(p - 2) + \varepsilon > p - 1.$$

---

[18] It does not work for the Heat Equation!

We have to stop, because the previous exponent $(k-1)(p-2) + \varepsilon$ has to obey the rule not to become larger than 1. This way we can reach that

$$v \in L^1(\Omega_{T'}),$$

with a $T' < T$, which will do to proceed. In fact, adjusting we can reach any exponent strictly below $p-1$, but the passage over the exponent $p-1$ requires a special device. Since only a finite number of steps were involved, we can take T' as close to T as we wish.

We use inequality (18) in the form

$$\frac{1}{2} \int_\Omega v_j(x,t)^2 \, dx \le \int_0^\tau \int_\Omega |\nabla v_j|^p \, dx \, dt + \frac{1}{2} \int_\Omega v_j(x,\tau)^2 \, dx, \qquad (21)$$

where $t < \tau$. For $t_1 < \tau < T$ it follows that

$$\operatorname*{ess\,sup}_{0<t<t_1} \int_\Omega v_j(x,t)^2 \, dx \le 2 \int_0^\tau \int_\Omega |\nabla v_j|^p \, dx \, dt + \int_\Omega v_j(x,\tau)^2 \, dx$$

$$\le 2j \int_0^\tau \int_\Omega |\nabla v_1|^p \, dx \, dt + 2j \int_\Omega v_j(x,\tau) \, dx$$

$$\le 2j \int_0^T \int_\Omega |\nabla v_1|^p \, dx \, dt + 2j \int_\Omega v_j(x,\tau) \, dx,$$

where the second step is from Lemma 27. We integrate the resulting inequality with respect to τ over the interval $[t_1, T]$, which affects only the last integral. Upon division by $T - t_1$, the last term is replaced by

$$\frac{2j}{T - t_1} \int_{t_1}^T \int_\Omega v_j \, dx \, dt.$$

We can combine this and the earlier estimate

$$\int_0^{t_1} \int_\Omega |\nabla v_j|^p \, dx \, dt \le j \int_0^T \int_\Omega |\nabla v_1|^p \, dx \, dt + \frac{j^{2-\varepsilon}}{T - t_1} \int_{t_1}^T \int_\Omega v^\varepsilon \, dx \, dt,$$

taking $\varepsilon = 1$, so that we finally arrive at

$$\int_0^{t_1} \int_\Omega |\nabla v_j|^p \, dx \, dt + \operatorname*{ess\,sup}_{0<t<t_1} \int_\Omega v_j(x,t)^2 \, dx$$

$$\le 3j \int_0^T \int_\Omega |\nabla v_1|^p \, dx \, dt + \frac{3j}{T - t_1} \int_{t_1}^T \int_\Omega v \, dx \, dt. \qquad (22)$$

The majorant is now $O(j)$, which is of the right order, as the following lemma shows with its sharp exponents.

Lemma 29. *If*

$$\int_0^T \int_\Omega |\nabla v_j|^p \, dx \, dt + \mathop{ess\,sup}_{0<t<T} \int_\Omega v_j(x,t)^2 \, dx \leq jK$$

when $j = 1, 2, \ldots$ *then*

$$v \in L^q(\Omega_T) \quad whenever \quad 0 < q < p - 1 + \frac{p}{n},$$

$$\nabla v \in L^q(\Omega_T) \quad whenever \quad 0 < q < p - 1 + \frac{1}{n+1}.$$

Proof. The first part is a repetition of the proof of Lemma 28. Denote again

$$E_j = \{(x,t) | j \leq v(x,t) \leq 2j\}, \quad \kappa = 1 + \frac{2}{n}.$$

We have as before

$$j^{\kappa p}|E_j| \leq \iint_{E_j} v_{2j}^{\kappa p} \, dx \, dt \leq \int_0^T \int_\Omega v_{2j}^{\kappa p} \, dx \, dt$$

$$\leq C \int_0^T \int_\Omega |\nabla v_{2j}|^p \, dx \, dt \cdot \left(\mathop{ess\,sup}_{0<t<T} \int_\Omega v_{2j}^2 \, dx \right)^{\frac{p}{n}} \leq CK^{1+\frac{p}{n}}(2j)^{1+\frac{p}{n}}.$$

It follows that

$$|E_j| \leq \text{Constant} \times j^{1-p-\frac{p}{n}}.$$

We estimate the L^q-norm using the subdivision of the domain. Thus

$$\int_0^T \int_\Omega v^q \, dx \, dt \leq T|\Omega| + \sum_{j=1}^\infty \iint_{E_{2^{j-1}}} v^q \, dx \, dt$$

$$\leq T|\Omega| + \sum_{j=1}^\infty 2^{jq}|E_{2^{j-1}}|$$

$$\leq T|\Omega| + C \sum_{j=1}^\infty 2^{j(q+1-p-\frac{p}{n})},$$

which converges in the desired range for q. Thus the first part is proved.

For the summability of the gradient, we use the bound on the measure of the level sets E_j and also the growth assumed for the energy of the truncated functions. Fix a large index k and write, using that $|\nabla v_k| \leq |\nabla v_{2^j}|$ on $E_{2^{j-1}}$:

$$\int_0^T \int_\Omega |\nabla v_k|^q \, dx \, dt \lesssim \sum_{j=1}^\infty \int\int_{E_{2^{j-1}}} |\nabla v_k|^q \, dx \, dt$$

$$\leq \sum_{j=1}^\infty \left(\int\int_{E_{2^{j-1}}} |\nabla v_k|^p \, dx \, dt \right)^{\frac{q}{p}} |E_{2^{j-1}}|^{1-\frac{q}{p}}$$

$$\leq \sum_{j=1}^\infty \left(\int_0^T \int_\Omega |\nabla v_{2^j}|^p \, dx \, dt \right)^{\frac{q}{p}} 2^{(j-1)(1-\frac{q}{p})(1-p-\frac{p}{n})}$$

$$\leq \sum_{j=1}^\infty 2^{(j-1)(1-\frac{q}{p})(1-p-\frac{p}{n})} (2^j K)^{\frac{q}{p}},$$

where the geometric series converges provided that $q < p - 1 + 1/(n+1)$. Strictly speaking, the "first term" $K^{q/p}(T|\Omega|)^{1-q/p}$ ought to be added to the sum, since the integral over the set $\{0 < v < 1\}$ was missing. Now we may let k go to infinity. □

3.4   Reduction to Zero Boundary Values

It remains to reduce the situation to a p-supercaloric function with zero boundary values and to estimate the p-energy of the first truncated function so that the majorant in Lemma 27 is finite. The constructions have to be performed only on the truncated functions v_k and so we are dealing only with bounded functions. A first reduction comes from the fact that if v is a nonnegative p-supercaloric function, so is the function obtained by redefining $v(x,t)$ as 0 when $t \leq \delta$. Thus the initial values can be regarded as zero. Given v so that $v \geq 0$ in, say, the closure of $B_{2r} \times (0,T)$, we first redefine $v(x,t)$ as 0 when $t \leq \delta$, where δ is small. Then we want to define the modified function

$$w = \begin{cases} v(x,t), & \text{if } (x,t) \in \overline{B_r} \times (\delta, T) \\ h(x,t), & \text{if } (x,t) \in (B_{2r} \setminus \overline{B_r}) \times (\delta, T) \\ 0, & \text{if } t \leq \delta \end{cases}$$

where h is the p-caloric function in the domain $(B_{2r} \setminus \overline{B_r}) \times (0,T)$ having

Outer boundary values 0 (this is on $\partial B_{2r} \times (0,T)$),
Inner boundary values v (this is on $\partial B_r \times (0,T)$),
Initial values 0 (this is when $t = 0$).

By the maximum principle, $h(x,t) = 0$ when $t < \delta$. Thus also $w(x,t) = 0$ when $t < \delta$.

This construction is easily done for $v \in C(\Omega_T) \cap L^p(0,T;W^{1,p}_{loc}(\Omega)$ It is sufficient to do this modification for each of the truncated functions appearing in Sect. 3.3. Thus we may regard v as bounded and hence as belonging to the required Sobolev space (Theorem 25). Finally, to reach the case of a discontinuous v, we first perform the above construction on the infimal convolution v_ε and we may arrange it in advance[19] so that this is a viscosity supersolution in the whole domain $B_{2r} \times (0,T)$. Now $0 \leq v_\varepsilon \leq v$ so that $v_\varepsilon(x,t) = 0$ when $t \leq \delta$. The corresponding h_ε takes the given initial and boundary values and it is easily seen that w_ε obeys the comparison principle. For the verification it is decisive to use the fact that v_ε is defined also outside the inner cylinder. By Theorem 25 w_ε is a weak supersolution. We need a uniform bound as $\varepsilon \to 0$. To this end, choose a test function $\zeta = \zeta(x)$ so that $\zeta = 0$ on ∂B_r and $\zeta = 1$ on $B_{2r} \setminus B_{3r/2}$. Taking into account that h_ε has outer boundary values zero, we get a Caccioppoli estimate of the form

$$\int_0^T \int_{B_{2r} \setminus B_{3r/2}} |\nabla h_\varepsilon|^p \, dx \, dt \leq C \int_0^T \int_{B_{2r} \setminus B_r} h_\varepsilon^p |\nabla \zeta|^p \, dx \, dt + \int_{B_{2r} \setminus B_r} h_\varepsilon^2(x,T) \, dx.$$

Indeed, to see this use the test function $\zeta^p h_\varepsilon$ and take into account that we deal with a solution. The Caccioppoli estimate for w_ε can be written as

$$\int_0^T \int_{B_{3r/2}} |\nabla w_\varepsilon|^p \, dx \, dt \leq C \left(\max(w_\varepsilon)^p r^{n-p} + \max(w_\varepsilon)^2 r^n \right).$$

Adding the estimates we get an integral over the whole $B_{2r} \times (0,T)$, and hence

$$\int_0^T \int_{B_{2r}} |\nabla w_\varepsilon|^p \, dx \, dt \leq C \left(\|v\|_\infty^p r^{n-p} + \|v\|_\infty^2 r^n \right).$$

Now it follows that, as $\varepsilon \to 0$, w_ε approaches a weak supersolution w. Indeed, w is semicontinuous and being the limit of an increasing sequence it satisfies the comparison principle. Since w is also bounded, Theorem 25 applies again. By the weak lower semicontinuity of the integral also

$$\int_0^T \int_{B_{2r}} |\nabla w|^p \, dx \, dt \leq C \left(\|v\|_\infty^p r^{n-p} + \|v\|_\infty^2 r^n \right).$$

Moreover, $w \in L^p(0,T;W_0^{1,p}(B_{2r}))$ and $w(x,0) = 0$ when $x \in B_{2r}$, provided, of course, that $w \in L^p(0,T;W^{1,p}_{loc}(\Omega))$. Performing the above reduction for the function[20] v_1, which is v cut at level 1, we obtain

---

[19]Start with a nonnegative v in a larger domain, say $B_{3r} \times (-\delta, T)$ and put $v = 0$ up to time $t = \delta$.

[20]To clarify the notation, this is not v_ε for $\varepsilon = 1$.

$$\int_0^T \int_{B_{2r}} |\nabla w_1|^p \, dw \, dt \le C(r^{n-p} + r^n)$$

as required in Lemma 27. We know that $w_1 = v_1$ in the smaller set.

The procedure is to perform the previous reduction with v replaced by each of the truncated functions v_k. Repeating the whole iteration of the previous section, we obtain the results for the functions w_k and thereby also for v_k, although in the smaller set where the functions are unaffected by the reduction. This is enough for our *local* results.

4 Weak Supersolutions are Semicontinuous

Are the weak supersolutions p-supercaloric functions (=viscosity supersolutions)? To qualify they have to obey the comparison principle and to be semicontinuous. The comparison principle is rather immediate. The semicontinuity is a delicate issue. For a weak supersolution defined in the classical way with test functions under the integral sign (Definition 16) the Sobolev derivative is assumed to exist, but the semicontinuity, which now is not assumed, has to be established. The proof requires parts of the classical regularity theory.[21] We will use a variant of the Moser iteration, for practical reasons worked out for weak subsolutions bounded from below. Our proof of the theorem below is essentially the same as in [18], but we avoid the use of infinitely stretched infinitesimal space-time cylinders.

Theorem 30. *Suppose that $v = v(x,t)$ is a weak supersolution of the Evolutionary p-Laplace equation. Then it is locally bounded from below and at almost every point (x_0, t_0) it holds that*

$$v(x_0, t_0) = \operatorname{ess} \liminf_{(x,t) \to (x_0,t_0)} v(x,t).$$

In particular, v is lower semicontinuous after a redefinition in a set of measure zero.

Functions like $\operatorname{ess\,lim\,inf} v(x,t)$ are lower semicontinuous, if they are bounded from below. Thus the problem is the formula. The hardest part of the proof is to establish that the supremum norm of a non-negative weak subsolution is 1^0 bounded (Lemma 34) and 2^0 bounded in terms of quantities that can carry information from the Lebesgue points (Theorem 36). With such estimates the proof follows easily (at the end of this section). Before entering into the semicontinuity proof we address the comparison principle.

---

[21] The preface of Giuseppe Mingione's work [25] is worth reading as an enlightenment.

Proposition 31 (Comparison Principle). *Suppose that v is a weak supersolution and u a weak subsolution, $u, v \in L^p(0, T; W^{1,p}(\Omega))$, satisfying*

$$\liminf v \geq \limsup u$$

on the parabolic boundary. Then $v \geq u$ in the whole domain Ω_T.

Proof. This is well-known and we only give a formal proof. For a non-negative test function $\varphi \in C_0^\infty(\Omega_T)$ the equations

$$\int_0^T \int_\Omega (-v\varphi_t + \langle |\nabla v|^{p-2}\nabla v, \nabla\varphi\rangle) \, dx \, dt \geq 0$$

$$\int_0^T \int_\Omega (+u\varphi_t - \langle |\nabla u|^{p-2}\nabla u, \nabla\varphi\rangle) \, dx \, dt \geq 0$$

can be added. Thus

$$\int_0^T \int_\Omega \left((u-v)\varphi_t + \langle |\nabla v|^{p-2}\nabla v - |\nabla u|^{p-2}\nabla u, \nabla\varphi\rangle\right) dx \, dt \geq 0.$$

These equations remain true if v is replaced by $v + \varepsilon$, where ε is any constant. To complete the proof we choose (formally) the test function to be

$$\varphi = (u - v - \varepsilon)_+ \eta,$$

where $\eta = \eta(t)$ is a cut-off function; even $\eta(t) = T - t$ will do here. We arrive at

$$\int_0^T \int_{u \geq v + \varepsilon} \eta(\langle |\nabla v|^{p-2}\nabla v - |\nabla u|^{p-2}\nabla u, \nabla v - \nabla u\rangle) \, dx \, dt$$

$$\leq \int_0^T \int_\Omega (u - v - \varepsilon)_+^2 \eta' \, dx \, dt + \frac{1}{2} \int_0^T \int_\Omega \eta \frac{\partial}{\partial t}(u - v - \varepsilon)_+^2 \, dx \, dt$$

$$= \frac{1}{2} \int_0^T \int_\Omega (u - v - \varepsilon)_+^2 \eta' \, dx \, dt$$

$$= -\frac{1}{2} \int_0^T \int_\Omega (u - v - \varepsilon)_+^2 \, dx \, dt \leq 0.$$

Since the first integral is non-negative by the structural inequality (22), the last integral is, in fact, zero. Hence the integrand $(u - v - \varepsilon)_+^2 = 0$ almost everywhere. But this means that

$$u \leq v + \varepsilon$$

almost everywhere. Since $\varepsilon > 0$ we have the desired inequality $v \geq u$ a.e. $\qquad\square$

We need some estimates for the semicontinuity and begin with the well-known Caccioppoli estimates, which are extracted directly from the differential equation.

Lemma 32 (Caccioppoli Estimates). *For a non-negative weak subsolution u in $\Omega \times (t_1, t_2)$ we have the estimates*

$$\operatorname*{ess\,sup}_{t_1 < t < t_2} \int_\Omega \zeta^p u^{\beta+1}\, dx \le \int_{t_1}^{t_2} \int_\Omega u^{\beta+1} \left| \frac{\partial}{\partial t} \zeta^p \right| dx\, dt$$

$$+ 2^{p-1}\beta^{2-p} \int_{t_1}^{t_2} \int_\Omega u^{p-1+\beta} |\nabla \zeta|^p\, dx\, dt$$

and

$$\int_{t_1}^{t_2} \int_\Omega \left| \nabla (\zeta u^{\frac{p-1+\beta}{p}}) \right|^p dx\, dt \le C\beta^{p-2} \int_{t_1}^{t_2} \int_\Omega u^{\beta+1} \left| \frac{\partial}{\partial t} \zeta^p \right| dx\, dt$$

$$+ C \int_{t_1}^{t_2} \int_\Omega u^{p-1+\beta} |\nabla \zeta|^p\, dx\, dt,$$

where the exponent $\beta \ge 1$, $C = C(p)$, and $\zeta \in C^\infty(\Omega \times [t_1, t_2])$, $\zeta(x, t_1) = 0$, $\zeta \ge 0$.

Proof. Use the test function $\varphi = u^\beta \zeta^p$ in the equation

$$\int_{t_1}^{\tau} \int_\Omega \left(-u\varphi_t + \langle |\nabla u|^{p-2}\nabla u, \nabla \varphi \rangle \right) dx\, dt$$

$$+ \int_\Omega u(x, \tau)\varphi(x, \tau)\, dx \le \int_\Omega u(x, t_1)\varphi(x, t_1)\, dx = 0,$$

where $t_1 < \tau \le t_2$. (The intermediate τ is needed to match the supremum in the first estimate.) Strictly speaking, the "forbidden" time derivative u_t is required at the intermediate steps. This can be handled through a regularization, which we omit. Proceeding, integration by parts leads to

$$\int_{t_1}^{\tau} \int_\Omega -u\varphi_t\, dx\, dt + \int_\Omega u(x, \tau)\varphi(x, \tau)\, dx$$

$$= \frac{1}{\beta + 1} \int_\Omega \zeta(x, \tau)^p u(x, \tau)^{\beta+1}\, dx - \frac{1}{\beta + 1} \int_{t_1}^{\tau} \int_\Omega u^{\beta+1} \left| \frac{\partial}{\partial t} \zeta^p \right| dx\, dt$$

valid for a.e. τ. To treat the "elliptic term", we use

$$\nabla \varphi = \beta \zeta^p u^{\beta-1} \nabla u + p \zeta^{p-1} u^\beta \nabla \zeta$$

and obtain

$$\frac{1}{\beta+1}\int_\Omega \zeta(x,\tau)^p u(x,\tau)^{\beta+1}\, dx + \beta \int_{t_1}^\tau \int_\Omega \zeta^p u^{\beta-1}|\nabla u|^p\, dx\, dt$$

$$\leq \frac{1}{\beta+1}\int_{t_1}^\tau \int_\Omega u^{\beta+1}\left|\frac{\partial}{\partial t}\zeta^p\right|\, dx\, dt + p\int_{t_1}^\tau \int_\Omega \zeta^{p-1} u^\beta |\nabla u|^{p-1}|\nabla \zeta|\, dx\, dt.$$

As much as possible of the last integral must be absorbed by the double integral in the left-hand member. It is convenient to employ Young's inequality

$$ab \leq \frac{a^q}{q} + \frac{b^p}{p}$$

to achieve the splitting

$$\zeta^{p-1} u^\beta |\nabla u|^{p-1}|\nabla \zeta|$$

$$= \overbrace{\left(\frac{\beta}{p}\right)^{\frac{p-1}{p}}\zeta^{p-1} u^{(\beta-1)\frac{p-1}{p}}|\nabla u|^{p-1}}^{a} \times \overbrace{\left(\frac{p}{\beta}\right)^{\frac{p-1}{p}} u^{\frac{p-1+\beta}{p}}|\nabla \zeta|}^{b}$$

$$\leq \frac{p-1}{p}\left(\frac{\beta}{p}\right)\zeta^p u^{\beta-1}|\nabla u|^p + \frac{1}{p}\left(\frac{p}{\beta}\right)^{p-1} u^{p-1+\beta}|\nabla \zeta|^p,$$

which has to be multiplied by p and integrated. Absorbing one integral into the left-hand member, we arrive at the fundamental estimate

$$\frac{1}{\beta+1}\int_\Omega \zeta(x,\tau)^p u(x,\tau)^{\beta+1}\, dx + \frac{\beta}{p}\int_{t_1}^\tau \int_\Omega \zeta^p u^{\beta-1}|\nabla u|^p\, dx\, dt$$

$$\leq \frac{1}{\beta+1}\int_{t_1}^\tau \int_\Omega u^{\beta+1}\left|\frac{\partial}{\partial t}\zeta^p\right|\, dx\, dt + \left(\frac{p}{\beta}\right)^{p-1}\int_{t_1}^\tau \int_\Omega u^{p-1+\beta}|\nabla \zeta|^p\, dx\, dt.$$

Since the integrands are positive it follows that

$$\frac{1}{\beta+1}\int_\Omega \zeta(x,\tau)^p u(x,\tau)^{\beta+1}\, dx$$

$$\leq \frac{1}{\beta+1}\int_{t_1}^{t_2}\int_\Omega u^{\beta+1}\left|\frac{\partial}{\partial t}\zeta^p\right|\, dx\, dt$$

$$+ \left(\frac{p}{\beta}\right)^{p-1}\int_{t_1}^{t_2}\int_\Omega u^{p-1+\beta}|\nabla \zeta|^p\, dx\, dt,$$

where the majorant now is free from τ. Taking the supremum over τ we obtain the first Caccioppoli inequality.

To derive the second Caccioppoli inequality, we start from

$$\frac{\beta}{p} \int_{t_1}^{t_2} \int_\Omega \zeta^p u^{\beta-1} |\nabla u|^p \, dx \, dt$$

$$\leq \frac{1}{\beta+1} \int_{t_1}^{t_2} \int_\Omega u^{\beta+1} \left| \frac{\partial}{\partial t} \zeta^p \right| \, dx \, dt$$

$$+ \left(\frac{p}{\beta} \right)^{p-1} \int_{t_1}^{t_2} \int_\Omega u^{p-1+\beta} |\nabla \zeta|^p \, dx \, dt$$

and notice that

$$\zeta^p u^{\beta-1} |\nabla u|^p = \left(\frac{p}{p-1+\beta} \right)^p \left| \zeta \nabla u^{\frac{p-1+\beta}{p}} \right|^p.$$

Then the triangle inequality

$$\left| \nabla (\zeta u^{\frac{p-1+\beta}{p}}) \right| \leq \left| \zeta \nabla u^{\frac{p-1+\beta}{p}} \right| + \left| u^{\frac{p-1+\beta}{p}} \nabla \zeta \right|$$

and a simple calculation yield the desired result.   □

In the following version of Sobolev's inequality the exponents are adjusted to our need. For a proof, see [6, Chap. 1].

Proposition 33 (Sobolev). *For* $\zeta \in C^\infty(\Omega_T)$ *vanishing on the lateral boundary* $\partial\Omega \times [0, T]$ *we have*

$$\int_0^T \int_\Omega \zeta^{p\gamma} |u|^{p-2+(\beta+1)\gamma} \, dx \, dt$$

$$\leq S \int_0^T \int_\Omega \left| \nabla (\zeta |u|^{\frac{p-1+\beta}{p}}) \right|^p \, dx \, dt \left\{ \operatorname*{ess\,sup}_{0<t<T} \int_\Omega \zeta^p |u|^{\beta+1} \, dx \right\}^{\frac{p}{n}},$$

where $\gamma = 1 + \frac{p}{n}$.

Now we can control the right-hand member in the Sobolev inequality by the quantities in the Caccioppoli estimates. Thus

$$\left(\int_{t_1}^{t_2} \int_\Omega \zeta^{p\gamma} u^{p-2+(\beta+1)\gamma} \, dx \, dt \right)^{\frac{1}{\gamma}}$$

$$\leq C \beta^{\frac{(2-p)p}{n+p}} \left(\beta^{p-2} \int_{t_1}^{t_2} \int_\Omega u^{\beta+1} \left| \frac{\partial}{\partial t} \zeta^p \right| \, dx \, dt + \int_{t_1}^{t_2} \int_\Omega u^{p-1+\beta} |\nabla \zeta|^p \, dx \, dt \right).$$

We select the test function ζ so that it is equal to 1 in the cylinder $B_{R-\Delta R} \times (T + \Delta T, t_2)$, $\zeta(x, T) = 0$, and so that $\zeta(x, t) = 0$ when x is outside B_R.

Then we can write

$$\left(\int_{T+\Delta T}^{t_2} \int_{B_{R-\Delta R}} u^{p-2+(\beta+1)\gamma} \, dx \, dt \right)^{\frac{1}{\gamma}}$$

$$\leq C\beta^{\frac{(2-p)p}{n+p}} \left(\frac{\beta^{p-2}}{\Delta T} \int_T^{t_2} \int_{B_R} u^{\beta+1} \, dx \, dt + \frac{1}{(\Delta R)^p} \int_T^{t_2} \int_{B_R} u^{p-1+\beta} \, dx \, dt \right),$$

where C is a new constant. Recall that $\gamma > 1$. This is a *reverse Hölder inequality*, which is most transparent for $p = 2$. It will be important to keep $\Delta T = (\Delta R)^p$. This is the basic inequality for the celebrated *Moser iteration*, which we will employ. The power of u increases to $p - 2 + (\beta + 1)\gamma$, but the integral is taken over a smaller cylinder. In order to iterate over a chain of shrinking cylinders $U_k = B(x_0, R_k) \times (T_k, t_2)$, starting with

$$U_0 = B(x_0, 2R) \times \left(\frac{T}{2}, t_2 \right)$$

and ending up with an estimate over the cylinder

$$U_\infty = B(x_0, R) \times (T, t_2),$$

we introduce the quantities

$$R_k = R + \frac{R}{2^k}, \qquad\qquad R_k - R_{k+1} = \frac{R}{2^{k+1}}$$

$$T_k = T - \frac{T}{2^{kp+1}}, \qquad\qquad T_{k+1} - T_k = \frac{T}{2^{(k+1)p}} s,$$

$$\omega = \frac{R^p}{Ts} = \frac{(\Delta R_k)^p}{\Delta T_k} \qquad\qquad s = \frac{2^{p-1} - 1}{2}.$$

We remark that ω *is independent of the index* k. Further, we write $\alpha = \beta + 1$, so that $\alpha \geq 2$. Thus

$$\left(\int\!\!\int_{U_{k+1}} u^{p-2+\alpha\gamma} \, dx \, dt \right)^{\frac{1}{\gamma}}$$

$$\leq C \frac{2^{(k+1)p} \beta^{\frac{(2-p)p}{n+p}}}{R^p} \left(\beta^{p-2} \omega \int\!\!\int_{U_k} u^\alpha \, dx \, dt + \int\!\!\int_{U_k} u^{p-2+\alpha} \, dx \, dt \right). \quad (23)$$

It is inconvenient to deal with *two* different integrals in the majorant. For simplicity we will perform two iteration procedures, depending on which integral is dominating. For the **first procedure** we assume that

$$\omega \leq u^{p-2}.$$

Then we have the simpler expression

$$\left(\iint_{U_{k+1}} u^{p-2+\alpha\gamma} \, dx \, dt\right)^{\frac{1}{\gamma}} \leq C_1 \frac{2^{kp}\alpha^{\frac{(p-2)}{\gamma}}}{R^p} \iint_{U_k} u^{p-2+\alpha} \, dx \, dt.$$

We start the iteration with $\alpha = 2$ and $k = 0$. Thus

$$\left(\iint_{U_1} u^{p-2+2\gamma} \, dx \, dt\right)^{\frac{1}{\gamma}} \leq C_1 \frac{2^{0p}2^{\frac{(p-2)}{\gamma}}}{R^p} \iint_{U_0} u^p \, dx \, dt.$$

Then take $\alpha = 2\gamma$ and $k = 1$ so that

$$\left(\iint_{U_2} u^{p-2+2\gamma^2} \, dx \, dt\right)^{\frac{1}{\gamma^2}} \leq \left(C_1 \frac{2^{1p}(2\gamma)^{\frac{(p-2)}{\gamma}}}{R^p} \iint_{U_1} u^{p-2+2\gamma} \, dx \, dt\right)^{\frac{1}{\gamma}}$$

$$\leq \left(C_1 \frac{2^{1p}(2\gamma)^{\frac{(p-2)}{\gamma}}}{R^p}\right)^{\frac{1}{\gamma}}$$

$$\times C_1 \frac{2^{0p}2^{\frac{(p-2)}{\gamma}}}{R^p} \iint_{U_0} u^p \, dx \, dt.$$

The result of the next step is

$$\left(\iint_{U_3} u^{p-2+2\gamma^3} \, dx \, dt\right)^{\frac{1}{\gamma^3}}$$

$$\leq \left(\frac{C_1 2^{\frac{p-2}{\gamma}}}{R^p}\right)^{1+\frac{1}{\gamma}+\frac{1}{\gamma^2}} 2^{p(\frac{1}{\gamma}+\frac{2}{\gamma^2})}\gamma^{(p-2)(\frac{1}{\gamma^2}+\frac{2}{\gamma^3})} \iint_{U_0} u^p \, dx \, dt.$$

Continuing the chain and noticing that the geometric series

$$1 + \frac{1}{\gamma} + \frac{1}{\gamma^2} + \frac{1}{\gamma^3} + \cdots = 1 + \frac{n}{p}$$

and the series $\sum k\gamma^{-k}$ appearing in the exponents converge, since $\gamma > 1$, we arrive at

$$\left(\iint_{U_{k+1}} u^{p-2+2\gamma^{k+1}}\, dx\, dt\right)^{\frac{1}{\gamma^{k+1}}} \le K R^{-p(1+\frac{1}{\gamma}+\frac{1}{\gamma^2}+\cdots+\frac{1}{\gamma^k})} \iint_{U_0} u^p\, dx\, dt.$$

Here K is a numerical constant. As $k \to \infty$, we obtain the final estimate

$$\operatorname*{ess\,sup}_{B_R \times (T, t_2)} (u^2) \le \frac{K}{R^{n+p}} \int_{\frac{T}{2}}^{t_2} \int_{B_{2R}} u^p\, dx\, dt = \frac{K}{\omega s T R^n} \int_{\frac{T}{2}}^{t_2} \int_{B_{2R}} u^p\, dx\, dt,$$

where the square came from the factor 2 in $2\gamma^{k+1}$. The sum of the geometric series determined the power of R.

Finally, if the assumption $\omega \le u^{p-2}$ is relaxed to $u \ge 0$, we can apply the previous estimate to the function

$$u(x,t) + \omega^{\frac{1}{p-2}} = u(x,t) + \left(\frac{R^p}{Ts}\right)^{\frac{1}{p-2}}.$$

A simple calculation gives us the bound in the next lemma.

Lemma 34. *Suppose that $u \ge 0$ is a weak supersolution in the cylinder $B_{2R} \times (\frac{T}{2}, t_2)$. Then*

$$\operatorname*{ess\,sup}_{B_R \times (T, t_2)} \{u^2\} \le C\left\{ \left(\frac{R^p}{T}\right)^{\frac{2}{p-2}} + \frac{T}{R^p}\left(\frac{1}{TR^n} \int_{\frac{T}{2}}^{t_2} \int_{B_{2R}} u^p\, dx\, dt\right)\right\},$$

where $C = C(n,p)$.

We can extract the following piece of information.

Corollary 35. *A weak supersolution that is bounded from above, is locally bounded from below.*

Proof. Use $u(x,t) = L - v(x,t)$.   □

The estimate in the lemma suffers from the defect that it is not sharp when $u \approx 0$ because of the presence of the constant term. Our remedy is a **second iteration procedure**, this time under the assumption that

$$0 \le u \le j,$$

where we take j so large that also

$$j^{p-2} \ge \omega.$$

Read j^{p-2} as $\max\{\omega, j^{p-2}\}$. The previous lemma shows that j is finite, but the point now is that u is not bounded away from zero. Then the first integral in the majorant of (23) is dominating and we can begin with the bound

$$\left(\int\!\!\!\int_{U_{k+1}} u^{p-2+\alpha\gamma}\,dx\,dt\right)^{\frac{1}{\gamma}} \le C\,j^{p-2}\frac{2^{kp}\alpha^{\frac{(p-2)}{\gamma}}}{R^p}\int\!\!\!\int_{U_k} u^\alpha\,dx\,dt.$$

We start the iteration with $\alpha = p$ and $k = 0$. Thus

$$\left(\int\!\!\!\int_{U_1} u^{p-2+p\gamma}\,dx\,dt\right)^{\frac{1}{\gamma}} \le C j^{p-2}\frac{2^{0p}p^{\frac{(p-2)}{\gamma}}}{R^p}\int\!\!\!\int_{U_0} u^p\,dx\,dt.$$

Then take $\alpha = p - 2 + p\gamma$, which is $< n\gamma^2$, and $k = 1$ so that

$$\left(\int\!\!\!\int_{U_2} u^{(p-2)(1+\gamma)+p\gamma^2}\,dx\,dt\right)^{\frac{1}{\gamma^2}} \le \left(C j^{p-2}\frac{2^{1p}(n\gamma^2)^{\frac{(p-2)}{\gamma}}}{R^p}\int\!\!\!\int_{U_1} u^{p-2+p\gamma}\,dx\,dt\right)^{\frac{1}{\gamma}}$$

$$\le \left(C j^{p-2}\frac{2^{1p}(n\gamma^2)^{\frac{(p-2)}{\gamma}}}{R^p}\right)^{\frac{1}{\gamma}}$$

$$\times\, C j^{p-2}\frac{2^{0p}(n\gamma)^{\frac{(p-2)}{\gamma}}}{R^p}\int\!\!\!\int_{U_0} u^p\,dx\,dt.$$

At the next step $\alpha = (p - 2)(1 + \gamma) + p\gamma^2 < n\gamma^3$ and $k = 2$. The result is

$$\left(\int\!\!\!\int_{U_3} u^{(p-2)(1+\gamma+\gamma^2)+p\gamma^3}\,dx\,dt\right)^{\frac{1}{\gamma^3}}$$

$$\le \left(\frac{C j^{p-2}n^{\frac{p-2}{\gamma}}}{R^p}\right)^{1+\frac{1}{\gamma}+\frac{1}{\gamma^2}} 2^{p(\frac{1}{\gamma}+\frac{2}{\gamma^2})}\gamma^{(p-2)(\frac{1}{\gamma}+\frac{2}{\gamma^2}+\frac{3}{\gamma^3})}\int\!\!\!\int_{U_0} u^p\,dx\,dt.$$

Continuing like this we end up with an estimate integrated over U_{k+1} with the power $\alpha_{k+1} = p - 2 + \alpha_k\gamma$, where

$$\alpha_k = (p - 2)(1 + \gamma + \gamma^2 + \cdots + \gamma^{k-1}) + p\gamma^k$$

$$= \frac{n(p - 2)}{p}(\gamma^k - 1) + p\gamma^k \approx \left(n + p - \frac{2n}{p}\right)\gamma^k$$

and $\alpha_k < n\gamma^{k+1}$. As $k \to \infty$ we find that

$$\underset{B_R \times (T,t_2)}{\text{ess sup}} \{u^{n+p-\frac{2n}{p}}\} \leq C \frac{j^{\frac{(p-2)(n+p)}{p}}}{R^{n+p}} \int_{\frac{T}{2}}^{t_2} \int_{B_{2R}} u^p \, dx \, dt.$$

We can summarize the result.

Theorem 36. *A weak supersolution u that is non-negative in the cylinder $U = B(x_0, 2R) \times (t_0 - 3T/2, t_0 + T)$ has the bound*

$$\underset{B_R \times (t_0-T,t_0+T)}{\text{ess sup}} \{u^{n+p-\frac{2n}{p}}\} \leq K \frac{\left(\frac{R^p}{T} + \|u\|_\infty^{p-2}\right)^{1+\frac{n}{p}}}{TR^n} \int_{t_0-\frac{3T}{2}}^{t_0+T} \int_{B_{2R}} u^p \, dx \, dt,$$

(24)

where $0 \leq u \leq \|u\|_\infty$ in U.

We need the fact that **the positive part $(u)_+$ of a weak subsolution is again a weak subsolution.** Here the proof has to avoid the comparison principle, which is not yet available. It reduces to the following lemma.

Lemma 37. *If v is a weak supersolution, so is $v_L = \min\{v, L\}$.*

Proof. Formally, the test function[22]

$$\varphi = \min\{k(L-v)_+, 1\}\zeta = \chi_k \zeta$$

inserted into

$$\int_0^T \int_\Omega \left(-v\varphi_t + \langle |\nabla v|^{p-2} \nabla v, \nabla \varphi \rangle \right) dx \, dt \geq 0$$

implies the desired inequality

$$\int_0^T \int_\Omega \left(-v_L \zeta_t + \langle |\nabla v_L|^{p-2} \nabla v_L, \nabla \zeta \rangle \right) dx \, dt \geq 0$$

at the limit $k = \infty$. As usual, $\zeta \in C_0^\infty(\Omega_T)$, $\zeta \geq 0$. The explanation is that $\lim \chi_k = $ the characteristic function of the set $\{v < L\}$. Under the assumption that the "forbidden" time derivative u_t is available at the intermediate steps we have

$$\int_0^T \int_\Omega \chi_k \left(-v\zeta_t + \langle |\nabla v|^{p-2} \nabla v, \nabla \zeta \rangle \right) dx \, dt$$

$$\geq k \iint_{L-\frac{1}{k}<v<L} \zeta |\nabla v|^p \, dx \, dt + \int_0^T \int_\Omega v\zeta \frac{\partial}{\partial t} \chi_k \, dx \, dt$$

---

[22]This is from Lemma 2.109 on page 122 of [23].

$$\geq \int_0^T \int_\Omega v\zeta \frac{\partial}{\partial t} \chi_k \, dx \, dt = -\frac{1}{2k} \int_0^T \int_\Omega \zeta \frac{\partial}{\partial t} (\chi_k)^2 \, dx \, dt$$

$$= +\frac{1}{2k} \int_0^T \int_\Omega (\chi_k)^2 \zeta_t \, dx \, dt \longrightarrow 0.$$

The formula $\partial \chi_k / \partial t = -vk$ or $= 0$ was used. The result follows.

Finally, to handle the problem with the time derivative, one has first to regularize the equation and then to use the test function

$$\varphi^\varepsilon = \min\{k(L - v^\varepsilon)_+, 1\}\zeta = \chi_k \zeta,$$

where v^ε is the convolution in (19). The term

$$\int_0^T \int_\Omega -v^\varepsilon \frac{\partial \varphi^\varepsilon}{\partial t} dx \, dt$$

can be written so that the derivative $\partial v^\varepsilon / \partial t$ disappears. Then one may safely let $\varepsilon \to 0$. The result follows as before. $\square$

Proof of Theorem 30. Let (x_0, t_0) be a Lebesgue point for the weak supersolution v. Then

$$\lim_{TR^n \to 0} \frac{1}{TR^n} \int_{t_0 - 2T}^{t_0 + T} \int_{B_{2R}} |v(x_0, t_0) - v(x, t)|^p \, dx \, dt = 0.$$

A fortiori

$$\lim_{TR^n \to 0} \frac{1}{TR^n} \int_{t_0 - 2T}^{t_0 + T} \int_{B_{2R}} (v(x_0, t_0) - v(x, t))_+^p \, dx \, dt = 0. \tag{25}$$

We claim that

$$v(x_0, t_0) \leq \text{ess} \liminf_{(x,t) \to (x_0, t_0)} v(x, t). \tag{26}$$

It is sufficient to establish that

$$\text{ess} \limsup_{(x,t) \to (x_0, t_0)} (v(x_0, t_0) - v(x, t))_+ = 0,$$

since those points where $v(x, t) \geq v(x_0, t_0)$ can do no harm to inequality (26).

To this end, notice that the function $v(x_0, t_0) - v(x, t)$ is a weak subsolution and so is its positive part, the function

$$u(x, t) = (v(x_0, t_0) - v(x, t))_+$$

by Lemma 37. It is locally bounded according to Lemma 34. Thus the essliminf is $> -\infty$ in (26). Use Theorem 36 and let $TR^n \to 0$, keeping $R^p/T \leq$ Constant. In virtue of (24) it follows that

$$\text{ess}\lim_{(x,t)\to(x_0,t_0)}\sup\left\{u(x,t)^{n+p-\frac{2n}{p}}\right\} = 0$$

and the exponent can be erased. This proves the claim (26) at the given Lebesgue point.

Furthermore, the Lebesgue points have the property that

$$v(x_0,t_0) \leq \text{ess}\lim_{(x,t)\to(x_0,t_0)}\inf v(x,t)$$

$$\leq \lim_{TR^n\to 0}\frac{1}{TR^n}\int_{t_0-2T}^{t_0+T}\int_{B_{2R}} v(x,t)\,dx\,dt = v(x_0,t_0).$$

Since almost every point is a Lebesgue point, we have established that

$$v(x_0,t_0) = \text{ess}\lim_{(x,t)\to(x_0,t_0)}\inf v(x,t)$$

almost everywhere. The right-hand member is a semicontinuous function. $\square$

5 The Equation with Measure Data

There is a close connexion between supersolutions and equations where the right-hand side is a Radon measure. The Barenblatt solution has the Dirac measure (multiplied by a suitable constant) as the right-hand side, and hence it is, indeed, a *solution* to an equation. The equation

$$\frac{\partial v}{\partial t} - \nabla \cdot (|\nabla v|^{p-2}\nabla v) = \mu$$

with a Radon measure μ has been much studied. For example, in [1] a summability result is given for the spatial gradient ∇v of the solution. There the starting point was the given measure and the above equation. However, we can do the opposite and **produce the measure**. Indeed, every viscosity supersolution (or p-supercaloric function) induces a Radon measure $\mu \geq 0$. This follows from our summability theorem, combined with the Riesz Representation Theorem for linear functionals.

Theorem 38. *Let v be a viscosity solution in $\Omega \times (0,T)$. Then there exists a non-negative Radon measure μ such that*

$$\int_0^T\int_\Omega\left(-v\frac{\partial\varphi}{\partial t} + \langle|\nabla v|^{p-2}\nabla v, \nabla\varphi\rangle\right)dx\,dt = \int_{\Omega\times(0,T)}\varphi\,d\mu$$

for all $\varphi \in C_0^\infty(\Omega \times (0,T))$.

Proof. We already know that $v, \nabla v \in L_{loc}^{p-1}(\Omega \times (0,T))$. In order to use Riesz's Representation Theorem we define the linear functional

$$\Lambda_v : C_0^\infty(\Omega \times (0,T)) \longrightarrow \mathbb{R},$$

$$\Lambda_v(\varphi) = \int_0^T \int_\Omega \left(-v\frac{\partial\varphi}{\partial t} + \langle |\nabla v|^{p-2}\nabla v, \nabla\varphi \rangle \right) dx\, dt.$$

Now $\Lambda_v(\varphi) \geq 0$ for $\varphi \geq 0$ according to Theorem 2. Thus the functional is positive and the existence of the Radon measure follows from Riesz's theorem, cf. [7, Sect. 1.8]. □

Some further results can be found in [16].

6   Pointwise Behaviour

The viscosity supersolutions are defined at each point, not only almost everywhere. Actually, the results in this section imply that two viscosity supersolutions that coincide almost everywhere do so at each point.

6.1   The Stationary Equation

We begin with the stationary case. At each point a p-superharmonic function v satisfies

$$v(x) \leq \liminf_{y\to x} v(y) \leq \operatorname*{ess\,lim\,inf}_{y\to x} v(y)$$

by lower semicontinuity. *Essential limes inferior* means that sets of Lebesgue measure zero be neglected in the calculation of the lower limit. The reverse inequalities also hold. To see this, we start by a lemma, which requires a pedantic formulation.

Lemma 39. *Suppose that v is p-superharmonic in the domain Ω. If $v(x) \leq \lambda$ at each point x in Ω and if $v(x) = \lambda$ at almost every point x in Ω, then $v(x) = \lambda$ at each point x in Ω.*

Proof. The proof is trivial for continuous functions and the idea is that v is everywhere equal to a p-harmonic function, which, of course, must coincide with the constant λ. We approximate v by the infimal convolutions v_ε. We can assume that the function v is bounded also from below in a given ball B_{2r}, strictly interior in Ω. We may even take $0 \leq v \leq \lambda$ by adding a constant. We approximate v by the infimal convolutions v_ε. Replace v_ε in B_r by the

p-harmonic function h_ε having boundary values v_ε. Thus we have the function

$$w_\varepsilon = \begin{cases} h_\varepsilon \text{ in } B_r \\ v_\varepsilon \text{ in } B_{2r} \backslash B_r \end{cases}$$

As we have seen before, also w_ε is p-superharmonic. By comparison

$$w_\varepsilon \leq v_\varepsilon \leq v$$

pointwise in B_{2r}. As ε approaches zero via a decreasing sequence, say $1, 1/2, 1/3, \cdots$, the h_ε's converge to a p-harmonic function h, which is automatically continuous because the family is uniformly equicontinuous so that Ascoli's theorem applies. The equicontinuity is included in the Hölder estimate (8), because $0 \leq h_\varepsilon \leq \lambda$. Thus

$$h \leq v \leq \lambda$$

at *each* point in B_r. Since $\lambda - v_\varepsilon \geq \lambda - v \geq 0$, the Caccioppoli estimate

$$\int_{B_r} |\nabla h_\varepsilon|^p \, dx \leq \int_{B_r} |\nabla v_\varepsilon|^p \, dx$$
$$\leq p^p \int_{B_{2r}} (\lambda - v_\varepsilon)^p |\nabla \zeta|^p \, dx$$
$$\leq Cr^{-p} \int_{B_{2r}} (\lambda - v_\varepsilon)^p \, dx$$

is valid. The weak lower semicontinuity of the integral implies that

$$\int_{B_r} |\nabla h|^p \, dx \leq \lim_{\varepsilon \to 0} \int_{B_r} |\nabla h_\varepsilon|^p \, dx \leq Cr^{-p} \int_{B_{2r}} (\lambda - v)^p \, dx = 0.$$

The conclusion is that h is constant almost everywhere, and hence everywhere by continuity. The constant must be λ, because it has boundary values λ in Sobolev's sense. We have proved that also $v(x) = \lambda$ at *each* point in the ball B_r. The result follows. $\square$

Lemma 40. *If v is p-superharmonic in Ω and if $v(x) > \lambda$ for a.e. x in Ω, then $v(x) \geq \lambda$ for every x in Ω.*

   Proof. If $\lambda = -\infty$, there is nothing to prove. Applying the previous lemma to the p-superharmonic function defined by

$$\min\{v(x), \lambda\}$$

we obtain the result in the case $\lambda > -\infty$. $\square$

Theorem 41. *At each point a p-superharmonic function v satisfies*

$$v(x) = \operatorname*{ess\,lim\,inf}_{y \to x} v(y).$$

Proof. Fix an arbitrary point $x \in \Omega$. We must show only that

$$\lambda = \operatorname*{ess\,lim\,inf}_{y \to x} v(y) \le v(x),$$

since the opposite inequality was clear. Given any $\varepsilon > 0$, there is a δ such that $v(y) > \lambda - \varepsilon$ for a.e. $y \in B(x, \delta)$. By the lemma $v(y) \ge \lambda - \varepsilon$ for *each* such y. In particular, $v(x) \ge \lambda - \varepsilon$. Because ε was arbitrary, we have established that $v(x) \ge \lambda$. $\qquad\square$

6.2 The Evolutionary Equation

We turn to the pointwise behaviour for the Evolutionary p-Laplacian Equation. At each point in its domain a lower semicontinuous function satisfies

$$v(x,t) \le \operatorname*{lim\,inf}_{(y,\tau) \to (x,t)} v(y,\tau) \le \operatorname*{ess\,lim\,inf}_{(y,\tau) \to (x,t)} v(y,\tau) \le \operatorname*{ess\,lim\,inf}_{\substack{(y,\tau) \to (x,t) \\ \tau < t}} v(y,\tau).$$

We show that for a viscosity supersolution also the reverse inequalities hold, thus establishing Theorem 3 in the Introduction. In principle, the proof is similar to the stationary case, but now a delicate issue of regularization arises. We first consider a non-positive viscosity supersolution $v = v(x,t)$ which is equal to zero at *almost* each point and, again, we show that locally it coincides with the viscosity solution having the same boundary values, now in a space-time cylinder. Then one has to conclude that v was identically zero.

We seize the opportunity to describe a useful procedure of **regularizing** by taking the convolution[23]

$$u^\star(x,t) = \frac{1}{\sigma} \int_0^t e^{(s-t)/\sigma} u(x,s)\, ds, \quad \sigma > 0.$$

The notation hides the dependence on the parameter σ. For continuous and for bounded semicontinuous functions u the averaged function $u^\star$ is defined at each point. We will stay within this framework. Observe that

$$\sigma \frac{\partial u^\star}{\partial t} + u^\star = u.$$

Some of its properties are listed in the next lemma.

---

[23] The origin of this function is unknown to me. In connexion with the Laplace transform it would be the convolution of u and $\sigma^{-1} e^{-t/\sigma}$.

Lemma 42.   *(i) If $u \in L^p(D_T)$, then*

$$\|u^\star\|_{L^p(D_T)} \leq \|u\|_{L^p(D_T)}$$

and

$$\frac{\partial u^\star}{\partial t} = \frac{u - u^\star}{\sigma} \in L^p(D_T).$$

Moreover, $u^\star \to u$ in $L^p(D_T)$ as $\sigma \to 0$.
(ii) If, in addition, $\nabla u \in L^p(D_T)$, then $\nabla(u^\star) = (\nabla u)^\star$ componentwise,

$$\|\nabla u^\star\|_{L^p(D_T)} \leq \|\nabla u\|_{L^p(D_T)},$$

and $\nabla u^\star \to \nabla u$ in $L^p(D_T)$ as $\sigma \to 0$.
(iii) Furthermore, if $u_k \to u$ in $L^p(D_T)$ then also

$$u_k^\star \to u^\star \quad and \quad \frac{\partial u_k^\star}{\partial t} \to \frac{\partial u^\star}{\partial t}$$

in $L^p(D_T)$.
  1. *(iv) If $\nabla u_k \to \nabla u$ in $L^p(D_T)$, then $\nabla u_k^\star \to \nabla u^\star$ in $L^p(D_T)$.*
  2. *(v) Finally, if $\varphi \in C(\overline{D_T})$, then*

$$\varphi^\star(x,t) + e^{-t/\sigma}\varphi(x,0) \to \varphi(x,t)$$

uniformly in D_T as $\sigma \to 0$.

Proof. We leave this as an exercise. (Some details are worked out on page 7 of [15].)   □

The *averaged equation* for a weak supersolution u in D_T reads as follows:

$$\int_0^T \int_D \left(\langle (|\nabla u|^{p-2}\nabla u)^\star, \nabla\varphi \rangle - u^\star \frac{\partial\varphi}{\partial t} \right) dx\, dt + \int_D u^\star(x,T)\varphi(x,T)\, dx$$

$$\geq \int_D u(x,0) \left(\frac{1}{\sigma} \int_0^T \varphi(x,s)e^{-s/\sigma}\, ds \right) dx$$

valid for all test functions $\varphi \geq 0$ vanishing on the lateral boundary $\partial D \times [0,T]$ of the space-time cylinder. For solutions one has equality. Notice the typical difficulty with obtaining $(|\nabla u|^{p-2}\nabla u)^\star$ and not $|\nabla u^\star|^{p-2}\nabla u^\star$, except in the linear case. The averaged equation follows from the equation for the retarded supersolution $u(x, t - s)$, where $0 \leq s \leq T$:

$$\int_s^T \int_D \Big(\langle |\nabla u(x,t-s)|^{p-2} \nabla u(x,t-s), \nabla \varphi(x,t) \rangle - u(x,t-s) \frac{\partial \varphi}{\partial t}(x,t) \Big) dx\, dt$$

$$+ \int_D u(x,T-s)\varphi(x,T)\, dx \geq \int_D u(x,0)\varphi(x,s)\, dx.$$

Notice that $(x, t - s) \in \overline{D_T}$ when $0 \leq s \leq t \leq T$. Multiply by $\sigma^{-1} e^{-s/\sigma}$, integrate over $[0,T]$ with respect to s, and, finally, interchange the order of integration between s and t. This yields the averaged equation above.

The advantage of this procedure over more conventional convolutions is that no values outside the original space-time cylinder are evoked.

We begin with a simple situation.

Lemma 43. *Suppose that v is a viscosity supersolution in a domain containing the closure of $B_T = B \times (0,T)$. If*

(i) $v \leq 0$ at each point in B_T and
(ii) $v = 0$ at almost every point in B_T,

then $v = 0$ at each point in $B \times (0,T]$.

Proof. We may assume that v is bounded. Construct the infimal convolution v_ε with respect to a larger domain than B_T. Fix a small time $t' > 0$ and let h^ε be the p-caloric function with boundary values induced by v_ε on the parabolic boundary of the cylinder $B \times (t', T)$ and define the function

$$w_\varepsilon = \begin{cases} h^\varepsilon, & \text{in } B \times (t', T] \\ v_\varepsilon & \text{otherwise.} \end{cases}$$

To be on the safe side concerning the validity at the terminal time T we may solve the boundary value problem in a slightly larger domain with terminal time $T' > T$. Also w_ε is a viscosity supersolution. By comparison

$$w_\varepsilon \leq v_\varepsilon \leq 0 \quad \text{pointwise in} \quad B_T.$$

We let ε go to zero through a monotone sequence, say $1, \frac{1}{2}, \frac{1}{3}, \cdots$. Then the limit

$$h = \lim_{\varepsilon \to 0} h^\varepsilon$$

exists pointwise and it follows from the uniform Hölder estimates (14) that this h is continuous without any correction made in a set of measure zero. It is important to preserve the information at *each* point. Thus h is a p-caloric function. The so obtained function

$$w = \begin{cases} h, & \text{in } B \times (t', T') \\ v & \text{otherwise} \end{cases}$$

is a viscosity supersolution. For the verification of the semicontinuity and the comparison principle, which proves this, the fact that $h \leq v$ is essential.

We know that $w \leq v \leq 0$ everywhere in a domain containing $B \times (0, T)$. In particular,

$$h \leq v \leq 0 \quad \text{everywhere in} \quad B \times (0, T).$$

We claim that $h = 0$ at each point. The claim immediately implies that $v = 0$ at each point in $B \times (0, T)$. Concerning the statement at the terminal time T, we notice that $v \geq h$ and

$$v(x, T) \geq h(x, T) = \lim_{t \to T-} h(x, t) = 0,$$

since h is continuous. On the other hand $v(x, T) \leq 0$ by the lower semicontinuity. Thus also $v(x, T) = 0$.

Therefore it is sufficient to prove the claim. To conclude that h is identically zero we use the averaged equation for $w^\star$ and write

$$\int_0^T \int_B \left(\langle (|\nabla w|^{p-2} \nabla w)^\star, \nabla \varphi \rangle + \varphi \frac{\partial w^\star}{\partial t} \right) dx \, dt$$

$$\geq \int_B w(x, 0) \left(\frac{1}{\sigma} \int_0^T \varphi(x, s) e^{-s/\sigma} \, ds \right) dx,$$

where the test function vanishes on the parabolic boundary (an integration by parts has been made with respect to time.) Select the test function $\varphi = (v_\varepsilon - w_\varepsilon)^\star$ and let ε approach zero. Taking into account that $\varphi = 0$ when $t < t'$, we arrive at

$$\int_{t'}^T \int_B \left(\langle (|\nabla h|^{p-2} \nabla h)^\star, \nabla v^\star - \nabla h^\star \rangle + (v^\star - h^\star) \frac{\partial h^\star}{\partial t} \right) dx \, dt$$

$$\geq \int_B v(x, 0) \left(\frac{1}{\sigma} \int_{t'}^T (v^\star(x, s) - h^\star(x, s)) e^{-s/\sigma} \, ds \right) dx.$$

The last integral (which could be negative) approaches zero as the regularization parameter σ goes to zero, because $t' > 0$, so that the exponential decays. Integrating

$$(v^\star - h^\star) \frac{\partial h^\star}{\partial t} = -(v^\star - h^\star) \frac{\partial (v^\star - h^\star)}{\partial t} + (v^\star - h^\star) \frac{\partial v^\star}{\partial t}$$

we obtain

$$\int_{t'}^{T}\int_{B}(v^{\star}-h^{\star})\tfrac{\partial h^{\star}}{\partial t}\,dx\,dt$$
$$=-\tfrac{1}{2}\int_{B}(v^{\star}(x,T)-h^{\star}(x,T))^2\,dx$$
$$+\int_{t'}^{T}\int_{D}(v^{\star}-h^{\star})\tfrac{\partial v^{\star}}{\partial t}\,dx\,dt.$$

because $v^{\star}(x,t')-h^{\star}(x,t')=0$. The last integral is zero because $v^{\star}$ and $\tfrac{\partial v^{\star}}{\partial t}$ are zero almost everywhere according to property (i) in Lemma 42. Erasing this integral and letting the regularization parameter σ go to zero (so that the $\star$'s disappear) we finally obtain

$$\int_{t'}^{T}\int_{B}|\nabla h|^{p}\,dx\,dt+\frac{1}{2}\int_{B}h^{2}(x,T)\,dx\le 0\quad\text{i.e.}\quad=0.$$

In fact,[24] the proof guarantees this only for almost all values of T in the range $t' < T < T'$. From this it is not difficult to conclude that h is identically zero. Thus our claim has been proved. □

Lemma 44. *Suppose that v is a viscosity supersolution in a domain containing $B_T = B \times (0,T)$. If $v(x,t) > \lambda$ for almost every $(x,t) \in B_T$, then $v(x,t) \ge \lambda$ for every $(x,t) \in B \times (0,T]$.*

Proof. The auxiliary function

$$u(x,t) = \min\{v(x,t),\lambda\} - \lambda$$

in place of v satisfies the assumptions in the previous lemma. Hence $u = 0$ everywhere in $B \times (0,T]$. This is equivalent to the assertion. □

Proof of Theorem 3. Denote

$$\lambda = \operatorname*{ess\,lim\,inf}_{\substack{(x,t)\to(x_0,t_0)\\t<t_0}} v(x,t).$$

According to the discussion in the beginning of this section, it is sufficient to prove that $\lambda \le v(x_0,t_0)$. Thus we can assume that $\lambda > -\infty$.

---

[24] It is the validity of

$$\lim_{\sigma\to 0}\frac{1}{2}\int_{B}(v^{\star}(x,T)-h^{\star}(x,T))^2\,dx=\frac{1}{2}\int_{B}h^2(x,T)\,dx$$

that requires some caution. We know that $v^{\star}$ is zero almost everywhere but with respect to the $(n+1)$-dimensional measure.

First, we consider the case $\lambda < \infty$. Given $\varepsilon > 0$, we can find a $\delta > 0$ and a ball B with centre x_0 such that the closure of $B \times (t_0 - \delta, t_0)$ is comprised in the domain and

$$v(x,t) > \lambda - \varepsilon$$

for almost every $(x,t) \in B \times (t_0 - \delta, t_0)$. According to the previous lemma

$$v(x,t) \geq \lambda - \varepsilon$$

for *every* $(x,t) \in B \times (t_0 - \delta, t_0]$. In particular, we can take $(x,t) = (x_0, t_0)$. Hence $v(x_0, t_0) \geq \lambda - \varepsilon$. Since ε was arbitrary, we have proved that $\lambda \leq v(x_0, t_0)$, as desired.

Second, the case $\lambda = \infty$ is easily reached via the truncated functions $v_k = \min\{v(x,t), k\}$, $k = 1, 2, \cdots$. Indeed,

$$v(x_0, t_0) \geq v_k(x_0, t_0) \geq \min\{\infty, k\} = k,$$

in view of the previous case. This concludes the proof of Theorem 3. $\square$

References

1. L. Boccardo, A. Dall'Aglio, T. Gallouët, L. Orsina, Nonlinear parabolic equations with measure data. J. Funct. Anal. **147**, 237–258 (1997)
2. M. Brelot, Éléments de la Théorie Classique du potential, 3e édition, Les cours de Sorbonne, 3e cycle, Centre de Documentation Universitaire, Paris, 1965
3. M. Crandall, J. Zhang, Another way to say harmonic. Trans. Am. Math. Soc. **355**, 241–263 (2003)
4. H.-J. Choe, A regularity theory for a more general class of quasilinear parabolic partial differential equations and variational inequalities. Differ. Integr. Equat. **5**, 915–944 (1992)
5. M. Crandall, H. Ishii, P.-L. Lions, User's guide to viscosity solutions of second order partial differential equations. Bull. Am. Math. Soc. **27**, 1–67 (1992)
6. E. DiBenedetto, *Degenerate Parabolic Equations* (Springer, Berlin, 1993)
7. L. Evans, R. Gariepy, *Measure Theory and Fine Propeties of Functions* (CRC Press, Boca Raton, 1992)
8. E. Giusti, *Metodi diretti nel calcolo delle variazioni* (Unione Matematica Italiana, Bologna, 1994)
9. E. Giusti, *Direct Methods in the Calculus of Variations* (World Scientific, Singapore, 2003)
10. P. Juutinen, P. Lindqvist, J. Manfredi, On the equivalence of viscosity solutions and weak solutions for a quasi-linear equation. SIAM J. Math. Anal. **33**, 699–717 (2001)
11. S. Kamin, J.L. Vasquez, Fundamental solutions and asymptotic behaviour for the p-Laplacian equation. Revista Mathematica Iberoamericana **4**, 339–354 (1988)
12. T. Kilpeläinen, P. Lindqvist, On the Dirichlet boundary value problem for a degenerate parabolic equation, SIAM J. Math. Anal. **27**, 661–683 (1996)
13. T. Kilpeläinen, J. Malý, Degenerate elliptic equations with measure data and nonlinear potentials. Annali della Scuola Normale Superiore di Pisa Cl. Sci. (4) **19**, 591–613 (1992)

14. J. Kinnunen, P. Lindqvist, Summability of semicontinuous supersolutions to a quasi-linear parabolic equation. Annali della Scuola Normale Superiore di Pisa Cl. Sci. (5) **4**, 59–78 (2005)

15. J. Kinnunen, P. Lindqvist, Pointwise behaviour of semicontinuous supersolutions to a quasilinear parabolic equation. Annali di Matematica Pura ed Applicata (4) **185**, 411–435 (2006)

16. J. Kinnunen, T. Lukkari, M. Parviainen, An existence result for superparabolic functions. J. Funct. Anal. **258**, 713–728 (2010)

17. S. Koike, in *A Beginner's Guide to the Theory of Viscosity Solutions*. MSJ Memoirs, vol. 13 (Mathematical Society of Japan, Tokyo, 2004)

18. T. Kuusi, Lower semicontinuity of weak supersolutions to nonlinear parabolic equations. Differ. Integr. Equat. **22**, 1211–1222 (2009)

19. J. Lewis, On very weak solutions of certain elliptic systems. Comm. Part. Differ. Equat. **18**, 1517–1537 (1993)

20. P. Lindqvist, On the definition and properties of p-superharmonic functions. Journal für die Reine und Angewandte Mathematik (Crelles Journal) **365**, 67–79 (1986)

21. P. Lindqvist, Notes on the p-Laplace equation. University of Jyväskylä, Report 102 (2006)

22. P. Lindqvist, J. Manfredi, Viscosity supersolutions of the evolutionary p-Laplace equation. Differ. Integr. Equat. **20**, 1303–1319 (2007)

23. J. Maly, W. Ziemer, in *Fine Regularity of Solutions of Elliptic Partial Differential Equations*. Math. Surveys Monogr., vol. 51 (AMS, Providence, 1998)

24. J. Michael, W. Ziemer, Interior regularity for solutions to obstacle problems. Nonlinear Anal. **10**, 1427–1448 (1986)

25. G. Mingione, Regularity of minima: An invitation to the dark side of the calculus of variations. Appl. Math. **51**, 355–426 (2006)

26. J. Serrin, Pathological solutions of elliptic differential equations. Annali della Scuola Normale Superiore di Pisa Cl. Sci. (3) **18**, 385–387 (1964)

27. N. Trudinger, On Harnack type inequalities and their application to quasilinear elliptic equations. Comm. Pure Appl. Math. **20**, 721–747 (1967)

28. N. Trudinger, Pointwise estimates and quasilinear parabolic equations. Comm. Pure Appl. Math. **21**, 205–226 (1968)

29. J. Urbano, *The Method of Intrinsic Scaling: A Systematic Approach to Regularity for Degenerate and Singular PDEs*. Lecture Notes in Mathematics, vol. 1930 (Springer, Berlin, 2008)

30. Z. Wu, J. Zhao, J. Yin, H. Li, *Non-linear Diffusion Equations* (World Scientific, Singapore, 2001)

31. C. Yazhe, Hölder continuity of the gradient of the solutions of certain degenerate parabolic equations. Chin. Ann. Math. Ser. B **8**(3), 343–356 (1987)

Introduction to Random Tug-of-War Games and PDEs

Juan J. Manfredi

1 Introduction

The fundamental contributions of Kolmogorov, Ito, Kakutani, Doob, Hunt, Lévy, and many others have shown the profound and powerful connection between classical linear potential theory and probability theory. The idea behind the classical interplay is that harmonic functions and martingales share a common cancelation property that can be expressed by using mean value properties. In these lectures, we will see how this approach turns out to very useful in the nonlinear theory as well.

The objective of this course is to provide an introduction to the connection between the theory of stochastic tug-of-war games and non-linear equations of p-Laplacian type in the Euclidean and discrete cases. These notes will provide the student with background to read [16] and [17]. Most of the material e based on the joint papers [11–13] with Mikko Parviainen and Julio Rossi, and on the 2010 doctoral thesis of Alexander Sviridov [18]. I am grateful to Alex for correcting many misprints and for suggesting changes that have improved the readability of the manuscript.

2 Probability Background

We present a quick introduction to Doob's Optimal Sampling Theorem and to Kolmogorov's construction of infinite product of measures. We follow the presentation in the book [21]. We refer to [21] for the basic probability definitions and the proofs. However, we have chosen to emphasize certain

J.J. Manfredi (✉)
Department of Mathematics, University of Pittsburgh, Pittsburgh, PA 15260
e-mail: manfredi@pitt.edu

J. Lewis et al., *Regularity Estimates for Nonlinear Elliptic and Parabolic Problems*, Lecture Notes in Mathematics 2045, DOI 10.1007/978-3-642-27145-8_3, © Springer-Verlag Berlin Heidelberg 2012

details in the proofs that are likely to help student of analysis going through this material for the first time.

We are given a set Ω endowed with a σ-algebra $\mathcal{F}$ and probability measure $\mathbb{P}$ on $\mathcal{F}$. The triplet $(\Omega, \mathcal{F}, \mathbb{P})$ is a probability space.

Definition 2.1. If $\mathcal{G} \subset \mathcal{F}$ is a sub-σ-algebra of $\mathcal{F}$ and $f \in L^1(\Omega, \mathbb{P})$ the conditional expectation of f given $\mathcal{G}$ is the only $\mathcal{G}$-measurable function $g = \mathbb{E}[f|\mathcal{G}]$ such that

$$\int_A f \, d\mathbb{P} = \int_A g \, d\mathbb{P}$$

for all sets $A \in \mathcal{G}$.

Example 2.2. Suppose that we have a finite partition $\Omega = A_1 \cup A_2 \cup \cdots \cup A_n$ where all the $A_i \in \mathcal{F}$ have positive probability $\mathbb{P}(A_i) > 0$. Let $\mathcal{G}$ the σ-algebra generated by this partition. For $f \in L^1(\Omega, \mathbb{P})$, the function $g = \mathbb{E}[f|\mathcal{G}]$ has a constant value g_i on each A_i given by

$$g_i = \fint_{A_i} f \, d\mathbb{P}$$

The notion of (discrete) martingale will be key to our developments later on.

Definition 2.3. Let $\mathcal{F}_i \subset \mathcal{F}_{i+1} \subset \cdots \subset \mathcal{F}$ be a filtration of σ algebras of $\mathcal{F}$ and $X_i \colon \Omega \mapsto \mathbb{F}$ be an $\mathcal{F}_i$-measurable random variable (or $\mathcal{F}_i$-measurable function.)

1. The sequence of random variables $\{X_i\}$ is a martingale if $X_i = \mathbb{E}[X_{i+1}|\mathcal{F}_i]$.
2. The sequence of random variables $\{X_i\}$ is a submartingale if $X_i \leq \mathbb{E}[X_{i+1}|\mathcal{F}_i]$.
3. The sequence of random variables $\{X_i\}$ is a supermartingale if $X_i \geq \mathbb{E}[X_{i+1}|\mathcal{F}_i]$.

These relations are supposed to hold a.e. with respect to $\mathbb{P}$.

When needed we will make explicit the σ-algebras by writing $\{X_i, \mathcal{F}_i\}$. If $\{X_i\}$ is a martingale, we have $\mathbb{E}[X_{i+1}] = \mathbb{E}[X_i] = \cdots = \mathbb{E}[X_1] = c$. The random variables $Y_{i+1} = X_{i+1} - X_i$ are the martingales differences. We clearly have $\mathbb{E}[Y_i] = 0$.

Example 2.4. If $X \in L^1(\Omega, \mathbb{P})$, the sequence $X_i = [X|\mathcal{F}_i]$ is a martingale. It turns that this is the most general L^1-martingale (see [21].)

Example 2.5. Let $\{Y_i\}$ be a collection of independent random variables with mean zero $\mathbb{E}[Y_i] = 0$. Set $\mathcal{F}_i = \sigma(Y_1, Y_2, \ldots, Y_i)$. Then, the sequence

$$X_i = c + \sum_{j=1}^i Y_j$$

is a martingale.

The following lemma is a consequence of Jensen's inequality for conditional expectations.

Lemma 2.6. *Let $\{X_i, \mathcal{F}_i\}$ be a martingale and $1 \leq p < \infty$. Then $\{|X_i|^p, \mathcal{F}_i\}$ is a submartingale provided that $\mathbb{E}[|X_i|^p] < \infty$.*

We present Doob's weak type 1-1 inequality for finite martingales. The simple proof contains all the ingredients of the more general cases.

Theorem 2.7. *Let $\{X_i\}_{i=1}^n$ be a martingale and $l > 0$. Then we have*

$$\mathbb{P}\left(\omega : \sup_{1 \leq i \leq n} |X_i(\omega)| \geq l\right) \leq \frac{1}{l} \int_{\{\sup |X_i| > l\}} |X_n|\, d\mathbb{P}$$

$$\leq \frac{1}{l} \int_\Omega |X_n|\, d\mathbb{P}.$$

Proof. The maximum random variable S is

$$S(\omega) = \sup\{|X_i(\omega)| : 1 \leq i \leq n\}.$$

For each $1 \leq i \leq n$ define the set describing the first time we go above l,

$$E_i = \{\omega : |X_1(\omega)| < l, |X_1(\omega)| < l, \ldots, |X_{i-1}(\omega)| < l, |X_i(\omega)| \geq l\}$$

and the set when the maximum is above l

$$E = \{\omega : S(\omega) \geq l\}.$$

We have a disjoint union

$$E = \bigcup_1^n E_i$$

and the basic estimate

$$\mathbb{P}(E_i) \leq \frac{1}{l} \int_{E_i} |X_i|\, d\mathbb{P}. \tag{1}$$

Since $|X_i|$ is a *submartingale* we have $\mathbb{E}\left[|X_n| \mid \mathcal{F}_i\right] \geq |X_i|$ for a.e. ω. Since $E_i \in \mathcal{F}_i$ we then have

$$\mathbb{E}\left[\chi_{E_i} |X_n| \mid \mathcal{F}_i\right] = \chi_{E_i} \mathbb{E}\left[|X_n| \mid \mathcal{F}_i\right] \geq \chi_{E_i} |X_i|.$$

Taking expectations we get

$$\int_{E_i} |X_n|\, d\mathbb{P} \geq \int_{E_i} |X_i|\, d\mathbb{P}$$

and using (1) we get

$$\mathbb{P}(E_i) \leq \frac{1}{l} \int_{E_i} |X_n| \, d\mathbb{P}.$$

A well-known argument using distribution functions (see [21]) gives the maximal theorem

Corollary 2.8. *(Doob's maximal theorem) Let $\{X_i\}_{i=1}^n$ be a martingale and $1 < p < \infty$. For $S = \sup\{|X_i| : 1 \leq i \leq n\}$ we have*

$$\mathbb{E}[S^p] \leq (\frac{p}{p-1})^p \mathbb{E}[|X_n|^p].$$

In addition to martingales, stopping times play a crucial role in stochastic games.

Definition 2.9. The random variable $\tau \colon \Omega \mapsto \mathbb{N} \cup \{0, \infty\}$ is a stopping time with respect to the filtration $\{\mathcal{F}_i\}$ if

$$\{\omega \colon \tau(\omega) \leq n\} \in \mathcal{F}_n \text{ for all } n \geq 0.$$

Example 2.10. The following are examples of stopping times. The reader would benefit from checking in detail that they are indeed stopping times.

1. $\tau(\omega) = k$, where k is constant.
2. If τ is a stopping time and f is monotone such that $f(t) \geq t$, then $\tau' = f \circ \tau$ is a stopping time.
3. If τ_1 and τ_2 are stopping times so are $\max\{\tau_1, \tau_2\}$ and $\min\{\tau_1, \tau_2\}$.

We see then that the truncated stopping time $\tau_n = \min\{\tau, n\}$ is also a stopping time, so that we have the following useful fact.

Corollary 2.11. *Every stopping time is the limit of an increasing sequence of bounded stopping times*

We next formalize the notion of information available up to time τ.

Definition 2.12. Let τ be a stopping time respect to the filtration $\{\mathcal{F}_i\}$. Set

$$\mathcal{F}_\tau = \{A \colon A \in \mathcal{F} \text{ and } A \cap \{\omega \colon \tau(\omega) \leq n\} \in \mathcal{F}_n \text{ for all } n\}$$

Lemma 2.13. *(Basic properties of $\mathcal{F}_\tau$)*

1. $\mathcal{F}_\tau$ is a σ-field.
2. If $\tau(\omega) = k$ constant, then $\mathcal{F}_\tau = \mathcal{F}_k$.
3. $\tau_1 \leq \tau_2 \implies \mathcal{F}_{\tau_1} \subset \mathcal{F}_{\tau_2}$.
4. τ is $\mathcal{F}_\tau$ measurable.
5. $\{\omega \colon \tau(\omega) < \infty\} \in \mathcal{F}_\tau$.

The last part follows from $\{\omega \colon \tau(\omega) < \infty\} = \bigcup_n \{\omega \colon \tau(\omega) \leq n\}$. We stop a martingale by defining X_τ as follows

$$X_\tau(\omega) = X_{\tau(\omega)}(\omega).$$

Lemma 2.14. *The stopped martingale X_τ is $\mathcal{F}_\tau$ measurable on the set where τ is finite $\{\omega \colon \tau(\omega) < \infty\}$.*

Proof. We need to check that for $\lambda \in \mathbb{R}$ we have

$$E_\lambda = \{\omega \colon \tau(\omega) < \infty, X_{\tau_\omega}(\omega) > \lambda\} \in \mathcal{F}_\tau,$$

which follows from the expression

$$E_\lambda = \bigcup_n \Big(\{X_n(\omega) > \lambda\} \cap \{\tau(\omega) = n\} \Big)$$

The cornerstone of the applications of martingale to partial differential equations is DOOB'S OPTIONAL SAMPLING THEOREM.

Theorem 2.15. *Let $\{X_n\}$ be a submartingale with respect to the filtration $\{\mathcal{F}_n\}$. Let $0 \leq \tau_1 \leq \tau_2$ be bounded stopping times. Then we have*

$$\mathbb{E}\left[X_{\tau_2} \mid \mathcal{F}_{\tau_1}\right] \geq X_{\tau_1}.$$

If $\{X_n\}$ is supermartingale we get instead

$$\mathbb{E}\left[X_{\tau_2} \mid \mathcal{F}_{\tau_1}\right] \leq X_{\tau_1},$$

and If $\{X_n\}$ is martingale we get the equality

$$\mathbb{E}\left[X_{\tau_2} \mid \mathcal{F}_{\tau_1}\right] = X_{\tau_1},$$

By making τ_1 equals to zero we get the following corollary.

Corollary 2.16. *Let $\{X_n\}$ be a martingale. For any stopping time we have $\mathbb{E}[X_\tau] = \mathbb{E}[X_0]$.*

Proof. Let us prove the theorem in the martingale case with $\tau_1 = \tau$ and $\tau_2 = k$. We need to establish that $\mathbb{E}[X_k \mid \mathcal{F}_\tau] = X_\tau$, or equivalently that for all $A \in \mathcal{F}_\tau$ we have

$$\int_A X_k \, d\mathbb{P} = \int_A X_\tau \, d\mathbb{P}. \tag{2}$$

Set $E_i = \{\omega \colon \tau(\omega) = i\}$ and decompose $\Omega = \cup_1^k E_i$ as a disjoint union. Note that if $A \in \mathcal{F}_\tau$ then $A \cap E_i \in \mathcal{F}_i$. We then have

$$\int_{A \cap E_i} X_k \, d\mathbb{P} = \int_{A \cap E_i} X_i \, d\mathbb{P} = \int_{A \cap E_i} X_\tau \, d\mathbb{P},$$

where the first equality follows by the martingale property and the second by the definition of E_i. By adding in i we obtain the required (2).

We now change gears and consider the product of probability spaces. Suppose that we have two probability spaces $(\Omega_1, \mathcal{B}_1, \mathbb{P}_1)$ and $(\Omega_2, \mathcal{B}_2, \mathbb{P}_2)$. A *rectangle* is a set $A = A_1 \times A_2$ where $A_1 \in \mathcal{B}_1$ and $A_2 \in \mathcal{B}_2$. We denote by $\mathcal{F}$ the field of all finite disjoint unions of rectangles. The product σ-algebra is the σ algebra generated by $\mathcal{F}$

$$\mathcal{B} = \mathcal{B}_1 \times \mathcal{B}_2 = \sigma(\mathcal{F}).$$

We define the product probability $\mathbb{P} = \mathbb{P}_1 \times \mathbb{P}_2$ on $\mathcal{B}$ as follows:

 (i) $\mathbb{P}(A_1 \times A_2) = \mathbb{P}_1(A_1) \cdot \mathbb{P}_2(A_2)$ for rectangles $A_1 \times A_2$.
 (ii) Extend $\mathbb{P}$ to $\mathcal{F}$ as finitely additive measure.
(iii) $\mathbb{P}$ is in fact countably additive on $\mathcal{F}$.
(iv) Extend $\mathbb{P}$ to $\mathcal{B}$ by using Caratheodory's theorem.

A finite product of probability spaces is defined similarly. We next consider infinite products. We are given probability measures $\mathbb{P}_n$ on $(\mathbb{R}^n, \mathcal{B}(\mathbb{R}^n))$, where $\mathcal{B}(\mathbb{R}^n))$ is the Borel σ-algebra in $\mathbb{R}^n$. The projection $\pi_n \colon \mathbb{R}^{n+1} \mapsto \mathbb{R}^n$ is given by

$$\pi_n(x_1, x_2, \ldots, x_n, x_{n+1}) = (x_1, x_2, \ldots, x_n).$$

The family $\{\mathbb{P}_n\}_n$ is *consistent* if for all n all rectangles $A_1 \times \ldots \times A_n$ we have

$$\mathbb{P}_{n+1}(A_1 \times \ldots \times A_n \times \mathbb{R}) = \mathbb{P}_n(A_1 \times \ldots \times A_n).$$

We also write this equation as $\mathbb{P}_{n+1}\pi_n^{-1} = \mathbb{P}_n$ and say that the marginal probability of $\mathbb{P}^{n+1}$ on $\mathbb{R}^n$ is $\mathbb{P}_n$.

From now on Ω will be the infinite cartesian product

$$\Omega = \mathbb{R}^\infty = \{\omega \colon \omega = (x_n)_{n \in \mathbb{N}}\}.$$

A *cylinder* with base $A \in \mathcal{B}(\mathbb{R}^n)$ is a set of the form

$$C = \{\omega \colon (x_1, x_2, \ldots, x_n) \in A\}.$$

The set of all cylinders form a field $\mathcal{F}$. They generate a σ-algebra

$$\Sigma = \sigma(\mathcal{F}).$$

Theorem 2.17. *(Kolmogorov) Let $\{\mathbb{P}_n\}$ be a consistent family of probabilities in $(\mathbb{R}^n, \mathcal{B}(\mathbb{R}^n))$. Then, there exists a unique probability in (Ω, Σ) such that*

$$\mathbb{P} \, \pi_n^{-1} = \mathbb{P}_n.$$

That is, for all rectangles $A_1 \times A_2 \times \ldots \times A_n$ we have

$$\mathbb{P}(A_1 \times A_2 \times \ldots A_n \times \mathbb{R} \ldots \times \mathbb{R} \times \ldots) = \mathbb{P}_n(A_1 \times A_2 \times \ldots A_n).$$

Proof. For a cylinder $C \in \mathcal{F}$ with base $A \in \mathcal{B}(\mathbb{R}^n)$ define

$$\mathbb{P}(C) = \mathbb{P}_n(A).$$

The first observation is that $\mathbb{P}$ is well-defined in $\mathcal{F}$ by the consistency hypothesis. The key point is to establish the $\mathbb{P}$ is countably additive on $\mathcal{F}$. Then we can extend $\mathbb{P}$ to Σ by the Caratheodory's procedure.

We need to show that if $B_n \in \mathcal{F}$, $B_{n+1} \subset B_n$, and $\cap_n B_n = \emptyset$, then

$$\lim_{n \to \infty} \mathbb{P}(B_n) = 0.$$

The proof is by contradiction. Suppose that for some $\delta > 0$ we have that $\mathbb{P}(B_n) > \delta$ for all $n \in \mathbb{N}$. Since B_n is a cylinder we write $B_n = \pi_n^{-1}(A_n)$. Suppose for the moment that all basis A_n are compact and write

$$B_n = A_1^n \times A_2^n \times \ldots \times A_n^n \times \mathbb{R} \times \mathbb{R} \ldots$$
$$B_{n+1} = A_1^{n+1} \times A_2^{n+1} \times \ldots \times A_n^{n+1} \times A_{n+1}^{n+1} \times \mathbb{R} \times \mathbb{R} \ldots$$

We see that for each j we have a nested sequence of compact subsets $A_j^{n+1} \subset A_j^n$. By the finite intersection property we get

$$A_j^\infty = \bigcap_{n=1}^\infty A_j^n \neq \emptyset.$$

But then we would get a contradiction since

$$\bigcap_{n=1}^\infty B_n = A_1^\infty \times A_2^\infty \times \ldots \neq \emptyset.$$

In the general case A_n is only a Borel set. For each n select a compact set $K_n \subset A_n$ such that

$$\mathbb{P}_n(A_n \setminus K_n) \leq \frac{\delta}{2^{n+2}}.$$

Write $D_n = \pi_n^{-1}(K_n) \subset B_n$ and observe that $E_n = \bigcap_{j=1}^{n} D_j = \pi_n^{-1}(F_n)$ for some compact set F_n. We then have

$$\mathbb{P}(B_n \setminus E_n) = \mathbb{P}\left(B_n \setminus \cap_{j=1}^{n} D_j\right)$$

$$= \mathbb{P}\left(\bigcup_{j=1}^{n}(B_n \setminus D_j)\right)$$

$$\leq \mathbb{P}\left(\bigcup_{j=1}^{n}(B_j \setminus D_j)\right)$$

$$\leq \sum_{j=1}^{n} \mathbb{P}\left(B_j \setminus D_j\right))$$

$$\leq \sum_{j=1}^{n} \mathbb{P}\left(A_j \setminus K_j\right))$$

$$\leq \sum_{j=1}^{n} \frac{\delta}{2^{j+2}}$$

$$\leq \frac{\delta}{4}.$$

From which we deduce that

$$\mathbb{P}(E_n) \geq \mathbb{P}(B_n) - \mathbb{P}(B_n \setminus E_n)$$

$$\geq \delta - \frac{\delta}{4}$$

$$\geq \frac{\delta}{2},$$

reducing the problem to the compact case.

Let see how Kolmogorov's theorem can be used to give a quick construction of the Lebesgue measure.

Consider the simplest case of ternary trees. We follow the formalism developed in [5]. A directed tree T with regular 3-branching consists of the empty set $\emptyset$ as the top vertex, 3 sequences of length 1 with terms chosen from the set $X = \{0, 1, 2\}$, 9 sequences of length 2 with terms chosen from the set $X^2 = \{0, 1, 2\}$,..., 3^r sequences of length r with terms chosen from the et $X^r = \{0, 1, 2\}$ and so on. A vertex b_r at level r is labeled by a sequence of digits $d_1 d_2 \ldots d_r$, where $d_j \in X$ for all $1 \leq j \leq r$. A *branch* of T is an infinite sequence of vertices, each followed by one of its immediate successors. We denote a branch $\mathbf{b}$ starting at the vertex b_1 as follows $\mathbf{b} = (b_1, b_2, \ldots, b_r, \ldots)$.

The collection of all branches forms the boundary of the tree T and is denoted by ∂T. A branch $\mathbf{b}$ determines a real number in the interval $[0, 1]$ by means of the ternary expansion

$$g(\mathbf{b}) = \sum_{k=1}^{\infty} \frac{b_k}{3^k}. \tag{3}$$

Note that the set of all branches that start at a given vertex $b = d_1 d_2 \ldots d_r$ is the ternary interval $I_b = [0.d_1 d_2 \ldots d_r, 0.d_1 d_2 \ldots d_r + 3^{-r}]$ of length 3^{-r}, where the expansions are in base 3. Note also that the *classical Cantor set* $\mathcal{C}$ is the subset of ∂T formed by branches that don't go through any vertex labeled 1.

Let $\mathbb{P}_n$ be the uniform probability in X^n. The family $\{\mathbb{P}_n\}$ is a consistent family of probabilities, so that by Theorem 2.17 there exists a unique probability in X^∞ such that

$$\mathbb{P}(A_1 \times A_2 \times \ldots A_n \times \mathbb{X} \ldots \times \mathbb{X} \times \ldots) = \mathbb{P}_n(A_1 \times A_2 \times \ldots A_n).$$

If we take $A_1 = \{d_1\}$, $A_2 = \{d_1\}$,..., $A_r = \{d_r\}$ we get

$$\mathbb{P}(\{b\} \times X \ldots \times X \ldots) = \mathbb{P}_r(\{b\}) = 3^{-r} = |I_b|$$

Corollary 2.18. *Consider the probability spaces $(X^\infty, \Sigma, \mathbb{P})$ and $([0,1], \mathcal{B}, \mathcal{L})$, where $\mathcal{L}$ is the Lebesgue measure in the interval $[0, 1]$. Let g be the ternary expansion mapping (3). Then we have*

$$\mathbb{P} g^{-1} = \mathcal{L}.$$

3 The p-Laplacian Gambling House

Start with a set $\mathfrak{X}$ endowed with a σ-algebra $\mathcal{B}$. Decompose

$$\mathfrak{X} = X \cup Y$$

as a disjoint union of two non-empty sets X and Y. We shall call X the interior and Y the boundary. For each point $x \in \mathfrak{X}$ we have a nonempty set $S(x) \subset \mathfrak{X}$ of successors of x. For points $y \in Y$ we require that $S(y) = \{y\}$. Moreover, the set $S(x)$ comes equipped with a probability measure supported in $S(x)$ denoted by $\mu(x)$. For points $y \in Y$ on the boundary we have that $\mu(y) = \delta_y$.

We are given non-negative numbers α and β so that $\alpha + \beta = 1$ and a pay-off function $F \colon Y \mapsto \mathbb{R}$.

At every point $x \in \mathfrak{X}$ we have a family of probability measures $\Gamma(x)$ in $(\mathfrak{X}, \mathcal{B})$ given by

$$\Gamma(x) = \left\{ \frac{\alpha}{2} \left(\delta_{x_I} + \delta_{x_{II}} \right) + \beta \, \mu(x) : x_I, x_{II} \in S(x) \right\} \tag{4}$$

To play a Tug-of-War game with noise starting at a point $x_0 \in \mathfrak{X}$, choose a probability $\gamma_0[x_0] \in \Gamma(x_0)$. The next position $x_1 \in S(x_0)$ is selected according to $\gamma_0[x_0]$. Once x_0 and x_1 are chosen, we pick a probability $\gamma_1[x_0, x_1] \in \Gamma(x_1)$ to determine the next game position $x_2 \in S(x_1)$. In this manner we determine a particular history

$$\mathbf{x} = (x_0, x_1, x_2, \dots) \in \mathfrak{X} \times \mathfrak{X} \times \cdots \times \mathfrak{X} \times \cdots = \mathfrak{X}^\infty.$$

The game ends when we reach the boundary Y since once $x_j \in Y$ we have $x_{j+1} \in S(x_j) = \{x_j\}$. We write

$$\tau(\mathbf{x}) = \inf\{k : x_k \in Y\}$$

for the first time we hit the boundary with the understanding that $\tau(\mathbf{x}) = \infty$ if the boundary is never reached. If the game ends at a point $y \in Y$ the pay-off value is $F(y)$.

We now apply a variant of the Kolmogorov's construction. Let us denote by $\mathcal{B}^j$ the product σ-algebra in $\mathfrak{X}^j$ and by $\mathcal{B}^\infty$ the σ-algebra in $\mathfrak{X}^\infty$ generated by the cylinder sets

$$A_0 \times A_1 \times \cdots \times A_j \times \mathfrak{X} \times \mathfrak{X} \times \cdots,$$

where $A_k \in \mathcal{B}_k$ for $k = 0, 1, \dots, j$.

We define a sequence of probability measures $\mathbb{P}_\sigma^{x_0, k}$ on $(\mathfrak{X}^k, \mathcal{B}^k)$ uniquely determined by the following properties:

(i) $\mathbb{P}_\sigma^{x_0, 1} = \gamma_0[x_0]$,

(ii) $\mathbb{P}_\sigma^{x_0, k+1}$ has marginal probability $\mathbb{P}_\sigma^{x_0, k}$ on $(\mathfrak{X}^k, \mathcal{B}^k)$, and

(iii) $\mathbb{P}_\sigma^{x_0, k+1}$ has conditional probabilities $\gamma_k[x_0, x_1, \dots, x_{k-1}]$ on the fibers $(x_0, x_1, \dots, x_{k-1}) \times \mathfrak{X}$; that is, for every rectangle $(A_0 \times A_1 \times \dots \times A_k)$ in $(\mathfrak{X}^{k+1}, \mathcal{B}^{k+1})$ we have

$$\mathbb{P}_\sigma^{x_0, k+1}(A_0 \times A_1 \times \dots \times A_k) = \int_{(A_0 \times A_1 \times \dots \times A_{k-1})} \gamma_k[x_0, x_1, \dots, x_{k-1}](A_k) \, d\mathbb{P}_\sigma^{x_0, k}$$

Under these conditions an extension of Kolmogorov's construction due to Tulcea [19, 20] shows that there exists a unique probability measure $\mathbb{P}_\sigma^{x_0} = \lim_{k \to \infty} \mathbb{P}_\sigma^{x_0, k}$ in $(\mathfrak{X}^\infty, \mathcal{B}^\infty)$ with transition probabilities

$$\mathbb{P}_\sigma^{x_0}(\{x_{j+1} \in A\} \mid \mathcal{B}_{j+1}) = \gamma_j[x_0, x_1, \dots, x_j] \tag{5}$$

See Chap. 4 in [21] for more details.

We call the collection of probability measures

$$\sigma = (\gamma_0[x_0], \gamma_1[x_0, x_1] \ldots, \gamma_k[x_0, x_1 \ldots x_k], \ldots)$$

a *strategy*.

This formalism, coming from [14], is equivalent to the presentation in [16]. In this paper a strategy S is a collection of mappings $\sigma_j \colon \mathfrak{X}^{j+1} \mapsto \mathfrak{X}$ indicating the next move $x_{j+1} = \sigma_j(x_0, x_1, \ldots, x_j)$ given the partial history $(x_0, x_1, \ldots, x_j)$. A pair of strategies S_I and S_{II} and a starting point determine a family of measures

$$\{\mathbb{P}^{x_0}_{S_I, S_{II}}\}_{x_0 \in \mathfrak{X}}$$

that describe the game played under this pair of strategies. That is, the players choose either x_I or x_{II} to move there in case they win the coin toss. Their choices determine the probability measures $\gamma[x_0, x_1, \ldots, x_k]$ given $(x_0, x_1, \ldots, x_{k-1})$ and vice versa. Player I will try to choose points x_I to maximize the pay-off while player II will try to choose points x_{II} to minimize the pay-off. Each pair of strategies (S_I, S_{II}), S_I for player I and S_{II} for player II as in [16], determine a strategy in this sense and vice versa. We write

$$\sigma = (S_I, S_{II})$$

Having fixed a strategy σ and assuming, as we do from now on, that the game ends a.s.

$$\mathbb{P}^{x_0}_\sigma(\tau(\mathbf{x}) < \infty) = 1, \tag{6}$$

we average with respect to $\mathbb{P}^{x_0}_\sigma$ to obtain the expected pay-off for the Tug-of-War game starting at x_0

$$u_\sigma(x_0) = \mathbb{E}^{x_0}_\sigma[F(x_\tau)]. \tag{7}$$

To write down the mean value property satisfied by u_σ we condition on the first move using (5) with $j = 0$.

Lemma 3.1. *([14], Chap. 2) The value function $u_\sigma(x)$ satisfies the mean value property*

$$u_\sigma(x) = \frac{\alpha}{2}\left(u_{\sigma[x_I]}(x_I) + u_{\sigma[x_{II}]}(x_{II})\right) + \beta \int_{S(x_0)} u_{\sigma[y]}(y)\, d\mu(y) \tag{8}$$

Here the conditional strategy $\sigma[y_0]$ is defined as follows for $y_0 \in S(x_0)$

$$\sigma[y_0] = (\gamma_1[x_0, y_0], \gamma_2[x_0, y_0, y_1] \ldots, \gamma_k[x_0, y_0, y_1 \ldots y_k], \ldots)$$

so that $\mathbb{P}^{y_0}_{\sigma[y_0]}$ is the conditional distribution of $(x_2, x_3, \ldots)$ given that $x_1 = y_0$.

Let us stop and consider the particular case when $\alpha = 0$ and $\beta = 1$. In this case –the linear case– the strategies are irrelevant since $\Gamma(x)$ is always $\mu(x)$ so that there is only one family of measures $\{\mathbb{P}^{x_0}\}_{x_0 \in \mathfrak{X}}$. We recover the classical mean value formula

$$u(x) = \int_{S(x)} u(y)\, d\mu(y).$$

But the case of interest to us is when we have $\alpha \neq 0$. In this case the value function for player I is

$$u_I(x) = \sup_{S_I} \inf_{S_{II}} \mathbb{E}_\sigma^x[F(x_\tau)]$$

and for player II

$$u_{II}(x) = \sup_{S_{II}} \inf_{S_I} \mathbb{E}_\sigma^x[F(x_\tau)].$$

Player I lets Player II choose a strategy, presumably to decrease $\mathbb{E}_\sigma^{x_0}[F(x_\tau)]$, and then do as best a possible. Notice that we always have

$$u_I(x) \leq u_{II}(x) \quad \text{for all} \quad x \in \mathfrak{X}.$$

It turns out that in many cases the game has a value; that is

$$u_I(x) = u_{II}(x) \quad \text{for all} \quad x \in \mathfrak{X}, \tag{9}$$

and that this function satisfies a version of the Mean Value Property (3.1) given by

$$u(x) = \frac{\alpha}{2}\left(\sup_{y \in S(x)} u(y) + \inf_{y \in S(x)} u(y)\right) + \beta \int_{S(x)} u(y)\, d\mu(y). \tag{10}$$

Equation (10) is the Dynamic Programming Principle or DPP for short. Next, we will present two scenarios in which all the details above have been worked out.

4 p-harmonious Functions

Consider a bounded Lipschitz domain $\Omega \subset \mathbb{R}^n$ and fix $\varepsilon > 0$. To prescribe boundary values, let us denote the compact boundary strip of width ε by

$$\Gamma_\varepsilon = \{x \in \mathbb{R}^n \setminus \Omega : \text{dist}(x, \partial\Omega) \leq \varepsilon\}.$$

Let $\mathfrak{X} = \overline{\Omega}$ with the Borel σ-algebra, $X = \Omega \setminus \Gamma_\varepsilon$ and $Y = \Gamma_\varepsilon$. The successors of x are $S(x) = \overline{B}_\varepsilon(x) = \{y \in \mathbb{R}^n : |y - x| \leq \varepsilon\}$ and the measure $\mu(x)$ is just the Lebesgue measure restricted to $S(x)$ and normalized so that $\mu(x)(S(x)) = 1$. As it will be clear later on, we take α and β to be

$$\alpha = \frac{p-2}{p+n}, \quad \text{and} \quad \beta = \frac{2+n}{p+n}. \tag{11}$$

Notice that since $\alpha \geq 0$ we necessarily have $p \geq 2$.

We are given a bounded Borel pay-off function $F : \Gamma_\varepsilon \to \mathbb{R}$ and play the Tug-of-War game with parameters α and β and obtain value functions u_I^ε and u_{II}^ε, where we have chosen to emphasize the dependence on the step size ε. The following results are from [12]:

Theorem 4.1. *The value functions u_I^ε and u_{II}^ε are p-harmonious in Ω with boundary values $F : \Gamma_\varepsilon \to \mathbb{R}$; that is, they both satisfy*

$$u_\varepsilon(x) = \frac{\alpha}{2} \left\{ \sup_{\overline{B}_\varepsilon(x)} u_\varepsilon + \inf_{\overline{B}_\varepsilon(x)} u_\varepsilon \right\} + \beta \fint_{B_\varepsilon(x)} u_\varepsilon \, dy \qquad \textit{for every} \quad x \in \Omega, \tag{12}$$

and

$$u_\varepsilon(x) = F(x), \qquad \textit{for every} \quad x \in \Gamma_\varepsilon.$$

The *existence* of p-harmonious functions with given boundary values is obtained by playing the Tug-of-War games with noise. *Uniqueness* follows by using martingales, although the equation is not linear. This was first proved to the best of my knowledge in [16] for $p = \infty$.

For finite p whether the original Tug-of-War game with noise described in [17] has a value is an open problem. For our modified version of the p-game *we do have a value*. The key is to judiciously choose strategies so that we can bring martingales into play.

Lemma 4.2. (KEY LEMMA) *Let v_ε be p-harmonious such that $F \leq v_\varepsilon$ on Γ_ε. Player I chooses an arbitrary strategy S_I and player II chooses a strategy S_{II}^0 that almost minimizes v_ε,*

$$v_\varepsilon(x_k) \leq \inf_{y \in \overline{B}_\varepsilon(x_{k-1})} v_\varepsilon(y) + \eta 2^{-k}.$$

Then $M_k = v_\varepsilon(x_k) + \eta 2^{-k}$ is a supermartingale for any $\eta > 0$ and $u_I^\varepsilon \leq v_\varepsilon$.

We can now see how the inequality at the boundary literally walks into the interior by using Doob's optional stopping theorem for martingales

$$u_I^\varepsilon(x_0) = \sup_{S_I} \inf_{S_{II}} \mathbb{E}_{S_I,S_{II}}^x[F(x_\tau)]$$

$$\leq \sup_{S_I} \mathbb{E}_{S_I,S_{II}^0}^{x_0}[v_\varepsilon(x_\tau) + \eta 2^{-\tau}]$$

$$\leq \sup_{S_I} \mathbb{E}_{S_I,S_{II}^0}^{x_0}[M_\tau]$$

$$\leq \sup_{S_I} M_0 = v^\varepsilon(x_0) + \eta$$

An extension of the above technique gives the uniqueness of the value function.

Theorem 4.3. *[12] The game has a value. That is $u_I^\varepsilon = u_{II}^\varepsilon$.*

Most importantly for our purposes is the fact the p-harmonious functions satisfy the Strong Comparison Principle:

Theorem 4.4. *[12] Let $\Omega \subset \mathbb{R}^n$ be a bounded domain and let u_ε and v_ε be p-harmonious with boundary data $F_u \geq F_v$ in Γ_ε. Then if there exists a point $x_0 \in \Omega$ such that $u_\varepsilon(x_0) = v_\varepsilon(x_0)$, it follows that $u_\varepsilon = v_\varepsilon$ in Ω, and, moreover, the boundary values satisfy $F_u = F_v$ in Γ_ε.*

To prove that p-harmonious functions converge to the unique solution of the Dirichlet problem for the p-Laplacian in Ω with fixed continuous boundary values, we assume that Ω is bounded and satisfies the exterior cone condition.

Theorem 4.5. *[12] Consider the unique viscosity solution u to*

$$\begin{cases} \operatorname{div}(|\nabla u|^{p-2}\nabla u)(x) = 0, \, x \in \Omega \\ u(x) = F(x), \qquad\qquad x \in \partial\Omega, \end{cases} \tag{13}$$

and let u_ε be the unique p-harmonious function with boundary values F. Then

$$u_\varepsilon \to u \quad \text{uniformly in} \quad \overline{\Omega} \quad \text{as} \quad \varepsilon \to 0.$$

The above limit only depends on the values of F on $\partial\Omega$, and therefore any continuous extension of $F|_{\partial\Omega}$ to Γ_{ε_0} gives the same limit.

The key to prove this theorem is to pass from the discrete setting of p-harmonious functions to the continuous case of p-harmonic functions. This is done by means of a characterization of p-harmonic functions in terms of asymptotic mean value properties.

Theorem 4.6. *[11] Let $u \in C(\Omega)$ such that for all $x \in \Omega$ we have*

$$\frac{\alpha}{2}\left(\sup_{B_\varepsilon(x)} u + \inf_{B_\varepsilon(x)} u\right) + \beta \fint_{B_\varepsilon(x)} u = u(x) + o(\varepsilon^2), \qquad \text{as } \epsilon \to 0.$$

Then u is p-harmonic in Ω. Here α and β are chosen as in (11).

The converse of this theorem holds if we weaken the asymptotic expansion to hold only in the viscosity sense. See [11] for details. Another approach to pass from the discrete to the continuous for fully-nonlinear equations has been given by Kohn and Serfaty [6] by using a deterministic control theory approach.

5 Directed Trees

Let T be a directed tree with regular 3-branching as in Sect. 1. Let $u\colon T \mapsto \mathbb{R}$ be a real valued function. The *gradient* of u at the vertex v is the vector in $\mathbb{R}^3$

$$\nabla u(v) = (u(v_0) - u(v), u(v_1) - u(v), u(v_2) - u(v)).$$

The *divergence* of a vector $X = (x, y, z) \in \mathbb{R}^3$ is

$$\mathrm{div}(X) = x + y + z.$$

A function u is *harmonic* if it satisfies the Laplace equation

$$\mathrm{div}(\nabla u) = 0. \tag{14}$$

Observe that a function u is harmonic if and only if it satisfies the mean value property

$$u(v) = \frac{1}{3}(u(v_0) + u(v_1) + u(v_2))$$

Set $\mathfrak{X} = T \cup \partial T$, $X = T$ and $Y = \partial T$. The measure $\mu(v)$ is the normalized counting measure in $S(v)$

$$\mu(v) = \frac{1}{3}\left(\delta_{v_0} + \delta_{v_1} + \delta_{v_2}\right).$$

The pay-off function $F\colon \partial T \mapsto \mathbb{R}$ is defined on the unit interval $[0, 1]$. We are ready to play games in T.

Think of a random walk started at the top vertex $\emptyset$ and move downward by choosing successors at random with uniform probability. When you get at ∂T at the branch point b determined by the random walk, you get paid $f(b)$ dollars. Every time we run the game we get a sequence of vertices $v_1, v_2, \ldots, v_k, \ldots$ that determine a point on b the boundary ∂T. The set of all boundary points that start at a given vertex v_r at level r is a ternary interval of length 3^{-r} that we denote by I_{v_r}. Averaging out over all possible plays that start at v_r we obtain the value function

$$\mathbb{E}^{v_r}\left[f(t)\right] = u(v_r) = \frac{1}{|I_{v_r}|} \int_{I_{v_r}} f(b)\, db, \tag{15}$$

which is indeed harmonic in T. Therefore we have the well-known

Lemma 5.1. *Dirichlet Problem in Trees ($p=2$): Given a continuous (indeed in $L^1([0,1])$) function $f: [0,1] \mapsto \mathbb{R}$ the unique harmonic function $u: T \mapsto \mathbb{R}$ such that*

$$\lim_{r \to \infty} u(b_r) = f(b)$$

for every branch $b = (b_r) \in \partial T$ is given by (15).

Let us now play a Tug-of-War game with noise. Choose $\alpha \geq 0$, $\beta \geq 0$ such that $\alpha + \beta = 1$. Start at $\emptyset$. With probability α the players play Tug-of-War. With probability β move downward by choosing successors at random. When you get at ∂T at the point b player II pays $f(b)$ dollars to player I. The value function u verifies the *dynamic programming principle* or *mean value property*

$$u(v) = \frac{\alpha}{2}\left(\max_i\{u(v_i)\} + \min_i\{u(v_i)\}\right) + \beta\left(\frac{u(v_0) + u(v_1) + u(v_2)}{3}\right) \tag{16}$$

that we can interpret as a PDE on the tree by using the following formula for a generalized divergence depending on the parameters α and β.

Definition 5.2. Let $X = (x, y, z)$ be a vector in $\mathbb{R}^3$. The (α, β)-divergence of X is given by

$$\operatorname{div}_{\alpha,\beta}(X) = \frac{\alpha}{2}\left(\max\{x, y, z\} + \min\{x, y, z\}\right) + \beta\left(\frac{x + y + z}{3}\right).$$

Theorem 5.3. *[18] We have the equivalence*

$$\boldsymbol{DPP \approx MVP \approx PDE}$$

in the sense that the function u satisfies (16) in the tree T if and only if

$$\operatorname{div}_{\alpha,\beta}(\nabla u) = 0 \tag{17}$$

Some particular cases are:

(i) The Linear Case:

$\alpha = 0$, $\beta = 1$ that corresponds to the linear case $p = 2$ of harmonic functions (14).

(ii) The Discrete ∞-Laplacian:

$\alpha = 1$, $\beta = 0$ that corresponds to the case $p = \infty$. In this case the divergence is

$$\operatorname{div}_\infty(X) = \operatorname{div}_{1,0}(X) = \frac{1}{2}\left(\max\{x, y, z\} + \min\{x, y, z\}\right)$$

and the equation is the discrete ∞-Laplacian $\operatorname{div}_\infty(\nabla u) = 0$.

(iii) The Discrete p-Laplacian:

For $\alpha \neq 0$ and $\beta \neq 0$ we can select p as in (11), but the role of n is not intrinsically defined, to obtain the discrete p-Laplacian We remark that this is the non-divergence form of the p-Laplacian (17). A discrete version of the p-Laplacian in divergence form can be found in [5].

While the formula (15) for the solution to the Dirichlet problem for $p = 2$ is explicit, there are not such formulas to my knowledge for the case $p \neq 2$. However, the game theoretic interpretation allows us to find explicit formulas in some special, but interesting cases.

Suppose that f is monotonically increasing. In this case the best strategy $S_I^\star$ for player I is always to move right and the best strategy $S_{II}^\star$ for player II always to move left. Starting at the vertex v at level k

$$v = 0.b_1 b_2 \ldots b_k, \qquad b_j \in \{0, 1, 2\}$$

we always move either left (adding a 0) or right (adding a 1). In this case I_v is the Cantor-like set $I_v = \{0.b_1 b_2 \ldots b_k d_1 d_2 \ldots\}$, $d_j \in \{0, 2\}$.

Theorem 5.4. *[18] The (α, β)-harmonic function with boundary values f in the tree T is given by*

$$u(v) = \fint_{I_v} f(b) d\mathbb{P}_v^{\alpha,\beta} db,$$

where $\mathbb{P}_v^{\alpha,\beta}$ is a probability in $[0, 1]$.

Moreover in the case $\alpha = 0$, $\beta = 1$, which corresponds to $p = 2$ the measure $\mathbb{P}_v^{0,1}$ is just the Lebesgue measure, and in the case $\alpha = 1$, $\beta = 0$, which corresponds to the case $p = \infty$, the measure $\mathbb{P}_v^{1,0}$ is a Cantor measure supported in I_v.

To see why this theorem is true observe that

$$u(v) = \sup_{S_I} \inf_{S_{II}} \mathbb{E}_{S_I, S_{II}}^v [f(b)] = \mathbb{E}_{S_I^\star, S_{II}^\star}^v [f(b)].$$

Since the strategies used are always the same, we are indeed in a linear situation. All we need to do is to compute the probability $\mathbb{P}_{S_I^\star, S_{II}^\star}^v$.

6   Epilogue

We take the opportunity to state two open problems for p-harmonic functions in $\mathbb{R}^n$ that have challenged the mathematical community for more than thirty years.

Open Problem 6.1 (STRONG COMPARISON PRINCIPLE): *Suppose that u and v are p-harmonic functions in $B_R(x_0)$ such that*

$$\begin{cases} u(x) & \leq & v(x) \text{ for } x \in B_R(x_0), \text{ and} \\ u(x_0) & = & v(x_0). \end{cases} \tag{18}$$

Does it follow that $u \equiv v$ in $B_R(x_0)$?

Open Problem 6.2 (UNIQUE CONTINUATION PROPERTY): *Suppose that u is p-harmonic in $B_{2R}(x_0)$ and that*

$$u(x) = 0 \text{ for every } x \in B_R(x_0).$$

Does it follow that $u \equiv 0$ in $B_{2R}(x_0)$?

Of course, the answer to this problems is clearly yes in the linear case $p = 2$. The answer to both problems is also yes in the planar case $n = 2$ since complex methods are then available [1, 10].

It is natural to try to apply new techniques developed in Analysis to these problems in the hope of improving our understanding of them. When the viscosity theory was developed, it was first proved that the notions of Sobolev weak solution and viscosity solution agree [3] allowing us to study the p-harmonic equation not only by variational methods for divergence form equations but also using viscosity methods for non-divergence form equations. Progress in various problems followed: The ∞-eigenvalue problem [4], unexpected superposition principles [2] and [7], and various proofs were extended and simplified by using sup-convolutions [8].

The non-linear potential theory on trees of Kauffman and Wu [5] proved essential to settle a long standing conjecture of Martio on the lack of subadditivity of p-harmonic measures even at the zero level [9].

In the opinion of this author, the connection between the p-harmonic equation and discrete stochastic games discovered in the case $p = \infty$ first by Peres, Sheffield, Schramm, and Wilson [16] and for finite p by Peres and Sheffield [17] (see also [15]) that opens the door to the use of game theoretic and control theoretic methods to the study of the p-Laplace equation, will have substantial applications [15]. A distinguishing feature of the stochastic games approach is that, since it is based on discrete stochastic processes, it provides good discrete approximations to p-harmonic functions, making a direct connection with the analysis in trees of [5].

References

1. G. Alessandrini, Critical points of solutions of elliptic equations in two variables, Ann. Scuola Norm. Sup. Pisa Cl. Sci. (4) **14**(2), 229–256 (1988)
2. M. Crandall and J. Zhang, Another way to say harmonic, Trans. Am. Math. Soc. **355**, 241–263 (2002)
3. P. Juutinen, P. Lindqvist, J. Manfredi, On the equivalence of viscosity solutions and weak solutions for a quasi-linear elliptic equation, SIAM J. Math. Anal. **33**, 699–717 (2001)
4. P. Juutinen, P. Lindqvist, J. Manfredi, The ∞-eigenvalue problem, Arch. Ration. Mech. Anal. **148** 89–105 (1999)
5. R. Kaufman, J.G. Llorente, J.M. Wu, Nonlinear harmonic measures on trees, Ann. Acad. Sci. Fenn. Math. **28**, 279–302 (2003)
6. R. Kohn, S. Serfaty, A deterministic-control-based approach to fully nonlinear parabolic and elliptic equations, Comm. Pure Appl. Math. **63**(10), 1298–1350 (2010)
7. P. Lindqvist, J. Manfredi, Note on a remarkable superposition for a nonlinear equation, Proc. Am. Math. Soc. **136**, 229–248 (2008)
8. P. Lindqvist, J. Manfredi, Viscosity solutions of the evolutionary p-Laplace equation, Differ. Integr. Equat. **20**, 1303–1319 (2007)
9. J.G. Llorente, J. Manfredi, J.M. Wu, p-harmonic measure is not subadditive, Annalli de della Scuola Normale Superiore di Pisa, Classe di Scienze **IV**(5), 357–373 (2005)
10. J. Manfredi, p-harmonic functions in the plane, Proc. Am. Math. Soc. **103**(2), 473–479 (1988)
11. J.J. Manfredi, M. Parviainen, J.D. Rossi, An asymptotic mean value characterization for p-harmonic functions. Proc. Am. Math. Soc., **258**, 713–728 (2010)
12. J.J. Manfredi, M. Parviainen, J.D. Rossi, On the definition and properties of p-harmonious functions. To appear in Ann. Sc. Norm. Super. Pisa Cl. Sci.
13. J.J. Manfredi, M. Parviainen, J.D. Rossi, Dynamic programming principle for tug-of-war games with noise. ESAIM: Control, Optimisation and Calculus of Variations, **18**, 81–90 (2012)
14. A.P. Maitra, W.D. Sudderth, Discrete gambling and stochastic games. Appl. Math. **32**, Springer-Verlag, 1996
15. Y. Peres, G. Pete, S. Somersielle, Biased tug-of-war, the biased infinity Laplacian and comparison with exponential cones, Calc. Var. Partial Differential Equations **38**(3), 541–564 (2010)
16. Y. Peres, O. Schramm, S. Sheffield, D. Wilson, Tug-of-war and the infinity Laplacian, J. Am. Math. Soc. **22**, 167–210 (2009)
17. Y. Peres, S. Sheffield, Tug-of-war with noise: a game theoretic view of the p-Laplacian, Duke Math. J. **145**(1), 91–120 (2008)
18. A. Sviridov, Elliptic Equations in Graphs via Stochastic Games, University of Pittsburgh 2010 doctoral dissertation
19. D.W. Stroock, S.R.S. Varadan, Multidimensional Diffusion Processes, Grundlehren der Mathematische Wissenschaften 233, Springer-Verlag, 1979
20. C.T. Ionescu Tulcea, Measures dans les espaces produits, Atti. Acad. Naz. Lincei. Rend. Cl. Sci. Fis. Mat. Nat. **7**(8), 208–211 (1949)
21. S.R.S. Varadan, Probability Theory, Courant Lecture Notes in Mathematics 7, New York University, 2000

The Problems of the Obstacle in Lower Dimension and for the Fractional Laplacian

Sandro Salsa

1 Introduction

The *obstacle problem for a fractional power of the Laplace operator* appears in many contexts, such as in the study of anomalous diffusion [5], in the so called quasi-geostrophic flow problem [12], and in pricing of American options governed by assets evolving according to jump processes [13].

It can be stated in several ways. Given a smooth function $\varphi : \mathbb{R}^n \to \mathbb{R}$, $n > 1$, with bounded support (or at least rapidly decaying at infinity), we look for a continuous function u satisfying the following system:

$$\begin{cases} u \geq \varphi & \text{in } \mathbb{R}^n \\ (-\Delta)^s u \geq 0 & \text{in } \mathbb{R}^n \\ (-\Delta)^s u = 0 & \text{when } u > \varphi \\ u(x) \to 0 & \text{as } |x| \to +\infty. \end{cases} \tag{1}$$

Here we consider only the case $s \in (0,1)$. The definition and some of the main properties of $(-\Delta)^s$ are collected in Appendix A. The set $\Lambda(u) = \{u = \varphi\}$ is called *the contact or coincidence set*. The boundary of the set $\mathbb{R}^n \backslash \Lambda(u)$ is the *free boundary*, denoted by $F(u)$.

The main theoretical issues in a constrained minimization problem are *optimal regularity of the solution* and the *analysis of the free boundary*.

If $s = 1$ and $\mathbb{R}^n$ is replaced by a bounded domain Ω our problem corresponds to the usual obstacle problem for the Laplace operator. The existence of a unique solution satisfying some given boundary condition $u = g$

S. Salsa (✉)
Department of Mathematics, Politecnico di Milano, Piazza Leonardo Da Vinci 32, I-20133 Milano, Italy
e-mail: sandro.salsa@polimi.it

J. Lewis et al., *Regularity Estimates for Nonlinear Elliptic and Parabolic Problems*, Lecture Notes in Mathematics 2045, DOI 10.1007/978-3-642-27145-8_4, © Springer-Verlag Berlin Heidelberg 2012

can be obtained by minimizing the Dirichlet integral in $H^1(\Omega)$ under the constraint $u \geq \varphi$. The solution is the least superharmonic function greater than equal to φ in Ω, with $u \geq g$ on $\partial\Omega$, and inherits up to a certain level the regularity of φ [15]. In fact, even if φ is smooth, u is only $C^{1,1}_{loc}$, which is the optimal regularity. A classical reference for the obstacle problem, including the regularity and the complete analysis of the free boundary is [7].

Analogously, the existence of a solution u for problem (1) can be obtained by variational methods as the unique minimizer of the functional

$$J(v) = \int_{\mathbb{R}^n} \int_{\mathbb{R}^n} \frac{|v(x) - v(y)|^2}{|x - y|^{n+2s}} dx dy$$

over a suitable set of functions $v \geq \varphi$. Also we can obtain u via a Perron type method, as the least supersolution of $(-\Delta)^s$ such that $u \geq \varphi$. By analogy (see also later the Signorini problem), when φ is smooth, we expect the optimal regularity for u to be $C^{1,s}$. This is indeed true as it is shown in [24] when the contact set $\{u = \varphi\}$ is convex and in [11] in the general case.

The case $s = 1/2$ is strongly related to the so called *thin (or lower dimensional) obstacle problem* for the Laplace operator. To keep a better connection with the obstacle problem for $(-\Delta)^s$, it is better to work in $\mathbb{R}^{n+1}$, writing $X = (x, y) \in \mathbb{R}^n \times \mathbb{R}$. The thin obstacle problem concerns the case in which the obstacle is not anymore $n + 1$ dimensional, but supported instead on a smooth n-dimensional manifold $\mathcal{M}$ in $\mathbb{R}^{n+1}$. This problem and variation of it also arises in many applied contexts, such as flow through semi-permeable membranes (Fig. 1), elasticity (known as *Signorini problem*), boundary control temperature or heat problems (see [14]).

More precisely, let Ω be a domain in $\mathbb{R}^{n+1}$ divided into two parts Ω^+ and Ω^- by $\mathcal{M}$. Let $\varphi : \mathcal{M} \to \mathbb{R}$ be the (thin) obstacle and g be a given function on $\partial\Omega$ satisfying $g > \varphi$ on $\mathcal{M} \cap \partial\Omega$.

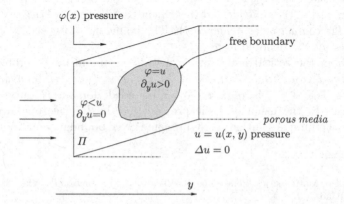

Fig. 1 Stationary flow through a semipermeable membrane of negligible thickness

The problem consists in the minimization of the Dirichlet integral

$$J(v) = \int_\Omega |\nabla v|^2$$

over the closed convex set

$$\mathbb{K} = \left\{ v \in H^1(\Omega) : v = g \text{ on } \partial\Omega \text{ and } v \geq \varphi \text{ on } \mathcal{M} \cap \partial\Omega \right\}.$$

Since we can perturb the solution u upwards and freely away from $\mathcal{M}$, it is apparent that u is superharmonic in Ω and harmonic in $\Omega \backslash \mathcal{M}$. One expects the continuity of the first derivatives along the directions tangential to $\mathcal{M}$, and the one sided continuity of normal derivatives [15]. In fact (see [8]), on $\mathcal{M}$, u satisfies the following complementary conditions

$$u \geq \varphi, \; u_{\nu+} + u_{\nu-} \leq 0, (u - \varphi)(u_{\nu+} + u_{\nu-}) = 0$$

where $\nu^{\pm}$ are the interior unit normals to $\mathcal{M}$ from the $\Omega^{\pm}$ side. The free boundary here is given by the boundary of the set $\Omega \backslash \Lambda(u)$ in the relative topology of $\mathcal{M}$, and in general, we expect it is a $(n-1)$-dimensional manifold.

A related problem is the *boundary thin obstacle problem (or Signorini problem)*, in which the manifold $\mathcal{M}$ is part of $\partial\Omega$ and one has to minimize the Dirichlet integral over the closed convex set

$$\mathbb{K} = \left\{ v \in H^1(\Omega) : v = g \text{ on } \partial\Omega \backslash \mathcal{M} \text{ and } v \geq \varphi \text{ on } \mathcal{M} \right\}.$$

In this case, u is harmonic in Ω and on $\mathcal{M}$ satisfies the complementary conditions

$$u \geq \varphi, \; u_{\nu+} \leq 0, (u - \varphi) u_{\nu+} = 0.$$

If $\mathcal{M}$ is a hyperplane (say $\{y = 0\}$) and Ω is symmetric with respect to $\mathcal{M}$, then the thin obstacle in Ω and the boundary obstacle problems in Ω^+ or Ω^- are equivalent.

It is this last case that is strongly related to the obstacle problem for $(-\Delta)^{1/2}$. This is clearly explained through the following remarks.

(a) *Reduction to a global problem.* Let $\Omega = B_1$ be the unit ball in $\mathbb{R}^{n+1}$ and $B_1' = B_1 \cap \{y = 0\}$. Let $\varphi : \mathbb{R}^n \to \mathbb{R}$ be a smooth obstacle, $\varphi < 0$ on $\partial B_1'$ and positive somewhere inside B_1'. Consider the following Signorini problem in $B_1^+ = B_1 \cap \{y > 0\}$:

$$\begin{cases} -\Delta u = 0 & \text{in } B_1^+ \\ u = 0 & \text{on } \partial B_1 \cap \{y > 0\} \\ u(x,0) \geq \varphi(x) & \text{in } \bar{B}_1' \\ u_y(x,0) \leq 0 & \text{in } \bar{B}_1' \\ u_y(x,0) = 0 & \text{when } u(x,0) > \varphi(x). \end{cases} \quad (2)$$

We want to convert the above problem in B_1 into a global one, that is in $\mathbb{R}^n \times (0, +\infty)$. To do this, let η be a radially symmetric cut-off function in B_1' such that

$$\{\varphi > 0\} \Subset \{\eta = 1\} \quad \text{and} \quad \text{supp}\,(\eta) \subset B_1'.$$

Extending ηu by zero outside B_1, we have $\eta(x)\,u(x,0) \geq \varphi(x)$ and also $(\eta u)_y\,(x,0) \leq 0$ for every $x \in \mathbb{R}^n$. Moreover, $(\eta u)_y\,(x,0) = 0$ if $\eta(x)\,u(x,0) > \varphi(x)$.

Let now v be the unique solution of the following Neumann problem in the upper half space, vanishing at infinity:

$$\begin{cases} \Delta v = \Delta\,(\eta u) & \text{in } \mathbb{R}^n \times \{y > 0\} \\ v_y\,(x,0) = 0 & \text{in } \mathbb{R}^n. \end{cases}$$

Then $w = \eta u - v$ is a solution of a global Signorini problem with $\varphi - v$ as the obstacle. Thus, the regularity of u in the local setting can be inferred from the regularity for the global problem.

The opposite statement is obvious.

(b) *Realization of* $(-\Delta)^{1/2}$ *as a Dirichlet-Neumann map.* Consider a smooth function $u_0 : \mathbb{R}^n \longrightarrow \mathbb{R}$ with rapid decay at infinity. Let $u : \mathbb{R}^n \times (0, +\infty) \longrightarrow \mathbb{R}$ be the unique solution of the Dirichlet problem

$$\begin{cases} \Delta u = 0 & \text{in } \mathbb{R}^n \times (0, +\infty) \\ u\,(x,0) = u_0\,(x) & \text{in } \mathbb{R}^n \end{cases}$$

vanishing at infinity. We call u the *harmonic extension of* u_0 to the upper half space.

Consider the *Dirichlet-Neumann map* $T : u_0\,(x) \longmapsto -u_y\,(x,0)$. We have:

$$\begin{aligned} (Tu_0, u_0) &= \int_{\mathbb{R}^n} -u_y\,(x,0)\,u\,(x,0)\,dx \\ &= \int_{\mathbb{R}^n \times (0,+\infty)} \left\{ \Delta u\,(X)\,u\,(X) + |\nabla u\,(X)|^2 \right\} dX \\ &= \int_{\mathbb{R}^n \times (0,+\infty)} |\nabla u\,(X)|^2\,dX \geq 0 \end{aligned}$$

so that T is a positive operator. Moreover, since u_0 is smooth and u_y is harmonic, we can write:

$$T \circ Tu_0 = -\partial_y(-\partial_y)u\,(x,0) = u_{yy}\,(x,0) = -\Delta u_0.$$

We conclude that

$$T = (-\Delta)^{1/2}.$$

As a consequence:

1. If $u = u(X)$ is a solution of the Signorini problem in $\mathbb{R}^n \times (0, +\infty)$, that is $\Delta u = 0$ in $\mathbb{R}^{n+1}$, $u(x, 0) \geq \varphi$, $u_y(x, 0) \leq 0$, and $(u - \varphi) u_y(x, 0) = 0$ in $\mathbb{R}^n$, then $u_0 = u(\cdot, 0)$ solves the obstacle problem for $(-\Delta)^{1/2}$.
2. If u_0 is a solution of the obstacle problem for $(-\Delta)^{1/2}$, then its harmonic extension to $\mathbb{R}^n \times (0, +\infty)$ solves the corresponding Signorini problem.

Therefore, the two problems are equivalent and any regularity result for one of them can be carried to the other one. More precisely, consider the optimal regularity for the solution u_0 of the obstacle problem for $(-\Delta)^{1/2}$, which is $C^{1,1/2}$. If we can prove a $C^{1,1/2}$ regularity of the solution u of the Signorini problem up to $y = 0$, then the same is true for u_0.

On the other hand, the $C^{1,\alpha}$ regularity of u_0 extends to u, via boundary estimates for the Neumann problem. Similarly, the analysis of the free boundary in the Signorini problem carries to the obstacle problem for $(-\Delta)^{1/2}$ as well and viceversa.

Although the two problem are equivalent, there is a clear advantage in favor of the Signorini type formulation. This is due to the possibility of avoiding the direct use of the non local pseudodifferential operator $(-\Delta)^{1/2}$, by localizing the problem and using local P.D.E. methods, such as monotonicity formulas and classification of blow-up profiles.

Concerning the Signorini problem, we will mainly refer to the papers [1, 3, 8, 18].

In [8] it is shown that u is $C^{1,\alpha}$ up to $y = 0$, for some $\alpha \leq 1/2$. For the zero-obstacle problem ($\varphi = 0$), the optimal regularity is shown in [1] while the analysis of the free boundary around a regular point is contained in [3]. A complete classification of singular points, i.e. points of $F(u)$ of vanishing density for $\Lambda(u)$, is developed in [18] (also for non zero obstacles).

At this point it is a natural question to ask whether there exists a P.D.E. realization of $(-\Delta)^s$ for every $s \in (0, 1)$, $s \neq \frac{1}{2}$.

The answer is positive as it is shown in [10]. Indeed using the results in Sect. C3, in a weak sense we have that

$$(-\Delta)^s u_0(x) = -\kappa_a \lim_{y \to 0+} y^a u_y(x, y)$$

for a suitable constant κ_a, where u is the solution of the problem

$$\begin{cases} L_a u = \operatorname{div}(y^a \nabla u) = 0 & \text{in } \mathbb{R}^n \times (0, +\infty) \\ u(x, 0) = u_0(x) & \text{in } \mathbb{R}^n \end{cases}$$

vanishing at infinity, where $a = 2s - 1$. Coherently, we call u the L_a-*harmonic extension of* u_0

Thus problem (1) is equivalent to the following Signorini problem for the operator $L_a = \operatorname{div}(y^a \nabla)$,

$$u(x,0) \geq \varphi \quad \text{in } \mathbb{R}^n \tag{3}$$

$$L_a u = \text{div}(y^a \nabla u) = 0 \quad \text{in } \mathbb{R}^n \times (0, +\infty) \tag{4}$$

$$\lim_{y \to 0+} y^a u_y(x,y) = 0 \quad \text{when } u(x,0) > \varphi(x) \tag{5}$$

$$\lim_{y \to 0+} y^a u_y(x,y) \leq 0 \quad \text{in } \mathbb{R}^n. \tag{6}$$

For $y > 0$, u is smooth so that (4) is understood in the classical sense. The equations at the boundary (5) and (6) should be understood in the weak sense (see Sect. A3). Since in [24] it is shown that $u(x,0) \in C^{1,\alpha}$ for every $\alpha < s$, for the range of values $2s - 1 < \alpha < s$, $\lim_{y \to 0+} y^a u_y(x,y)$ can be understood in the classical sense too.

The solution u of the above Signorini problem can be extended to the whole space by symmetrization, setting $u(x,-y) = u(x,y)$. Then, by the results in [10] (see Sect. A3), condition (5) holds if and only if the extended u is a solution of $L_a u = 0$ across $y = 0$, where $u(x,0) > \varphi(x)$. On the other hand, condition (6) is equivalent to $L_a u \leq 0$ in the sense of distributions. Thus, for the extended u, the Signorini problem translates into the following system:

$$\begin{cases} u(x,0) \geq \varphi(x) & \text{in } \mathbb{R}^n \\ u(x,-y) = u(x,y) & \text{in } \mathbb{R}^{n+1} \\ L_a u = 0 & \text{in } \mathbb{R}^{n+1} \setminus \{(x,0): \ u(x,0) = \varphi(x)\} \\ L_a u \leq 0 & \text{in } \mathbb{R}^{n+1}, \text{ in the sense of distributions.} \end{cases}$$

Again, we can exploit the advantages to analyze the obstacle problem for a nonlocal operator in P.D.E. form by considering a local version of it. Indeed, to study the optimal regularity properties of the solution we will focus on the following local version, where $\varphi : B' \longrightarrow \mathbb{R}$:

$$\begin{cases} u(x,0) \geq \varphi(x) & \text{in } B_1' \\ u(x,-y) = u(x,y) & \text{in } B_1 \\ L_a u = 0 & \text{in } B_1 \setminus \{(x,0): \ u(x,0) = \varphi(x)\} \\ L_a u \leq 0 & \text{in } B_1, \text{ in the sense of distributions.} \end{cases}$$

The above problem can be thought as the minimization of the weighted Dirichlet integral

$$J_a(v) = \int_{B_1} |y|^a |\nabla u(X)|^2 \, dX$$

over the set

$$\mathbb{K}_a = \{v \in W^{1,2}(B_1, |y|^a) : u(x,0) \geq \varphi(x)\}.$$

In a certain sense, this corresponds to an obstacle problem, where the obstacle is defined in a set of codimension $1 + a$, where a is not necessarily an integer.

The operator L_a is degenerate elliptic, with a degeneracy given by the weight $|y|^a$. This weight belongs to the Muckenhoupt class $A_2\left(\mathbb{R}^{n+1}\right)$.

A positive weight function $w = w(X)$ belongs to $A_2\left(\mathbb{R}^N\right)$ if

$$\left(\frac{1}{|B|}\int_B w\right)\left(\frac{1}{|B|}\int_B w^{-1}\right) \leq C$$

for every ball $B \subset \mathbb{R}^N$. For the class of degenerate elliptic operators of the form $Lu = \operatorname{div}(A(X)\nabla u)$, where

$$\lambda w(X)|\xi|^2 \leq A(X)\xi \cdot \xi \leq \Lambda w(X)|\xi|^2,$$

there is a well established potential theory for solutions in the weighted Sobolev space $W^{1,2}(\Omega, w)$, Ω bounded domain in $\mathbb{R}^N$, defined as the closure of $C^\infty(\bar{\Omega})$ in the norm

$$\left[\int_\Omega v^2 w + \int_\Omega |\nabla v|^2 w\right]^{1/2}$$

(see [16]). Since $w \in A_2$ the gradient of a function in $W^{1,2}(\Omega, w)$ is well defined in the sense of distributions[1] and belongs to the weighted space $L^2(\Omega, w)$.

The outline of the paper. We intend here to give a somewhat self-contained presentation of the results concerning the analysis of the solution and the free boundary of the thin obstacle problem and more generally of the obstacle for the fractional Laplacian, contained in the papers $[1, 3, 8, 10, 11, 18, 24]$.

We will start from the thin obstacle problem, considering the case of *zero obstacle*. In this case the main ideas and tools are clearly seen and developed without too many technicalities and in a somewhat self-contained fashion.

Afterwards, following the same strategy and using the results of [24] for the initial partial regularity of the solution, we extend the results on the optimal regularity and the analysis of the *regular part* of the free boundary to the general case for $(-\Delta)^s$.

Since all the main difficulties in the analysis appear in the behavior of the solution near $F(u)$, we *will always assume that the origin belongs to the free boundary.*

---

[1] In fact, if $\{\varphi_k\} \subset C_0^\infty(\Omega)$ and $\|\varphi_k\|_{L^2(\Omega, w)} \to 0$, $\|\nabla\varphi_k - \mathbf{v}\|_{L^2(\Omega, w)} \to 0$, then $\mathbf{v} = \mathbf{0}$.

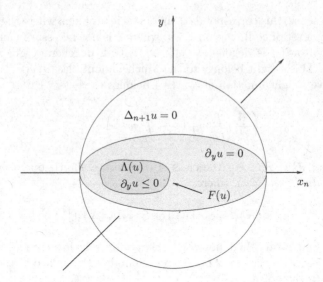

Fig. 2 Zero obstacle problem

2   The Zero Obstacle Problem

2.1   *Setting of the Problem*

In this section we give a rather complete and self-contained analysis of the thin obstacle problem. To better emphasize the ideas, we consider in this section the *zero obstacle problem*: *to minimize the Dirichlet integral* (see Fig. 2)

$$\int_{B_1} |\nabla v|^2 \, dX$$

over the set of function belonging to the closed convex set

$$\mathbb{K} = \left\{ v \in H^1(B_1) : v = g \text{ on } \partial B_1, \, v(x,0) \geq 0 \text{ in } B_1' \right\}$$

where g is a smooth function on ∂B_1.

It is easily seen that the minimizer u coincides with the least among all superharmonic functions v in B_1 with $v \geq g$ on ∂B_1, and $v(x,0) \geq 0$ in B_1'. Thus, $\{u > 0\}$ is open, and the contact set

$$\Lambda(u) = \{(x,0) : u(x,0) = 0, \, x \in B_1'\}$$

is closed. The free boundary is the boundary of the set $\{u(x,0) > 0\} \cap B_1$ relative to the hyperplane $y = 0$ that is:

$$F(u) = \partial \{x : u(x,0) > 0\} \cap B_1'.$$

In order to avoid trivial contact set and free boundary, we assume that g changes sign and that $g(\theta', 0) > 0$ for $\theta' \in \partial B_1'$.

Moreover, we set $\Lambda'(u) = \{x : u(x, 0) = 0\}$ and $\Omega' = B_1' \backslash \Lambda'(u)$.

Our analysis consists in proving the following steps:

1. Preliminary estimates: Lipschitz continuity and $C^{1,\alpha}$ local estimates for some $0 < \alpha < 1$.
2. Optimal regularity for tangentially convex global solutions.
3. Classification of asymptotic blow-up profiles around a free boundary point.
4. Almgren's frequency formula and optimal regularity (i.e. $C^{1,1/2}$) of the solution.
5. Analysis of the free boundary for non-degenerate profiles around stable points: Lipschitz continuity.
6. Boundary Harnack Principles and optimal regularity of the free boundary.
7. Structure of the subset of the non stable points of the free boundary at which $\Lambda(u)$ has vanishing density (*singular set*).

It is worthwhile to start with an observation of Hans Levy in $\mathbb{R}^2$ that gives a clue of why $C^{1,1/2}$ is the optimal regularity.

If u is a global solution of the zero obstacle problem, the complex function $w = u_x - iu_y$ is analytic outside $\Lambda(u)$, and therefore, there,

$$u_x u_y = \frac{1}{2} \operatorname{Im}(w^2)$$

is harmonic and vanishes on $y = 0$. Thus, $u_x u_y$ has an harmonic odd extension across $y = 0$ and w^2 has an analytic extension. Then $w \in C^{1/2}$ which implies $u \in C^{1,1/2}$.

An important example of global solution is $u = \rho^{3/2} \cos \frac{3}{2}\theta$; this is a typical profile that we call *nondegenerate* (Fig. 3). On the other hand, there are solutions like

$$\rho^{k+1/2} \cos \frac{2k+1}{2}\theta \quad \text{or} \quad \rho^{2k} \cos 2k\theta, k \geq 2,$$

vanishing of higher order at the origin. In these cases we cannot expect any regularity of the free boundary.

In fact, for the free boundary, the *regular (or nondegenerate or stable)* points are those for which a suitable asymptotic profile is nondegenerate in the above sense. Around this kind of points we expect regularity of $F(u)$.

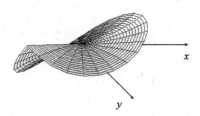

Fig. 3 Nondegenerate global profile

The non regular points of the free boundary can be divided in two classes: the set $\Sigma(u)$ at which $\Lambda(u)$ has a vanishing density (*singular points*), that is

$$\Sigma(u) = \left\{ (x_0, 0) \in F(u) : \lim_{r \to 0^+} \frac{\mathcal{H}^n(\Lambda(u) \cap B_r'(x_0))}{r^n} = 0 \right\},$$

and the set of *non regular, non singular points*. The following example (see [18]), given by the harmonic polynomial

$$p(x_1, x_2, y) = x_1^2 x_2^2 - (x_1^2 + x_2^2)y^2 + \frac{1}{3}y^4$$

shows that the entire free boundary could be composed by singular points. In fact $F(p) = \Lambda(p)$ is given by the union of the lines $x_1 = y = 0$ and $x_2 = y = 0$.

We shall see that, as in this example, the singular set is contained in the union of C^1-manifold of suitable dimension [18].

2.2   Lipschitz Continuity and Semiconvexity

We start by proving some preliminary estimate and partial regularity. Notice that, without loss of generality, we can assume that $u(x, -y) = u(x, y)$. Otherwise, we solve

$$\begin{cases} \Delta w = 0 \text{ in } B_1 \\ w = g^* \text{ on } B_1 \end{cases}$$

where $g^*(x, y) = (g(x, y) - g(x, -y))/2$. Then, $u - w$ is symmetric about $y = 0$ and solves the same kind of problem. Also, it is clear that u is harmonic outside $\Lambda(u)$ and $u(x, 0) > 0$ in $B_1' \backslash B_{1-d_0}'$ for some d_0 depending on g.

Lemma 2.2.1. Let u be a solution of the zero thin obstacle problem in B_1. Then:

(a) u^+ and u^- are subharmonic.

(b) There exist constants C such that:

$$\|u\|_{H^1(B_{1/2})} \leq C \|u\|_{L^2(B_1)}$$

$$\|u\|_{L^\infty(B_{1/2})} \leq C \|u\|_{L^2(B_1)}.$$

Proof. Since u is superharmonic, $\min\{u, 0\} = -u^-$ is superharmonic. To show that u^+ is subharmonic we perturb u as follows.

Let $\eta \in C_0^\infty(B_1)$, $\eta \geq 0$, and $f_\delta(s)$ be a δ approximation of the Heaviside function, that is

$$f_\delta(s) = \begin{cases} 0 & s \le 0 \\ s/\delta & 0 < s < \delta \\ 1 & s \ge \delta. \end{cases}$$

For $0 < \varepsilon < \delta$, small, the function $u_\varepsilon = u - \varepsilon \eta f_\delta(u)$ belongs to $\mathbb{K}$. Therefore we can write

$$\int_{B_1} |\nabla u_\varepsilon|^2 \ge \int_{B_1} |\nabla u|^2$$

which gives, letting $\varepsilon \to 0$, after a division by ε:

$$\int_{B_1} f_\delta(u) \nabla u \cdot \nabla \eta \le - \int_{\{0<u<\delta\}} \eta \nabla u \cdot \nabla f_\delta(u)$$

or, setting $F_\delta(u) = \int_0^u f_\delta(s)\, ds$,

$$\int_{B_1} \nabla F_\delta(u) \cdot \nabla \eta \le - \int_{\{0<u<\delta\}} \eta \nabla u \cdot \nabla f_\delta(u) = - \int_{\{0<u<\delta\}} \eta |\nabla u|^2 f_\delta'(u) \le 0.$$

Therefore, $F_\delta(u)$ is subharmonic. Now, if $\delta \to 0$, $F_\delta(u) \rightharpoonup u^+$ weakly in $H^1(B_1)$ and we deduce that u^+ is subharmonic too.

(b) Since $u^\pm$ are subharmonic, Caccioppoli inequality gives

$$\int_{B_{1/2}} |\nabla u^\pm|^2 \le C \int_{B_1} (u^\pm)^2$$

so that $\|u\|_{H^1(B_{1/2})} \le C \|u\|_{L^2(B_1)}$.

Moreover, since $(u^\pm)^2$ also are subharmonic, we can write

$$\sup_{B_{1/2}} (u^\pm)^2 \le \int_{B_1} (u^\pm)^2$$

from which $\|u\|_{L^\infty(B_{1/2})} \le C \|u\|_{L^2(B_1)}$. $\qquad\square$

Now we prove Lipschitz continuity and semi-convexity along tangential directions. By a *tangential direction* we mean any unit vector τ parallel to the hyperplane $\{y = 0\}$.

Theorem 2.2.2. *Let u be a solution of the zero thin obstacle problem in B_1. Then:*

(a) u *is Lipschitz. Moreover*

$$\|u\|_{\mathrm{Lip}(B_{1/2})} \le C \|u\|_{L^2(B_1)}.$$

(b)

$$\inf_{B_{1/2}} u_{\tau\tau} \ge -C \|u\|_{L^2(B_1)}.$$

Proof.   We enlarge the obstacle to $h_\varepsilon(x, y) = -\frac{y^2}{\varepsilon}$ and let u_ε be the corresponding solution of the thick obstacle problem. Notice that $u \leq u_\varepsilon$. Then:

i) $\|u_\varepsilon\|_{C^{1,1}(B_{2/3})} \leq C(\varepsilon)$.

ii) $\Lambda_\varepsilon = \{X : u_\varepsilon = h_\varepsilon\} \subset \{|y| < C\varepsilon\}$

Indeed, *i)* is well known (see [15]). To prove *ii)* let $m > 0$ such that

$$-m < \min_{\partial B_1} u_\varepsilon$$

and take $h \in C(\overline{B_1^+})$ such that

$$\begin{cases} \Delta h = 0 & \text{in } B_1^+ \\ h = 0 & \text{in } B_1' \\ h = -m & \text{on } \partial B_1^+ \cap \{y > 0\}. \end{cases}$$

Since h is Lipschitz up to B'_{1-d_0}, then $h > h_\varepsilon$ if $y > C\varepsilon$ and, since $u_\varepsilon \geq h$, *ii)* follows by symmetry.

Let us prove (a). Since ∇u_ε is harmonic outside Λ_ε and $\nabla u_\varepsilon = \nabla h_\varepsilon$ on Λ_ε, it follows from *ii)* and the maximum principle that, in B_1, $|\nabla u_\varepsilon| \leq C$, with $C = C(n, g)$ independent of ε. In particular, in $B_{2/3}$, we have $|\nabla u_\varepsilon| \leq c \|u_\varepsilon\|_{L^2(B_1)}$. Letting $\varepsilon \to 0$ and using $b)$ in Lemma 2.2.1, we deduce (a).

For (b) we give two proofs.

First proof. We use a penalization technique. Let $\beta_\delta = \beta_\delta(s)$ a family of smooth increasing, concave functions such that $\beta'_\delta(s) = \delta$ if $s > 0$ and as $\delta \to 0$, β_δ converges to the graph $\beta_0(s) = -\infty$ if $s < 0$, $\beta_0(s) = 0$ if $s > 0$.

Let now solve the following penalized problem:

$$\begin{cases} \Delta v_{\varepsilon,\delta} = \beta_\delta(v_{\varepsilon,\delta} - h_\varepsilon) & \text{in } B_1 \\ v_{\varepsilon,\delta} = g & \text{on } \partial B_1. \end{cases}$$

From [6], we know that a smooth unique solution $v_{\varepsilon,\delta}$ exists and $\|v_{\varepsilon,\delta}\|_{H^{2,p}} \leq c(\|g\|_{H^{2-1/p,p}})$. Differentiating twice the equation along the direction τ, we get, setting $w = \partial_{\tau\tau} v_{\varepsilon,\delta}$:

$$\Delta w = \beta''_\delta(v_{\varepsilon,\delta} - h_\varepsilon)[\partial_\tau v_{\varepsilon,\delta}]^2 + \beta'_\delta(v_{\varepsilon,\delta} - h_\varepsilon)w \leq \beta'_\delta(v_{\varepsilon,\delta} - h_\varepsilon)w.$$

Hence, since near ∂B_1 we have $v_{\varepsilon,\delta} > h_\varepsilon$ and $v_{\varepsilon,\delta}$ is bounded, we have $\Delta w = \delta w$ and w remains bounded.

On the other hand, if w has a global minimum at a point $x^* \in B_1$, then $\Delta w(x^*) \geq 0$. Since $\beta'_\delta \geq 0$, we infer $w \geq 0$. Using again the results in [6] and passing to the limit first as $\delta \to 0$ and then for $\varepsilon \to 0$, we conclude the first proof.

Second proof. For global solutions (b) follows simply from the observation that $u(X + h\tau)$ and $u(X - h\tau)$ are admissible nonnegative superharmonic functions and therefore

$$\frac{1}{2}\left[u(X + h\tau) + u(X - h\tau)\right] \geq u(X).$$

In the local case, we modify the argument in the following way. Consider the ring

$$S = \{X \in B_1 : 1 - d_0 < |X| < 1\}.$$

Inside S, u is harmonic. Then, from interior estimates, we get

$$\|u\|_{C^{2,\alpha}(S^*)} \leq C \|u\|_{L^2(B_1)}$$

where

$$S^* = \left\{x \in B_1 : 1 - \frac{2}{3}d_0 < |X| < 1 - \frac{1}{3}d_0\right\}.$$

Thus,

$$w_h(X) \equiv \frac{1}{2}\left[u(X + h\tau) + u(X - h\tau)\right] + Ch^2 \|u\|_{L^2(B_1)} \geq u(X)$$

for any X with $|X| = 1 - d_0/2$ and $-d_0/6 < h < d_0/6$.

Observe now that w_h is superharmonic in $B_{1-d_0/2}$ and $w_h(x,0) \geq 0$ in $B'_{1-d_0/2}$. On the other hand, u is the solution of the zero thin obstacle in $B_{1-d_0/2}$. Therefore, since w_h is admissible for comparison with u in $B_{1-d_0/2}$ we can write

$$w_h(X) \geq u(X)$$

in all $B_{1-d_0/2}$. This shows that $\inf_{B_{1/2}} u_{\tau\tau} \geq -C \|u\|_{L^2(B_1)}$.   □

Remark 2.2.1 As a consequence:

$$u_{yy} \leq C \text{ in } B_1 \backslash \Lambda(u).$$

In particular, the function $u_y - Cy$ is monotone and bounded. Thus we are allowed to define in B'_1 :

$$\sigma(x) = \lim_{y \to 0+} u_y(x,y).$$

Since

$$0 \geq \Delta u = 2u_y \mathcal{H}^n_{|\Lambda(u)}$$

in $\mathcal{D}'(B_1)$, we have $\sigma(x) \leq 0$ in $\Lambda(u)$ and by symmetry, $\sigma(x) = 0$ in $B'_1 \backslash \Lambda(u)$. We may summarize the property of the solution of a zero thin obstacle problem in complementary form as follows:

$$\begin{cases} \Delta u \leq 0, \ u\Delta u = 0 & \text{in } B_1 \\ \Delta u = 0 & \text{in } B_1 \backslash \Lambda(u) \\ u(x,0) \geq 0, \ \sigma(x) \leq 0, \ u(x,0)\sigma(x) = 0 & \text{in } B_1' \\ \sigma(x) = 0 & \text{in } B_1' \backslash \Lambda(u) \end{cases}$$

Let us point out some consequences of the above estimates.

Lemma 2.2.3. *Let u be a solution of the zero thin obstacle problem in B_1, normalized by $\|u\|_{L^2(B_1)} = 1$. Then, for some universal constant, in $B_{1/2}^+$:*

(a) $y \longmapsto u(x,y) - Cy^2$ *is concave and* $x \longmapsto u(x,y) + C|x|^2$ *is convex.* In particular:

$$u_y(x,t) - u_y(x,s) \leq 2C(t-s) \qquad (t > s) \qquad (7)$$

and

$$u(x,t) - u(x,0) \leq Ct^2. \qquad (8)$$

(b) If $u(x,t) \geq h$, then in the half ball (a ball if $\nabla_x u(x,t) = \mathbf{0}$)

$$HB_\rho'(x) = \{z : |x - z| \leq \rho, \ \langle z - x, \nabla_x u(x,t) \rangle \geq 0\}$$

we have $u(z,t) \geq h - C\rho^2$.

Proof. (a) follows directly from Theorem 2.2.2b. Then $u_y - 2Cy$ is decreasing and therefore

$$u_y(x,t) - 2Ct \leq u_y(x,s) - 2Cs$$

which is (7). Integrating $u_y(x,s) \leq Cs$ over $0, t$ gives (8).

(b) We have, by convexity:

$$u(z,t) + C|z|^2 \geq u(x,t) + C|x|^2 + \langle z - x, \nabla_x u(x,t) + 2Cx \rangle.$$

If $u(x,t) \geq h$, in $HB_\rho'(x)$ we have

$$u(z,t) \geq u(x,t) - C|x - z|^2 \geq h - C\rho^2. \qquad \square$$

2.3 Local $C^{1,\alpha}$ Estimate

Let u be a solution of the zero thin obstacle problem in B_1, normalized by $\|u\|_{L^2(B_1)} = 1$. We prove now a local $C^{1,\alpha}$ estimate. It is enough to show that $\sigma \in C^{0,\alpha}$ near the free boundary $F(u)$.

In fact, in the interior of $\Lambda'(u)$, $u(x,0)$ is smooth and so σ is. On the other hand, on $\Omega' = B'\backslash\Lambda'(u)$, $\sigma = 0$. Thus if we show that σ is $C^{0,\alpha}$ in a neighborhood of $F(u)$, then $u \in C^{1,\alpha}$ from both sides of the free boundary by standard estimates for the Neumann problem.

In particular it is enough to show uniform estimates around a free boundary point, say the origin.

We distinguish two steps:

Step 1: To show that near the free boundary we can locate large regions where $-\sigma$ grows at most linearly (estimates in measure of the oscillation of $-\sigma$).

Step 2: Using Poisson representation formula and concavity, we convert the estimate in average of the oscillation of $-\sigma$, done in step 1, into pointwise estimates, suitable for iteration.

$\Rightarrow$ *Step 1*: First a barrier.

Lemma 2.3.1. Let $x_0 \in \Omega' = B_1' \backslash \Lambda'(u)$. Define

$$h_{x_0}(X) = |x - x_0|^2 - ny^2.$$

Then, for any open set A such that $(x_0, 0) \in A \subset B_1$,

$$\sup_{\partial A}(u - h_{x_0}) \geq 0.$$

Proof. The function $w = u - h_{x_0}$ is harmonic in $A \backslash \Lambda(u)$ and $w(x_0, 0) \geq 0$. Since $x_0 \notin \Lambda'(u)$ we infer that

$$\sup_{\partial(A \backslash \Lambda)}(u - h_{x_0}) \geq 0.$$

On the other hand, on $\Lambda(u)$, $h_{x_0} > u$, and therefore the lemma follows. $\square$

The point in the above lemma is that the supremum of w is not attained at a point on $\Lambda(u)$.

Lemma 2.3.2. Let $x_0 \in \Omega' = B_1' \backslash \Lambda'(u)$ and define

$$E_\gamma = \{x : \sigma(x) \geq -\gamma\} \qquad (\gamma > 0).$$

Then, for suitable positive constants $C^*, \bar{C}$ and any sufficiently small γ, there exists a ball $B'_{C^* \gamma}(x^*)$ such that

$$B'_{C^* \gamma}(x^*) \subset B'_{\bar{C}\gamma}(x_0) \cap E_\gamma.$$

In particular

$$\left| B'_{\bar{C}\gamma}(x_0) \cap E_\gamma \right| \geq c\gamma^n.$$

Proof. We apply Lemma 2.3.1, choosing

$$A = B'_{C_1\gamma}(x_0) \times (-C_2\gamma, C_2\gamma) \qquad \text{with } C_1 \gg C_2.$$

The maximum of $u - h_{x_0}$ can be achieved only on the lateral side of the cylinder or on one of the two bases. By symmetry, it is enough to consider the base $B'_{C_1\gamma}(x_0) \times \{y = C_2\gamma\}$.

Case 1. $\sup_{\partial A}(u - h_{x_0})$ is attained at a point (ξ, t) on the lateral side. In this case:

$$u(\xi, t) \geq h_{x_0}(\xi, t) \geq \left(C_1^2 - nC_2^2\right)\gamma^2 \equiv C_3\gamma^2.$$

Apply now Lemma 2.2.3 (b) with $h = C_3\gamma^2$ and $\rho = C_4\gamma$, with C_4 to be chosen small enough. Then in the half ball

$$HB'_\rho(\xi) = \{z : |\xi - z| \leq \rho, \langle z - \xi, \nabla_x u(\xi, t)\rangle \geq 0\}$$

we have

$$u(z, t) \geq C_3\gamma^2 - CC_4\gamma^2 \equiv C_5\gamma^2$$

if C_4 is small enough. Notice that $HB'_\rho(\xi) \subset B'_{(C_1+C_4)\gamma}(x_0)$, if C_4 is small, and that $HB'_\rho(\xi)$ contains a ball $B_{\rho/2}(x^*)$.

To prove the theorem it is enough to show that if $z \in HB'_\rho(\xi)$ then $z \in E_\gamma$. If not, $\sigma(z) < -\gamma$. Then $u(z, 0) = 0$ so that, since $u_{yy} \leq C$:

$$u(z, t) = u(z, t) - u(z, 0) \leq t\sigma(z) + Ct^2 \leq -\gamma t + Ct^2$$
$$\leq (-C_2^{-1} + C)t^2 < 0$$

if C_2 is chosen small, since $t \leq C_2\gamma$.

This gives a contradiction and ends the proof in the first case, choosing $\bar{C} = C_1 + C_4$.

Case 2. $\sup_{\partial A}(u - h_{x_0})$ is attained at a point $(\xi, C_2\gamma)$ on the base of the cylinder. Then

$$u(\xi, C_2\gamma) \geq h_{x_0}(\xi, C_2\gamma) \geq -n(C_2\gamma)^2.$$

As before, apply Lemma 2.2.3 (b) with $h = -n(C_2\gamma)^2$ and $\rho = C_2\gamma$. Then in the half ball

$$HB'_\rho(\xi) = \{z : |\xi - z| \leq \rho, \langle z - \xi, \nabla_x u(\xi, t)\rangle \geq 0\}$$

we have

$$u(z, C_2\gamma) \geq -n(C_2\gamma)^2 - C(C_2\gamma)^2 \equiv -C_5(C_2\gamma)^2.$$

If $\sigma(z) < -\gamma$, we would have $u(z, 0) = 0$ and, using once more $u_{yy} \leq C$, we get

$$u\left(z, C_2\gamma\right) - u\left(z, 0\right) = C_2\gamma\sigma\left(z\right) + C\left(C_2\gamma\right)^2 \le -C_2\gamma^2 + C\left(C_2\gamma\right)^2 < -C_5\left(C_2\gamma\right)^2$$

if C_2 is small. Contradiction.

We complete the proof choosing $\overline{C} = C_1 + C_2$. $\qquad\square$

$\Rightarrow$ *Step 2.* We convert the above estimate into a pointwise estimate. We need the following remark, consequence of the Poisson formula.

Remark 2.3.1. Let v be a positive harmonic function in $B_1\left(x_0\right) \times \left(0, 1\right)$ continuous in $\overline{B_1'\left(x_0\right)} \times \left[0, 1\right]$. Assume that

$$v\left(x, 0\right) \ge 1 \quad \text{in } B_\delta'\left(x^*\right)$$

for some $B_\delta\left(x^*\right) \subset B_1\left(x_0\right)$. Then, by Poisson formula,

$$v\left(X\right) \ge \eta\left(\delta\right) > 0 \quad \text{in } \overline{B_{1/2}'\left(x_0\right)} \times \left[1/4, 3/4\right].$$

Theorem 2.3.3. *Let u be our normalized solution and $x_0 \in \Omega'$. Then there exist α, $0 < \alpha < 1$, and a constant C, depending only on n and the number d_0 (defined at the beginning of the section) such that*

$$\sigma\left(x\right) \ge -C\left|x - x_0\right|^\alpha \quad \text{for every } x \in B_{2/3}'.$$

As a consequence, $\sigma \in C^{0,\alpha}(B_{2/3}')$ and $u \in C^{1,\alpha}(B_{1/2})$. In particular

$$\|u\|_{C^{1,\alpha}\left(\overline{B_{1/2}^+}\right)} \le C.$$

Proof. Recall that u_y is bounded in $B_{2/3}$ and $u_{yy} \le C$ in $B_1 \backslash \Lambda$. Moreover, $\sigma \le 0$. Thus, it is enough to prove that

$$u_y\left(X\right) \ge -\beta^k \quad \text{in } B_{\gamma^k}'\left(x_0\right) \times \left[0, \gamma^k\right] \tag{9}$$

for some $0 < \gamma < 1, 0 < \beta < 1$, both to be chosen, and any $k \ge 0$.

The case $k = 0$ follows from the boundedness of u_y in $B_{2/3}$. Now, assume that, inductively, we have proved (9) up to k :

$$u_y\left(X\right) \ge -\beta^k \quad \text{in } B_{\gamma^k}'\left(x_0\right) \times \left[0, \gamma^k\right]$$

for $0 < \gamma \ll \beta < 1$.

Consider the function

$$w\left(X\right) = \frac{u_y\left(X\right) + \beta^k}{-C_0\mu\gamma^k + \beta^k} \quad \text{in } B_{\mu\gamma^k}'\left(x_0\right) \times \left[0, \mu\gamma^k\right].$$

If μ is fixed, sufficiently small, adjusting the constant[2] C_0, by Lemma 2.3.2 there is a ball

$$B'_{c^*\mu\gamma^k}(x^*) \subset B'_{\mu\gamma^k}(x_0) \cap E_{C_0\mu\gamma^k}.$$

Then, w is positive and harmonic in $B'_{\mu\gamma^k}(x_0) \times (0, \mu\gamma^k)$ and $w(x, 0) \geq 1$ in $B'_{c^*\mu\gamma^k}(x^*)$. Remark 2.3.1 gives

$$u_y(X) \geq -\beta^k + \eta\left(-C_0\mu\gamma^k + \beta^k\right) \geq -\beta^k + \frac{1}{2}\eta\beta^k \qquad \text{with } \eta = \eta(c^*).$$

for $(x, y) \in B'_{\frac{\mu}{2}\gamma^k}(x_0) \times \left[\frac{1}{4}\mu\gamma^k, \frac{3}{4}\mu\gamma^k\right]$, since $\gamma \ll \beta$. Note that μ is independent of γ and β.

We have to fill the gap from $y = 0$ to $y = \frac{1}{4}\mu\gamma^k$. Using $u_{yy} < C$, we obtain

$$u_y(x, y) - u_y\left(x, \frac{1}{4}\mu\gamma^k\right) \geq -\frac{C}{4}\mu\gamma^k$$

or

$$u_y(x, y) \geq -\beta^k + \frac{1}{2}\eta\beta^k - \frac{C}{4}\mu\gamma^k$$

for $(x, y) \in B_{\frac{\mu}{2}\gamma^k}(x_0) \times \left[0, \frac{3}{4}\mu\gamma^k\right]$.

We have to choose γ and β such that

$$-\beta^k + \frac{1}{2}\eta\beta^k - \frac{C}{4}\mu\gamma^k \geq -\beta^{k+1}.$$

Choose $\gamma < \mu/2$ and $\gamma \ll \beta$ so that the left hand side is larger than $-\left(1 - \frac{\eta}{4}\right)\beta^k$. If $\beta \leq \left(1 - \frac{\eta}{4}\right)$ then

$$u_y(x, y) \geq -\left(1 - \frac{\eta}{4}\right)\beta^k \geq -\beta^{k+1}$$

in $B_{\gamma^{k+1}}(x_0) \times \left[0, \gamma^{k+1}\right]$ and the inductive argument is complete. $\qquad\square$

2.4   Optimal Regularity for Tangentially Convex Global Solutions

In this section we consider global solution which are tangentially convex. Typically these solutions are obtained as a consequence of a blow up centered

---

[2] We apply the Lemma 2.3.2 with γ replaced by $C_0\mu\gamma^k$, $\bar{C}\gamma$ replaced by $\bar{C}C_0\mu\gamma^k$ and $C^*\gamma$ replaced by $C^*C_0\mu\gamma^k$. Take $C_0 = \bar{C}^{-1}$.

at a free boundary point. Notice that, in this case, the set $\{x : \sigma(x) < 0\}$ is convex. We also assume it is non empty.

We have already seen an explicit example, given by

$$u(X) = \operatorname{Re}(x_n + i\,|y|)^{3/2} = \rho^{3/2} \cos \frac{3}{2}\theta$$

where $\rho = \sqrt{x_n^2 + y^2}$ and $\tan\theta = y/x_n$, which once more displays optimal growth at the origin. In this example the zero set for $n = 1$ is given by the half line $x_n \leq 0$ and the free boundary is given by the origin.

Lemma 2.4.1. Let ∇_θ denote the surface gradient on the unit sphere ∂B_1. Set

$$\lambda_0 = \inf \left\{ \frac{\int_{\partial B_1^+} |\nabla_\theta w|^2\, dS}{\int_{\partial B_1^+} w^2 dS} : w \in H^{1/2}(\partial B_1^+) : w = 0 \text{ on } (\partial B_1')^- \right\}$$

where $(\partial B_1')^- = \{(x', x_n) \in \mathbb{R}^n, x_n < 0\}$. Then

$$\lambda_0 = \frac{2n-1}{4}.$$

Proof. Let $\rho = \sqrt{x_n^2 + y^2}$ and $\tan\psi = y/x_n$. Define, for $0 \leq \psi \leq \pi$,

$$w_0(x, y) = \rho^{1/2} \cos \frac{\psi}{2}.$$

Then, w_0 is the restriction to the surface of upper hemisphere of an harmonic function homogeneous of degree $1/2$ of the form[3]

$$w(x, y) = r^{1/2} h(\theta) \qquad r^2 = |x|^2 + y^2,\ \theta \in \partial B_1.$$

Indeed, h is an eigenfunction of the Laplace-Beltrami operator $-\Delta_\theta$ on ∂B_1. Computing $-\Delta_\theta h$ we get, since the eigenvalue corresponding to the homogeneity α is $\alpha(\alpha - 1) + n\alpha$:

$$-\Delta_\theta h = \left(-\frac{1}{4} + \frac{n}{2}\right) h = \frac{2n-1}{4} h.$$

Since h is nonnegative and if we reflect it even, it is an eigenfunction in ∂B_1 with homogeneous Dirichlet data on $(\partial B_1')^-$, it follows that h is (a multiple) of the first eigenfunction of the problem and therefore $\lambda_0 = \frac{2n-1}{4}$. $\qquad\square$

---

[3] For instance, if $n + 1 = 3$, $\theta = (\varphi, \psi)$, then: $h(\theta) = \left(\cos^2\psi\cos^2\varphi + \sin^2\psi\right)^{1/4} \cos\psi/2$.

Now, a monotonicity formula.

Lemma 2.4.2. *Let w be continuous in $\overline{B_r^+}$, harmonic in B_r^+, $w(0,0) = 0$, $w(x,0) \leq 0$, $w(x,0)\,w_y(x,0) = 0$ in B_r'. Assume that*

$$\{x \in B_r' : w(x,0) < 0\}$$

is a non empty and convex cone. Set

$$\beta(r) = \frac{1}{r} \int_{B_r^+} \frac{|\nabla w(X)|^2}{|X|^{n-1}} dX.$$

Then, $\beta(r)$ is bounded and increasing for $r \in (0, 1/2]$.

Proof. We have $\Delta w^2 = 2w\Delta w + 2|\nabla w|^2 = 2|\nabla w|^2$, so that

$$\beta(r) = \frac{1}{r} \int_{B_r^+} \frac{|\nabla w(X)|^2}{|X|^{n-1}} dX = \frac{1}{2r} \int_{B_r^+} \frac{\Delta w^2}{|X|^{n-1}} dX.$$

Now (letting $\partial B_r^+ = (\partial B_r \cap \{y \geq 0\})$):

$$\beta'(r) = -\frac{1}{2r^2} \int_{B_r^+} \frac{\Delta w^2}{|X|^{n-1}} dX + \frac{1}{r^n} \int_{\partial B_r^+} |\nabla w|^2 dS.$$

Since $w(0) = 0$ and $ww_y = 0$ on $y = 0$, we can write

$$\frac{1}{2r^2} \int_{B_r^+} \frac{\Delta w^2}{|X|^{n-1}} dX = \frac{1}{r^{n+1}} \int_{\partial B_r^+} ww_\nu dS - \frac{1}{2r^2} \int_{B_r^+} \nabla w^2 \cdot \nabla \left(\frac{1}{|X|^{n-1}} \right) dX$$

$$= \frac{1}{r^{n+1}} \int_{\partial B_r^+} ww_\nu dS + \frac{n-1}{2r^{n+2}} \int_{\partial B_r^+} w^2 dS$$

$$\leq \left(\frac{1}{2r^{n+2}} \int_{\partial B_r^+} w^2 dS \right)^{1/2} \left(\frac{2}{r^n} \int_{\partial B_r^+} w_\nu^2 dS \right)^{1/2} + \frac{n-1}{2r^{n+2}} \int_{\partial B_r^+} w^2 dS$$

$$\leq \frac{2n-1}{4r^{n+2}} \int_{\partial B_r^+} w^2 dS + \frac{1}{r^n} \int_{\partial B_r^+} w_\nu^2 dS.$$

On the other hand,

$$\int_{\partial B_r^+} |\nabla w|^2 dS = \int_{\partial B_r^+} |\nabla_\theta w|^2 dS + \int_{\partial B_r^+} w_\nu^2 dS$$

so that we obtain:

$$\beta'(r) \geq -\frac{2n-1}{4r^{n+2}} \int_{\partial B_r^+} w^2 \, dS + \frac{1}{r^n} \int_{\partial B_r^+} |\nabla_\theta w|^2 \, dS.$$

The convexity of $\{x \in B_r' : w(x,0) < 0\}$ implies that w vanishes at least on $(\partial B_1')^-$ so that, according to Lemma 2.4.1, the Rayleigh quotient

$$\frac{\int_{\partial B_1^+} |\nabla_\theta w|^2 \, dS}{\int_{\partial B_1^+} w^2 \, dS}$$

must be greater than $\lambda_0 = \frac{2n-1}{4}$. Thus we conclude that $\beta'(r) \geq 0$ and in particular, $\beta(r) \leq \varphi(1/2)$. $\qquad\square$

Theorem 2.4.3 (Optimal regularity for global solutions). *Let u be a global solution of a zero thin obstacle problem, which is convex along any tangential direction. Then $u \in C^{1,1/2}$ up to the coincidence set, from both sides.*

Proof. It is enough to prove that u_y goes to zero in $C^{1/2}$ fashion when X approaches a free boundary point, which we take to be the origin. Set $w = u_y$. Then, since $\Lambda(u)$ is convex, w satisfies all the hypotheses of Lemma 2.4.2. Therefore, if the coincidence set is non empty, for r small,

$$\frac{1}{r^n} \int_{B_r^+} |\nabla w(X)|^2 \, dX \leq \frac{1}{r} \int_{B_r^+} \frac{|\nabla w(X)|^2}{|X|^{n-1}} \, dX \leq \varphi(1/2).$$

On the other hand, since w vanishes on at least half of the ball B_r', by Poincaré inequality

$$\fint_{B_r^+} w^2 \leq Cr^2 \fint_{B_r^+} |\nabla w|^2 \leq C_0 r.$$

Finally, since w^2 is subharmonic across $y = 0$, we conclude that

$$w^2 \Big|_{B_{r/2}^+} \leq \fint_{B_r^+} w^2 \leq cr. \qquad\square$$

2.5 Almgren's Frequency Formula

Once the optimal regularity for global solutions has been established, a blow up analysis around a free boundary point allows to carry the same regularity to local solutions. A key tool is a variant of the so called frequency formula of Almgren. In the case of solutions to the zero thin obstacle problem, this kind of monotonicity formula takes the following form:

Theorem 2.5.1. *Let u be a continuous function in B_1, harmonic in $B_1^+ \backslash \Lambda(u)$ and $u \in C^{1,\alpha}(\overline{B_1^+}) \cap C^{1,\alpha}(\overline{B_1^-})$.*

Assume that $u(0,0) = 0$ and $u(x,0) u_y(x,0) = 0$ in B_1'. Define, for $0 < r < 1$:

$$\Phi(r;u) = r \frac{\int_{B_r} |\nabla u|^2}{\int_{\partial B_r} u^2 dS} \equiv r \frac{V(r;u)}{H(r;u)}.$$

Then, for $0 < r < 1$, $\Phi(r;u)$ is nondecreasing. Moreover, let

$$\mu = \lim_{r \to 0+} \Phi(r;u).$$

Then $\Phi(r;u) \equiv \mu$ in $(0,1)$ if and only if u is a homogeneous function of the form

$$u(X) = |X|^\mu g(\theta) \qquad \theta \in \partial B_1.$$

In particular, in this last case, if u is a solution of the zero thin obstacle problem (that is $u(x,0) \geq 0$), then

$$\mu \geq \frac{3}{2}.$$

Remark 2.5.1. Thus, u gives the homogeneity degree of u at the origin. We know that, if u is the restriction of a global solution of the zero obstacle problem, then $\mu \geq 3/2$, from optimal regularity (Theorem 2.4.3).

Finally, note the scaling property of $D(r;u)$: if $u_r(X) = u(rX)$ then:

$$\Phi(rR;u) = \Phi(R, u_r).$$

Proof of Theorem 2.5.1. We have

$$\log \Phi(r;u) = \log r + \log V(r;u) + \log H(r;u)$$

so that

$$\frac{d}{dr} \log \Phi(r;u) = \frac{1}{r} + \frac{V'(r;u)}{V(r;u)} - \frac{H'(r;u)}{H(r;u)}.$$

We have (ν is the exterior normal):

$$V'(r;u) = \int_{\partial B_r} |\nabla u|^2 dS \quad \text{and} \quad H'(r;u) = \frac{n}{r} H(r;u) + 2 \int_{\partial B_r} u u_\nu dS \quad (10)$$

and therefore we want to prove that

$$\frac{d}{dr} \log \Phi(r;u) = \frac{1-n}{r} + \frac{\int_{\partial B_r} |\nabla u|^2 dS}{\int_{B_r} |\nabla u|^2} - 2 \frac{\int_{\partial B_r} u u_\nu dS}{\int_{\partial B_r} u^2 dS} \geq 0. \qquad (11)$$

Now, since $u(x,0) u_y(x,0) = 0$ and $\Delta(u^2) = 2u\Delta u + 2|\nabla u|^2 = 2|\nabla u|^2$ in $B_r \backslash \Lambda(u)$, we can write

$$\int_{B_r} |\nabla u|^2 = \frac{1}{2} \int_{B_r} \Delta(u^2) = \int_{\partial B_r} u u_\nu dS. \tag{12}$$

To control $\int_{\partial B_r} |\nabla u|^2 dS$, we use the divergence theorem in $B_r \backslash \Lambda(u)$. Let

$$h(X) = \operatorname{div}\left[X|\nabla u|^2 - 2(X \cdot \nabla u)\nabla u\right].$$

Under our hypotheses, in $B_r \backslash \Lambda(u)$,

$$h(X) = (n-1)|\nabla u|^2.$$

On the other hand, since $(X \cdot \nabla u) u_y$ vanishes continuously on Λ, we can write

$$(n-1)\int_{B_r} |\nabla u|^2 = \int_{B_r} h = r \int_{\partial B_r} |\nabla u|^2 dS - 2r \int_{\partial B_r} (u_\nu)^2 dS$$

or

$$\int_{\partial B_r} |\nabla u|^2 dS = \frac{(n-1)}{r} \int_{B_r} |\nabla u|^2 + 2 \int_{\partial B_r} (u_\nu)^2 dS. \tag{13}$$

Inserting into (11), we get

$$\frac{d}{dr} \log \Phi(r;u) = 2 \frac{\int_{\partial B_r} (u_\nu)^2 dS}{\int_{\partial B_r} u u_\nu dS} - 2 \frac{\int_{\partial B_r} u u_\nu dS}{\int_{\partial B_r} u^2 dS}.$$

By Schwarz inequality

$$\left(\int_{\partial B_r} u u_\nu dS\right)^2 \leq \int_{\partial B_r} (u_\nu)^2 dS \int_{\partial B_r} u^2 dS$$

so that $\frac{d}{dr} \log \Phi(r;u) \geq 0$.

The equality sign $\Phi(r;w) \equiv \mu$ in $(0,1)$ holds if and only if w is proportional to w_ν on ∂B_r for every $0 < r < 1$, which implies that w must be of the form

$$w(X) = f(|X|)g(\theta) \qquad \theta \in \partial B_1.$$

From the radial formula for the Laplace operator, it must be $f(|X|) = |X|^\mu$. In fact, by unique continuation μ must be the same for all component of $B_1 \backslash \Lambda(u)$ where g has constant sign.

If w is a solution of the zero obstacle problem, then, being homogeneous, can be extended to a global solution and therefore, from Theorem 2.4.3 we infer $\mu \geq 3/2$.   □

Let us point out an important consequence of Theorem 2.5.1. First observe that if

$$\varphi(r) = \varphi(r; u) = \int_{\partial B_r^+} u^2.$$

then

$$\Phi(r; u) = \frac{r}{2} \frac{d}{dr} \log \varphi(r; u). \tag{14}$$

Corollary 2.5.2. *Let u be a solution of the zero obstacle problem and*

$$\mu = \lim_{r \to 0+} \Phi(r, u).$$

Then:

(a) The function $r \longmapsto r^{-2\mu}\varphi(r)$ is nondecreasing for $0 < r < 1$ and in particular

$$\varphi(r) \leq r^{2\mu}\varphi(1) \leq r^{2\mu} \sup_{B_1} |u|.$$

(b) $\frac{d}{dr}[r^{-2\mu}\varphi(r)] \equiv 0$ in $(0,1)$ if and only if

$$u(X) = |X|^\mu g(\theta) \qquad \theta \in \partial B_1$$

with $\mu \geq 3/2$.

(c) Let $0 < r < R < 1$. Given $\varepsilon > 0$, for every $r, R \leq r_0(\varepsilon)$,

$$\varphi(R) \leq \left(\frac{R}{r}\right)^{2(\mu+\varepsilon)} \varphi(r).$$

Proof. (a) and (b). We have:

$$\varphi'(r; u) = \frac{d}{dr} \int_{\partial B_r} u^2 dS = 2 \int_{\partial B_r} u u_\nu dS = \frac{2}{|\partial B_r|} \int_{B_r} |\nabla u|^2$$

so that

$$\frac{d}{dr}[r^{-2\mu}\varphi(r)] = -2\mu r^{-2\mu-1}\varphi(r) + 2r^{-2\mu} \frac{1}{|\partial B_r|} \int_{B_r} |\nabla u|^2$$

$$= \frac{2\mu r^{-2\mu-1}}{|\partial B_r|} \left\{ r \int_{B_r} |\nabla u|^2 - \mu \int_{\partial B_r} u^2 \right\} \geq 0$$

and $(a), (b)$ follow from Theorem 2.5.1.

(c) Let $r_0 = r_0(\varepsilon)$ such that $\Phi(r; u) \leq \mu + \varepsilon$ for $r, R \leq r_0$, that is (recall (14)):

$$\Phi(r; u) = \frac{r}{2} \frac{d}{dr} \log \varphi(r) \leq \mu + \varepsilon.$$

An integration over (r, R) gives (c). $\square$

Here is a first step towards the optimal regularity of the solution.

Corollary 2.5.3. *Let u be a solution of the zero obstacle problem, with $(0, 0) \in F(u)$. Then*

$$|u(X)| \leq r^\mu \sup_{B_1} |u| \qquad \forall X \in B_{r/2}. \tag{15}$$

Proof. Since $(u^+)^2$ is subharmonic, we have, for any $X \in B_{r/2}$

$$\left(u^+(X)\right)^2 \leq \fint_{\partial B_r(X)} \left(u^+\right)^2 dS \leq r^{2\mu} \sup_{B_1} \left(u^+\right)^2.$$

A similar inequality holds for $(u^-)^2$. $\square$

2.6 Asymptotic Profiles and Optimal Regularity

The last step to obtain optimal regularity requires the analysis of suitable blow-up sequences around a free boundary point. Also a careful classification of this kind of limiting profiles is crucial for the study of the free boundary regularity.

Given a solution u of our zero thin obstacle problem, we consider the blow-up family

$$v_r(X) = \frac{u(rX)}{\left[\varphi(r; u)\right]^{1/2}} \tag{16}$$

around $(0, 0) \in F(u)$. Notice that the "natural" rescaling $u(rX)/r^\mu$, where $\mu = \lim_{r \to 0+} \Phi(r, u)$, would not be appropriate, because on this kind of rescaling we have precise control of its behavior as $r \to 0$ merely from one-side. Rescaling by an average over smaller and smaller balls provides the necessary adjustments for controlling the oscillations of u around the origin.

Our purpose is to classify the limit of v_r as $r \to 0$. Observe that

$$\|v_r\|_{L^2(\partial B_1)} = 1. \tag{17}$$

Moreover,

$$\int_{B_R} |\nabla v_r|^2 dX = \frac{\Phi(R, v_r)}{R} \int_{\partial B_R} v_r^2 dS = |\partial B_1| R^{n-1} \Phi(rR, u) \frac{\varphi(rR; u)}{\varphi(r; u)}.$$

Thus, for every $R > 1$ and every r such that $rR \leq r_0\left(\varepsilon\right)$, we can write

$$\int_{B_R} |\nabla v_r|^2 \, dX \leq |\partial B_1| \left(\mu + \varepsilon\right) R^{n-1+2(\mu+\varepsilon)}$$

and by Poincaré inequality,

$$\int_{B_R} v_r^2 dX \leq |\partial B_1| \left(\mu + \varepsilon\right) R^{n+1+2(\mu+\varepsilon)}.$$

Therefore, v_r is equibounded in H^1_{loc} and by the $C^{1,\alpha}$ estimates of Sect. 2.3, also in $C^{1,\alpha}_{oc}$. Hence there exists a subsequence $\{v_j\}$, $v_j = v_{r_j}$, such that, in every compact subset of $\mathbb{R}^{n+1}$:

$$v_j \to v^* \quad \text{and} \quad \nabla v_j \to \nabla v^*, \text{ both uniformly.}$$

Because of (17), v^* is a nontrivial global solution of the zero thin obstacle problem. Moreover, since

$$\Phi\left(r_j; u\right) = \Phi\left(1; v_{r_j}\right) \to \Phi\left(1; v^*\right) = \mu, \tag{18}$$

as $r_j \to 0$, it follows from Theorem 2.5.1 that v^* is homogeneous, that is:

$$v^*\left(X\right) = |X|^\mu g\left(\theta\right) \qquad \theta \in \partial B_1$$

and from Theorem 2.4.3 that
$$\mu \geq \frac{3}{2}.$$

From (18) and Corollary 2.5.3 we infer:

Theorem 2.6.1. (Optimal regularity). *Let u be a solution of the zero obstacle problem, with $(0,0) \in F\left(u\right)$. Then u is $C^{1,1/2}$ at the origin, in the sense that*

$$|u\left(X\right)| \leq r^{3/2} \sup_{B_1} |u| \qquad \forall X \in B_{r/2}. \tag{19}$$

Thus, u is $C^{1,1/2}$ in both $\overline{B^+_{1/2}}$ and $\overline{B^-_{1/2}}$.

We are now ready for our classification theorem.

Theorem 2.6.2 (Blow-up limits). *Let u be a solution of our zero thin obstacle problem and*

$$\mu = \lim_{r \to 0+} \Phi\left(r, u\right). \tag{20}$$

We know that $\mu \geq 3/2$. Let v^* be the global solution defined above. Up to a multiplicative constant, in a suitable system of coordinates, the following hold:

(a) *Assume* $\mu < 2$. *Then* $\mu = \frac{3}{2}$ *and*

$$v^* (X) = \rho^{3/2} \cos \frac{3}{2} \psi$$

where $\rho^2 = x_n^2 + y^2$ and $\tan \psi = y/x_n$.

(b) *Assume* $\mu = 2$. *Then* v^* *is a quadratic polynomial:*

$$v^* (X) = \sum_{i \le n} a_i x_i^2 - cy^2, \qquad a_i \ge 0.$$

Proof. (a) From the tangential quasi-convexity of u, for every tangential direction τ, we have:

$$\partial_{\tau\tau} v_j \ge -c \frac{r_j^2}{[\varphi (r_j)]^{1/2}}. \tag{21}$$

Choose ε such that $\mu + \varepsilon < 2$ in Corollary 2.5.2 (c). Then

$$\frac{r_j^2}{[\varphi (r_j)]^{1/2}} \to 0 \text{ as } r_j \to 0$$

and we obtain $v^*_{\tau\tau} \ge 0$. Thus, v^* is tangentially convex and $\Lambda (v^*)$ is a thick convex cone (since v^* is nontrivial).

Claim: $w = v^*_{x_n}$ *is the first eigenfunction for the spherical Laplacian, vanishing on* $\partial B_1 \cap \Lambda (v^*)$.

Proof. Observe that on $\Lambda (v^*)$, $v^* = 0$, $v^*_y \le 0$ and for $y \ne 0$, $v^*_{yy} \le 0$.

This implies that $v^* (x, y) \le 0$ if $(x, 0) \in \Lambda (v^*)$. Assume that the vector $-e_n$ belongs to $(\Lambda (v^*))^0$. For any point X, consider the line $L_X = \{X + te_n\}$. For t large, negative, the function $v^* (X + te_n)$ becomes nonpositive, from the previous remark.

Since v^* is convex along L_X it follows that the derivative w cannot become negative anywhere on L_X, otherwise $v^* (X + te_n) \to +\infty$ as $t \to -\infty$. In particular, since X is arbitrary, w must be nonnegative in $\mathbb{R}^{n+1}$.

On the other hand, $w = 0$ on $\Lambda (v^*)$ and $w_y = 0$ on $\{y = 0\} \backslash \Lambda (v^*)$, by symmetry. Thus, the restriction of w to the unit sphere must be the first eigenfunction for the spherical Laplacian, with zero Dirichlet data on $\partial B_1 \cap \Lambda (v^*)$.

This proves the claim.

Now, since $\Lambda (v^*)$ is convex, it is strictly contained in a half plane. If it is not a half plane, from Lemma 2.4.1 we deduce that the homogeneity degree of w should be less than $1/2$, since $1/2$ corresponds to the case in which a half plane is removed. However this is a contradiction to $\mu \ge 3/2$.

We conclude that $\Lambda (v^*)$ is a half plane, say $\{x_n < 0, y = 0\}$, and $w (X) = \rho^{1/2} \cos \frac{1}{2} \psi$. This implies that, up to a multiplicative factor,

$$v^* (X) = \rho^{3/2} \cos \frac{3}{2} \psi.$$

(b) Let now $\mu = 2$. The limiting profile v^* is of the form $|X|^2 g(\theta)$, $\theta \in \partial B_1$, and $\Lambda(v^*)$ is a cone. Consider $w = v_y^*$. Then w is linearly homogeneous and $w = 0$ on $\{y = 0\} \setminus \Lambda(v^*)$.

We reflect evenly w across the hyperplane $y = 0$, defining

$$\tilde{w}(X) = \begin{cases} w(x,y) & y > 0 \\ w(x,-y) & y < 0 \end{cases}.$$

Claim 1: $\tilde{w}$ has constant sign. Suppose not. Then, since $\tilde{w}$ is harmonic on his support and $w(0,0) = 0$, we may apply the monotonicity formula of [4] to w^+ and w^- and, in particular, Corollary 12.4 in [9]. Accordingly, the homogeneity and the linear behavior of $\tilde{w}$ forces $\tilde{w}$ to be a two-planes solution with respect to a direction transversal to the plane $y = 0$, say e_n, due to the even symmetry of $\tilde{w}$. Therefore $\tilde{w}(X) = \alpha x_n^+ - \beta x_n^-$. This is a contradiction, since $w(x,0) \leq 0$ on $\Lambda(v^*)$ and zero outside $\Lambda(v^*)$. The claim is proved.

Claim 2: $\Lambda(v^*)$ has empty interior. Suppose not. Then $\tilde{w}$ is the first eigenfunction for the spherical Laplacian, with zero Dirichlet data on

$$[\{y = 0\} \setminus \Lambda(v^*)] \cap \partial B_1.$$

This forces a superlinear behavior of $\tilde{w}$ at the origin, since linear behavior corresponds to a half sphere, giving again a contradiction with point (a). Thus, $\Lambda(v^*)$ has empty interior.

Therefore, v^* is harmonic across $\Lambda(v^*)$ and must coincide with a quadratic polynomial.   □

Remark 2.6.1. Let $\mu = 3/2$. Observe that if $\tau = \alpha e_n + \beta e$, where e is tangential, $e \perp e_n$, and $\alpha^2 + \beta^2 = 1$, $\alpha > 0$, then, outside a η-strip $|y| < \eta$, we have in B_1:

$$v_\tau^* \geq C(\alpha, \eta) > 0. \tag{22}$$

2.7   Lipschitz Continuity of the Free Boundary at Stable Points

Form the knowledge of the optimal regularity of the solution and the classification of blow-up profiles we can start the analysis of the free boundary. First of all we distinguish the points of the free boundary according to the following definition.

Definition 2.7.1. *Let $X_0 \in F(u)$ and*

$$\mu(X_0) = r \frac{\int_{B_r(X_0)} |\nabla u|^2 \, dX}{\int_{\partial B_r(X_0)} u^2 \, dS}.$$

We say that X_0 is a regular or stable point if $\mu(X_0) = 3/2$.

The examples in the introductory Sect. 2.1 reveal that any hope to achieve regularity of $F(u)$ is restricted to a neighborhood of regular points. Therefore from now on we will focus on regular points. The first step is to prove that near a regular point (e.g. the origin) the free boundary is a Lipschitz graph. Precisely:

Theorem 2.7.1. *Let u be a solution of the zero thin obstacle problem in B_1. Assume that $(0,0) \in F(u)$ and is a regular point. Then, with respect to a suitable coordinate system, there exists a neighborhood of the origin B_ρ and a cone of tangential directions $\Gamma'(e_n, \theta)$ with axis e_n and opening $\theta \geq \pi/3$ (say), such that for every $\tau \in \Gamma'(e_n, \theta)$ we have*

$$D_\tau u \geq 0.$$

In particular, in that neighborhood, $F(u)$ is the graph of a Lipschitz function $x_n = f(x_1, ..., x_{n-1})$.

Proof. As in (16) let

$$v_{r_j}(X) = \frac{u(r_j X)}{[\varphi(r_j; u)]^{1/2}}.$$

From Theorem 2.6.2 we know that, for a suitable sequence $\{r_j\}$, $r_j \to 0$, $v_{r_j} \to v^*$ and $\nabla v_j \to \nabla v^*$, both uniformly in compact sets, where

$$v^*(X) = \rho^{3/2} \cos \frac{3}{2}\psi$$

($\rho^2 = x_n^2 + y^2$ and $\tan\psi = y/x_n$).

Let $\tau = \alpha e_n + \beta e$ be a tangential direction ($\alpha^2 + \beta^2 = 1$). If we fix $\alpha \geq \alpha_0 > 0$, so that $\beta \leq \beta_0 \equiv (1 - \alpha_0^2)^{1/2}$, then the set of all these directions fills a cone $\Gamma'(e_n, \theta)$ with $\theta = \arctan \beta_0/\alpha_0$. Thus, if α_0 is small then $\theta \geq \pi/3$.

We want to show that $D_\tau v_{r_j} \geq 0$ along every $\tau \in \Gamma'(e_n, \theta)$.

For σ small and $r_j \leq r_0(\sigma)$, we deduce from (22) that $D_\tau v_{r_j}$ enjoys the following properties in $B_{5/6}$:

(i) $D_\tau v_{r_j} \geq 0$ outside the strip $|y| \leq \sigma$,
(ii) $D_\tau v_{r_j} \geq c_0$ for $|y| \geq \frac{1}{2}$, and from optimal regularity,
(iii) $D_\tau v_{r_j} \geq -c\sigma^{1/2}$ inside the strip $|y| \leq \sigma$.

To get rid of the lack of positivity inside the strip $|y| \leq \sigma$ and to conclude the proof, we apply to $h = D_\tau v_{r_j}$ the following lemma.

Lemma 2.7.2. *Let u be as in Theorem 2.6.2. Assume h is a function with the following properties:*

(i) $\Delta h \leq 0$ *in* $B_1 \backslash \Lambda(u)$;
(ii) $h \geq 0$ *for* $|y| \geq \sigma$, $h = 0$ *on* $\Lambda(u)$, *with* $\sigma > 0$, *small;*
(iii) $h \geq c_0$ *for* $|y| \geq \frac{1}{8n}$;
(iv) $h \geq -\omega(\sigma)$, *for* $|y| \leq \sigma$, *where* ω *is the modulus of continuity of* h.

There exists $\sigma_0 = \sigma_0(n, c_0, \omega)$ *such that, if* $\sigma \leq \sigma_0$ *then* $h \geq 0$ *in* $B_{1/2}$.

Proof. Suppose there is $Z = (z, t) \in B_{1/2}$ such that $h(Z) < 0$. Let

$$Q = \left\{ (x, y) : |x - z| \leq \frac{1}{3}, \ |t - y| < \frac{1}{4n} \right\}$$

and

$$P(x, y) = |x - z|^2 - ny^2.$$

Define

$$v(X) = h(X) + \delta P(X)$$

where $\delta > 0$, to be chosen later. We have:

(a) $v(Z) = h(Z) - \delta nt^2 < 0$;
(b) $\Delta v \leq 0$ in $B_1 \backslash \Lambda(u)$;
(c) $v \geq 0$ on $\Lambda(u)$, since $h = 0$, $P \geq 0$ there.

Thus $v_{|Q}$ must have a negative minimum on ∂Q.

However, on $\partial Q \cap \{|y| \geq \frac{1}{8n}\}$, we have

$$v \geq c_0 - \frac{\delta}{16n} = 0$$

if $\delta = 16nc_0$.

On

$$\left\{ |x - z| = \frac{1}{3}, \ \sigma \leq |y| \leq \frac{1}{8n} \right\}$$

we have $h \geq 0$ so that

$$v \geq \delta \left[\frac{1}{9} - \frac{1}{64n} \right] \geq 0.$$

Finally, on

$$\left\{ |x - z| = \frac{1}{3}, \ |y| < \sigma \right\}$$

we have

$$v \geq -\omega(\sigma) + \delta \left[\frac{1}{9} - n\sigma^2 \right] \geq 0$$

if σ is chosen small depending on n, c_0, ω.

Hence $v \geq 0$ on ∂Q and we have reached a contradiction. Therefore $h \geq 0$ in $B_{1/2}$. $\square$

2.8 Boundary Harnack Principles and $C^{1,\alpha}$ Regularity of the Free Boundary at Stable Points

We are now in position to show that if the origin is a regular point, then nearby, the free boundary is a $C^{1,\alpha}$ graph. Precisely, we are going to prove the following result.

Theorem 2.8.1. Let u be a solution of the zero thin obstacle problem in B_1. Assume that $(0,0) \in F(u)$ and is a regular point. Then, with respect to a suitable coordinate system, there exists a neighborhood of the origin B_ρ in which $F(u)$ is the graph of a $C^{1,\alpha}$ function $x_n = f(x_1, ..., x_{n-1})$.

There are two ways to prove the above theorem. One way, that we only sketch (see for the details [2]) is to use a bi-lipschitz transformation to map a neighborhood of the origin in $B_1 \backslash \Lambda(u)$ onto a half ball, say,

$$B^+ = \{Z = (z, \eta) : |Z| < 1, \eta > 0\}.$$

The Laplace operator in transformed into a uniformly elliptic operator $\mathcal{L}$ in divergence form. Each tangential derivative $D_\tau u$ with τ belonging to the cone $\Gamma'(e_n, \theta)$, defined in Theorem 2.7.1, is mapped into a positive solution $w(Z; \tau)$ of $\mathcal{L}w = 0$ in B^+ vanishing on $\{\eta = 0\}$.

An application of Corollary 1 in [2] implies that the ratio

$$\frac{w(Z; \tau)}{w(Z; e_n)}$$

is Hölder continuous up to $\{\eta = 0\} \cap \overline{B^+_{2/3}}$. This in turn, implies that any quotient $D_\tau u / D_{e_n} u$, with $\tau \in \Gamma'(e_n, \theta)$, is Hölder continuous up to $F(u)$ in a neighborhood of the origin. This implies that the level sets of u are tangentially $C^{1,\alpha}$ surfaces and in particular that $F(u)$ is a $C^{1,\alpha}$ graph.

The second way is a somewhat more direct proof, based on the next theorem, which is a generalization of by now well known boundary Harnack principles (see e.g. [14, 37]) and which could be interesting in itself.

First, we define a class of domains, closely related to the Nontangentially Accessible Domains (briefly NTA) in [37].

Let D be a subdomain of B_1 and let $\partial^* D = \partial D \cap B_1$. Let $\Upsilon(X, Y)$ be the set of regular curves contained in D with end points X, Y. We denote by $d_g(X, Y)$ the *geodesic distance* in D of X and Y, that is

$$d_g(X, Y) = \inf\{l(\gamma) : \gamma \in \Upsilon(X, Y)\}.$$

Definition 2.8.1. *We say that D is a Δ- NTA domain if the following properties hold:*

(1) *For every $X, Y \in D$, $d_g(X, Y)$ is finite.*
(2) *Non tangential ball condition. Let Q be arbitrary on $\partial^* D$. There exist positive numbers $r_0 = r_0(Q, D)$ and $\eta = \eta(D)$ such that, for every $r \leq r_0$ there is a point $A_r(Q) \in B_r(Q)$ such that*

$$B_{\eta r}(A_r(Q)) \subset B_r(Q) \cap D.$$

(3) *Harnack chain condition. There is a constant $M = M(D)$ such that, for all $X, Y \in D$, $\varepsilon > 0$ and $k \in N$ satisfying*

$$d(X, \partial^* D) > \varepsilon, \ d(Y, \partial^* D) > \varepsilon, \ d_g(X, Y) < 2^k \varepsilon,$$

there exists a sequence of Mk balls $B_{r_1}(P_1), ..., B_{r_{Mk}}(P_{Mk})$ contained in D, such that

$$X \in B_{r_1}(P_1), \ Y \in B_{r_{Mk}}(P_{Mk}) \text{ and}$$

$$B_{r_j}(P_j) \cap B_{r_{j+1}}(P_{j+1}) \neq \emptyset \qquad (j = 1, ..., Mk - 1)$$

and

$$\frac{1}{2} r_j < d(B_{r_j}(P_j), \partial D^*) < 4 r_j \qquad (j = 1, ..., Mk - 1).$$

(4) *Uniform capacity condition. Let Q be an arbitrary point on $\partial^* D$. There exist positive numbers $r_0 = r_0(Q, D)$ and $\gamma = \gamma(D)$ such that, for every $r \leq r_0$,*

$$\text{cap}_\Delta\left[(B_r(Q) \setminus B_{r/2}(Q))' \cap \partial^* D\right] \geq \gamma r^{n-1}$$

where $\text{cap}_\Delta(K)$ is the capacity of K in B_1 with respect to the Laplace operator.[4]

Remark 2.8.1. Conditions (2) and (3) appears in the definition of NTA domains as given in [37]. Condition (4) replaces the exterior ball property in that definition. A simple example of $\Delta - NTA$ domains is an n-dimensional smooth manifold with Lipschitz boundary.

Remark 2.8.2. Let $\mathcal{L} = \text{div}(A(X)\nabla)$ be a uniformly elliptic operator with ellipticity constants λ, λ^{-1}, and bounded measurable coefficients. Since for any $K \subset B_1$, $\text{cap}_{\mathcal{L}}(K) \sim \text{cap.}(K)$ with equivalence constant depending only on n, λ, the notion of $\Delta - NTA$ domain refers actually to the above

---

[4] $\text{Cap}_\Delta(K) = \inf\left\{\|\nabla\varphi\|_{L^2} : \varphi \geq 1 \text{ on } K \text{ and } \varphi \in C_0^\infty(B_1)\right\}.$

class of operators. From now on, it is understood that the constants $c's$ and $C's$ appearing in the statements and proofs are understood depend only on some of the relevant parameters n, η, γ, M and λ.

According to Remark 2.8.2 we prove the following result.

Theorem 2.8.2 Let $D \subset B_1$ be a $\Delta - NTA$ domain. Suppose v and w are positive functions in D, continuously vanishing on $\partial^* D$ and satisfying $Lv = Lw = 0$ in D. Let $X_0 \in D \cap B_{2/3}$ such that $d(X_0, \partial^* D) = d_0 > 0$ and

$$v(X_0) = w(X_0) = 1.$$

Then:

(a) (Carleson estimate) For every $Q \in \partial^* D \cap B_{1/2}$:

$$\sup_{D \cap B_{1/3}(Q)} v \leq C$$

(b) (Boundary Harnack Principle)

$$\sup_{D \cap B_{1/3}(Q)} \frac{v}{w} \leq C.$$

(c) (Hölder continuity for quotients). The quotient $\frac{v}{w}$ is Hölder continuous in $B_{1/2} \cap D$ up to $\partial^* D$.

Proof. The proof follows by now standard lines (see for instance [9], Sect. 11.2, and [37]). We emphasize the main steps and differences.

(a) Fix $Q \in \partial^* D \cap B_{1/2}$ and let Y_0 be such that

$$v(Y_0) = \sup_{B_{1/3}(Q) \cap D} v \equiv N.$$

The interior ball condition and the Harnack chain condition plus the interior Harnack inequality imply that, if N is large then

$$d(Y_0, \partial^* D) \equiv |Y_0 - Q_0| \leq N^{-\delta}$$

with $Q_0 \in \partial^* D$ and $\delta = \delta(n, \lambda, d_0, M) > 0$.

Let $r_0 = |Y_0 - Q_0|$. Then, the uniform capacity condition implies that

$$\sup_{B_{2r_0}(Q_0) \cap D} v \equiv v(Y_1) \geq CN$$

with $C > 1$. Thus, for some $Q_1 \in \partial^* D$,

$$d(Y_1, \partial^* D) \equiv |Y_1 - Q_1| \leq (CN)^{-\delta}.$$

Iterating the process, we construct a sequence of points $\{Y_m\}$ satisfying the following conditions:

(i) $v(Y_m) \geq C^m N$

(ii) $d(Y_m, \partial^* D) \leq (C^m N)^{-\delta}$

(iii) $|Y_m - Y_{m+1}| \leq 4(C^m N)^{-\delta}$.

If N is large enough, we can make

$$\sum_{m \geq 0} |Y_m - Y_{m+1}| \leq \frac{1}{16}$$

and therefore $\{Y_m\} \subset D \cap B_{2/3}$. On the other hand, $d(Y_m, \partial^* D) \to 0$ and $v(Y_m) \to +\infty$ which leads to a contradiction, since $v(Y_m) \to 0$. This proves (a).

We prove (b). Let $P \in \partial^* D \cap B_{1/3}$ and $R_0 = |X_0 - P| + \frac{d_0}{2}$. Notice that $\frac{3}{2} d_0 \leq R_0 \leq 1$. Define

$$\Psi_{R_0}(P) = B_{R_0}(P) \cap D$$

and

$$\Sigma_0 = \partial B_{R_0}(P) \cap B_{R_0}(P), \; \Sigma_1 = \partial \Psi_{R_0}(P) \setminus \partial^* D$$

Observe that $\Sigma_0, \Sigma_1 \subset D$.

Let $\omega_{\mathcal{L}}^X$ be the $\mathcal{L}-$harmonic measure in $\Psi_{R_0}(P)$.

Claim: for every $X \in B_{R_0/8}(P) \cap D$,

$$\omega_{\mathcal{L}}^X(\Sigma_1) \leq C \omega_{\mathcal{L}}^X(\Sigma_0), \tag{23}$$

which is a doubling property for the $\mathcal{L}-$harmonic measure.

To prove the claim, introduce the Green's function $G(X, X_0)$ for the operator $\mathcal{L}$ in $\Psi_{R_0}(P)$ with pole at X_0. From Hölder continuity, on $\partial B_{d_0/3}(X_0)$, we have

$$\omega_{\mathcal{L}}^X(\Sigma_0) \geq c > 0$$

and by the estimates for the Green's function (we restrict for simplicity to the case $n + 1 > 2$).

$$G(X, X_0) \leq c d_0^{1-n}.$$

On the other hand, on $\partial \Psi_{R_0}(P)$, we have $G(X, X_0) = 0$ and $\omega_{\mathcal{L}}^X(\Sigma_0) \geq 0$. Therefore, by maximum principle, outside $B_{R_0/3}(X_0)$ we get

$$G(X, X_0) \leq c d_0^{1-n} \omega_{\mathcal{L}}^X(\Sigma_0). \tag{24}$$

We need a bound from below of G in terms of $\omega_{\mathcal{L}}^X(\Sigma_1)$. Let $\varphi \in C_0^\infty(B_{R_0}(P))$ such that

$$\varphi \equiv 0 \text{ in } B_{R_0/4}(P), \; \varphi \equiv 1 \text{ outside } B_{R_0/2}(P)$$

and $0 \leq \varphi \leq 1$ in $C_0(P) \equiv B_{R_0/2}(P) \backslash B_{R_0/4}(P)$. Note that $|\nabla \varphi| \leq c/R_0$. We have:

$$\omega_{\mathcal{L}}^X(\Sigma_1) \leq \int_{\partial \Psi_{R_0}(P)} \varphi d\omega_{\mathcal{L}}^X.$$

Fix $X \in B_{R_0/8}(P) \cap D$. Then we can write

$$0 = \varphi(X) = \int_{\partial \Psi_{R_0}(P)} \varphi d\omega_{\mathcal{L}}^X - \int_{C_0(P) \cap D} A(X) \nabla_Y G(X,Y) \cdot \nabla \varphi(Y) dY.$$

Therefore, from Caccioppoli estimate and Carleson estimate, we have, in $B_{R_0/8}(P) \cap D$,

$$\omega_{\mathcal{L}}^X(\Sigma_1) \leq CR_0^{(n-1)/2} \left(\int_{C_0(P) \cap D} \left| \nabla_Y G(X,Y)^2 \right| dY \right)^{1/2}$$

$$\leq CR_0^{(n-3)/2} \left(\int_{C_0(P) \cap D} G^2(X,Y) dY \right)^{1/2}$$

$$\leq CR_0^{n-1} G(X,X_0).$$

Comparing with (24) we conclude the proof of the claim.

With the doubling condition at hand the proof of (a) is quickly finished. In fact, let $Q \in \partial^* D \cap B_{1/2}$. by the Carleson estimate and maximum principle, in $D \cap B_{1/3}(Q)$ we get

$$v(X) \leq C\omega_{\mathcal{L}}^X(\Sigma_1).$$

On the other hand, the maximum principle, interior Harnack inequality and the doubling property (23) give, in $B_{R_0/8}(P) \cap D$,

$$v(X) \leq C\omega_{\mathcal{L}}^X(\Sigma_1) \leq C\omega_{\mathcal{L}}^X(\Sigma_0) \leq w(X).$$

Thus

$$\sup_{D \cap B_{R_0/8}(Q)} \frac{v}{w} \leq C.$$

Using a covering argument and the interior Harnack inequality one can easily conclude the proof.

At this point, the proof of (c) follows, for instance, as in [9, Sect. 11.2].$\square$

Proof of Theorem 2.8.1. We apply Theorem 2.8.2 with $D = B_1 \backslash \Lambda(u)$, $\partial^* D = \Lambda(u)$ and $v = D_\tau u$, $w = D_{e_n} u$ where $\tau \in \Gamma'(e_n, \theta)$. Since $F(u)$ is Lipschitz,

$$\text{cap}_\Lambda \left[(B_r(Q) \backslash B_{r/2}(Q)) \cap \Lambda(u) \right] \geq \gamma r^{n-1}.$$

We obtain, in particular, that on $\{y = 0\} \backslash \Lambda(u)$ the quotient $D_\tau u/D_{e_n} u$, is Hölder continuous up to $F(u)$ in a neighborhood of the origin. As already mentioned before, this implies that the level sets of u are tangentially $C^{1,\alpha}$ surfaces and in particular that $F(u)$ is a $C^{1,\alpha}$ graph in $B_{1/2}$. $\square$

2.9 Structure of the Singular Set

2.9.1 Main Statements and Strategy

As we have already seen, the non regular (or non stable) points of the free boundary can be divided in two classes: the set $\Sigma(u)$ at which $\Lambda(u)$ has a vanishing density (*singular points*), that is

$$\Sigma(u) = \left\{ (x_0,0) \in F(u) : \lim_{r \to 0^+} \frac{\mathcal{H}^n(\Lambda(u) \cap B_r'(x_0))}{r^n} = 0 \right\},$$

and the set of *non regular, non singular points*. In this section, we describe the results from [18] on the structure of the set of *singular* points of the free boundary. As we shall see, around these kind of points a precise analysis of the behavior of u and the structure of the free boundary can be carried out. The analysis of the free boundary around the other kind of points is still an open question.

It is convenient to classify a point on $F(u)$ according to the degree of homogeneity of u, given by the frequency formula centered at that point. In other words, set

$$\Phi^{X_0}(r;u) = r \frac{\int_{B_r(X_0)} |\nabla u|^2}{\int_{\partial B_r(X_0)} u^2 dS}$$

and define

$$F_\kappa(u) = \left\{ X_0 \in F(u) : \Phi^{X_0}(0+;u) = \kappa \right\}$$

$$\Sigma_\kappa(u) = \Sigma(u) \cap F_\kappa(u).$$

According to these notations, X_0 is a regular point if it belongs to $F_{3/2}(u)$. Since $r \longmapsto \Phi^{X_0}(r;u)$ is nondecreasing, it follows that the mapping

$$X_0 \longmapsto \Phi^{X_0}(0+;u)$$

is upper-semicontinuous. Moreover, since $\Phi^{X_0}(0+;u)$ misses all the values in the interval $(3/2,2)$, it follows that $F_{3/2}(u)$ is a relatively open subset of $F(u)$.

Before stating the structure theorems of $\Sigma(u)$, it is necessary to examine the asymptotic profiles obtained at a singular point from the rescalings (16). Note that, in terms of these rescalings, saying that $X_0 = (x_0,y) \in \Sigma(u)$ is equivalent to

$$\lim_{r \to 0^+} \mathcal{H}^n \left(\Lambda \left(v_r \right) \cap B_1' \left(x_0 \right) \right) = 0. \tag{25}$$

As we see immediately, this implies that any blow-up v^* at a singular point is harmonic in B_1. Moreover, it is possible to give a complete characterization of these blow-ups in terms of the value $\kappa = \Phi^{X_0} \left(r; u \right)$. In particular

$$\Sigma_\kappa \left(u \right) = F_\kappa \left(u \right) \qquad \text{for } \kappa = 2m, \ m \in \mathbb{N}.$$

Theorem 2.9.1 (Blow-ups at singular points). *Let* $(0,0) \in F_\kappa \left(u \right)$. *The following statements are equivalent:*

(i) $(0,0) \in \Sigma_\kappa \left(u \right)$.

(ii) *Any blow-up of* u *at the origin is a non zero homogeneous polynomial* p_κ *of degree* κ *satisfying*

$$\Delta p_\kappa = 0, \quad p_\kappa \left(x, 0 \right) \geq 0, \quad p_\kappa \left(x, -y \right) = p_\kappa \left(x, y \right).$$

(iii) $\kappa = 2m$ *for some* $m \in \mathbb{N}$.

Proof. $(i) \Longrightarrow (ii)$. Since u is harmonic in $B_1^\pm$, we have:

$$\Delta v_r = 2 (\partial_y v_r) \mathcal{H}^n_{|\Lambda(v_r)} \quad \text{in } \mathcal{D}' \left(B_1 \right). \tag{26}$$

From Sect. 2.5, v_r is equibounded in $H^1_{loc} \left(B_1 \right)$ and (25) says that

$$\mathcal{H}^n \left(\Lambda \left(v_r \right) \cap B_1' \right) \to 0$$

as $r \to 0$. Thus (26) implies that $\Delta v_r \to 0$ in $\mathcal{D}' \left(B_1 \right)$ and therefore any blow-up v^* must be harmonic in B_1.

From Sect. 2.6 we know that v^* is homogeneous and non trivial and therefore can be extended to a harmonic function in all $\mathbb{R}^{n+1}$. Being homogeneous, v^* has at most a polynomial growth at infinity, hence Liouville Theorem implies that v^* is a non trivial homogeneous harmonic polynomial p_κ of integer degree κ. The properties of u imply that $p_\kappa \left(x, 0 \right) \geq 0$, and $p_\kappa \left(x, -y \right) = p_\kappa \left(x, y \right)$ in $\mathbb{R}^{n+1}$.

$(ii) \Longrightarrow (iii)$. We must show that κ is an even integer. If κ is odd, the nonnegativity of p_κ on $y = 0$, implies that p_κ vanishes on the hyperplane $y=0$. On the other hand, from the even symmetry in y we infer that $\partial_y p_\kappa \left(x, 0 \right) \equiv 0$ in $\mathbb{R}^n$. Since p_κ is harmonic, the Cauchy-Kowaleskaya Theorem implies that $p_\kappa \equiv 0$ in $\mathbb{R}^{n+1}$. Thus $\kappa = 2m$, for some $m \in \mathbb{N}$.

$(ii) \Longrightarrow (i)$. Suppose $(0,0)$ is not a singular point so that there exists a sequence $r_j \to 0$ such that

$$\mathcal{H}^n \left(\Lambda \left(v_r \right) \cap B_1' \right) \geq \delta > 0.$$

We may assume that v_{r_j} converges to a blow-up p^*. We claim that

$$\mathcal{H}^n\left(\Lambda\left(p^*\right)\cap B_1'\right)\geq\delta>0.$$

Indeed, otherwise, there exists an open set $U\subset\mathbb{R}^n$ with $\mathcal{H}^n\left(U\right)<\delta$ such that $\Lambda\left(p^*\right)\cap\overline{B}_1'\subset U$. Then, for j large, we must have $\Lambda\left(v_r\right)\cap\overline{B}_1'\subset U$ which is a contradiction, since $\mathcal{H}^n(\Lambda\left(v_{r_j}\right)\cap\overline{B}_1')\geq\delta>\mathcal{H}^n\left(U\right)$.

This implies that $p^*\left(x,0\right)\equiv 0$ in $\mathbb{R}^n$ and consequently in $\mathbb{R}^{n+1}$, by the Cauchy-Kowaleskaya theorem. Contradiction to (ii).

$(iii)\Longrightarrow(ii)$. From Almgren's formula, any blow-up is a κ-homogeneous solution of the zero thin obstacle problem in $\mathbb{R}^{n+1}$. Then $\Delta v=2v_y\mathcal{H}^n_{|\Lambda(v)}$ in $\mathbb{R}^{n+1}$, with $v_y\leq 0$ on $y=0$. Since $\kappa=2m$, the following auxiliary lemma implies that $\Delta v=0$ in $\mathbb{R}^{n+1}$ and therefore v is a polynomial.   □

Lemma. Let $v\in H_{loc}^1\left(\mathbb{R}^{n+1}\right)$ satisfy $\Delta v\leq 0$ in $\mathbb{R}^{n+1}$ and $\Delta v=0$ in $\mathbb{R}^{n+1}\setminus\{y=0\}$. If v is homogeneous of degree $\kappa=2m$ then $\Delta v=0$ in $\mathbb{R}^{n+1}$.

Proof. By assumption, $\mu=\Delta v$ is a nonpositive measure, supported on $\{y=0\}$. We have to show that $\mu=0$.

Let q be a $2m$-homogeneous harmonic polynomial, which is positive on $\{y=0\}\setminus(0,0)$. For instance:

$$q\left(X\right)=\sum_{j=1}^n\mathrm{Re}(x_j+iy)^{2m}.$$

Take $\psi\in C_0^\infty\left(0,+\infty\right)$ such that $\psi\geq 0$ and let $\Psi\left(X\right)=\psi\left(|X|\right)$. Then, we have:

$$-\langle\mu,\Psi q\rangle=-\langle\Delta v,\Psi q\rangle=\int_{\mathbb{R}^{n+1}}\left(\Psi\nabla v\cdot\nabla q+q\nabla v\cdot\nabla\Psi\right)dX$$

$$=\int_{\mathbb{R}^{n+1}}\left(-\Psi v\Delta q-v\nabla q\cdot\nabla\Psi+q\nabla v\cdot\nabla\Psi\right)dX$$

$$=\int_{\mathbb{R}^{n+1}}[-\Psi v\Delta q-v\frac{\psi'\left(|X|\right)}{|X|}(X\cdot\nabla q)+q\frac{\psi'\left(|X|\right)}{|X|}(X\cdot\nabla v)]dX$$

$$=0$$

since $\Delta q=0$, $X\cdot\nabla q=2mq$, $X\cdot\nabla v=2mv$. This implies that μ is supported at $X=0$ that is $\mu=c\delta_{(0,0)}$. On the other hand, $\delta_{(0,0)}$ is homogeneous of degree $-\left(n+1\right)$ while μ is homogeneous of degree $2m-2$ and therefore $\mu=0$.   □

Definition 2.9.1: *We denote by P_κ the class of homogeneous harmonic polynomials of degree $\kappa=2m$, defined in Theorem 2.9.1, that is:*

$$P_\kappa=\{p_\kappa:\Delta p_\kappa=0,\nabla p_\kappa\cdot X=\kappa p_\kappa,p_\kappa\left(x,0\right)\geq 0,p_\kappa\left(x,-y\right)=p_\kappa\left(x,y\right)\}.\tag{27}$$

Via the Cauchy-Kovaleskaya Theorem, it is easily shown that the polynomials in P_κ can be uniquely determined from their restriction to the hyperplane $y = 0$. Thus, if $p_\kappa \in P_\kappa$ is not trivial, then also its restriction to $y = 0$ must be non trivial.

The next theorem gives an exact asymptotic behavior of u near a point $X_0 \in \Sigma_\kappa(u)$.

Theorem 2.9.2 (κ-differentiability at singular points). *Let* $X_0 \in \Sigma_\kappa(u)$, *with* $\kappa = 2m$, $m \in \mathbb{N}$. *Then there exists a non trivial* $p_\kappa^{X_0} \in P_\kappa$ *such that*

$$u(X) = p_\kappa^{X_0}(X - X_0) + o(|X - X_0|^\kappa). \tag{28}$$

Moreover, the mapping $X_0 \longmapsto p_\kappa^{X_0}$ *is continuous on* $\Sigma_\kappa(u)$.

The proof is given in Sect. 2.9.3. Note that, since P_κ is a convex subset of the space of the homogeneous polynomial of degree κ, all the norms on P_κ are equivalent. Thus, the continuity in Theorem 2.9.2 can be understood, for instance, in the $L^2(\partial B_1)$ norm.

The structure of $F(u)$ around a singular point X_0 depends on the *dimension* of the singular set at that point, as defined below in terms of the polynomial $p_\kappa^{X_0}$:

Definition 2.9.3 (Dimension at a singular point). Let $X_0 \in \Sigma_\kappa(u)$. The *dimension of* $\Sigma_\kappa(u)$ *at* X_0 is defined as the integer

$$d_\kappa^{X_0} = \dim \left\{ \xi \in \mathbb{R}^n : \xi \cdot \nabla_x p_\kappa^{X_0}(x,0) = 0 \text{ for all } x \in \mathbb{R}^n \right\}.$$

Since $p_\kappa^{X_0}(x,0)$ is *not* identically zero on $\mathbb{R}^n$, we have

$$0 \le d_\kappa^{X_0} \le n - 1.$$

For $d = 0, 1, ..., n - 1$ we define

$$\Sigma_\kappa^d(u) = \left\{ X_0 \in \Sigma_\kappa(u) : d_\kappa^{X_0} = d \right\}.$$

Here is the structure theorem:

Theorem 2.9.3 (Structure of the singular set). *Every set* $\Sigma_\kappa^d(u)$, $\kappa = 2m$, $m \in \mathbb{N}$, $d = 0, 1, ..., n-1$, *is contained in a countable union of* d- *dimensional* C^1 *manifolds.*

For the harmonic polynomial

$$p(x_1, x_2, y) = x_1^2 x_2^2 - (x_1^2 + x_2^2)y^2 + \frac{1}{3}y^4$$

considered in Sect. 2.1, it is easy to check that $(0,0) \in \Sigma_4^0(u)$ and the rest of the points on $F(u)$ belongs to $\Sigma_2^1(u)$.

As in the classical obstacle problem, the main difficulty in the analysis consists in establishing the uniqueness of the Taylor expansion (28), which in turn is equivalent to establish the uniqueness of the limiting profile obtained by the sequence of rescalings (16).

A couple of monotonicity formulas, strictly related to Almgren's formula and to formulas in [25] and [22], play a crucial role in circumventing these difficulties.

2.9.2 Monotonicity Formulas

We introduce here two main tools. We start with a one-parameter family of monotonicity formulas (see also [25]) based on the functional:

$$
W_\kappa^{X_0}(r; u) = \frac{1}{r^{n-1+2\kappa}} \int_{B_r(X_0)} |\nabla u|^2 \, dX - \frac{\kappa}{r^{n+2\kappa}} \int_{\partial B_r(X_0)} u^2 dS.
$$

where $\kappa \geq 0$. If $X_0 = (0,0)$ we simply write $W_\kappa(r; u)$.

The functionals $W_\kappa^{X_0}(r; u)$ and $\Phi^{X_0}(r; u)$ are strictly related. Indeed, taking for brevity $X_0 = (0,0)$, in terms of the notations in Sect. 2.5, we have:

$$
W_\kappa(r; u) = \frac{1}{r^{n-1+2\kappa}} V(r) - \frac{\kappa}{r^{n+2\kappa}} H(r) = \frac{H(r)}{r^{n+2\kappa}} [\Phi(r; u) - \kappa]. \tag{29}
$$

This formula shows that $W_\kappa^{X_0}(r; u)$ is particularly suited for the analysis of asymptotic profiles at points $X_0 \in F_\kappa(u)$. Moreover, for these points, since from Almgren's frequency formula we have $\Phi^{X_0}(r; u) \geq \Phi^{X_0}(0+; u) = \kappa$ we deduce that

$$
W_\kappa^{X_0}(r; u) \geq 0. \tag{30}
$$

The next theorem shows the main properties of $W_\kappa(r; u)$.

Theorem 2.9.4. (W-type monotonicity formula). *Let u be a solution of our zero obstacle problem in B_1. Then, for $0 < r < 1$,*

$$
\frac{d}{dr} W_\kappa(r; u) = \frac{1}{r^{n+2\kappa}} \int_{\partial B_r} (X \cdot \nabla u - \kappa u)^2 dS.
$$

As a consequence, the function $r \longmapsto W_\kappa(r; u)$ is nondecreasing in $(0,1)$. Moreover, $W_\kappa(\cdot; u)$ is constant if and only if u is homogeneous of degree κ.

Proof. Using the identities (10), (12) and (13), we get:

$$
\frac{d}{dr} W_\kappa(r; u)
$$

$$
= \frac{1}{r^{n-1+2\kappa}} \left\{ V'(r) - \frac{n-1+2\kappa}{r} V(r) - \frac{\kappa}{r} H'(r) + \frac{\kappa(n+2\kappa)}{r^2} H(r) \right\}
$$

$$
= \frac{1}{r^{n-1+2\kappa}} \left\{ 2 \int_{\partial B_r} u_\nu^2 dS - \frac{4\kappa}{r} \int_{\partial B_r} u u_\nu dS + \frac{2\kappa^2}{r^2} \int_{\partial B_r} u^2 dS \right\}
$$

$$
= \frac{2}{r^{n+1+2\kappa}} \int_{\partial B_r} (X \cdot \nabla u - \kappa u)^2 dS. \qquad \square
$$

The next one is a generalization of a formula in [22], based on the functional

$$M_\kappa^{X_0}(r; u, p_\kappa) = \frac{1}{r^{n+2\kappa}} \int_{\partial B_r(X_0)} (u - p_\kappa)^2 dS.$$

As usual we set $M_\kappa(r; u, p_\kappa) = M_\kappa^{(0,0)}(r; u, p_\kappa)$. Here $\kappa = 2m$ and p_κ is a polynomial in the class P_k defined in (27). Since it measures the distance of u from an homogeneous polynomial of even degree, it is apparent that $M_\kappa^{X_0}(r; u, p_\kappa)$ is particularly suited for the analysis of blow-up profiles at points $X_0 \in \Sigma_\kappa(u)$. We have:

Theorem 2.9.5. (*M*-type monotonicity formula). *Let u be a solution of our zero obstacle problem in B_1. Assume $(0, 0) \in \Sigma_\kappa(u)$, $\kappa = 2m$, $m \in \mathbb{N}$. Then, for $0 < r < 1$, the function $r \longmapsto M_\kappa(r; u, p_\kappa)$ is nondecreasing in $(0, 1)$.*

Proof. We show that

$$\frac{d}{dr} M_\kappa(r; u, p_\kappa) \geq \frac{2}{r} W_\kappa(r; u) \geq 0.$$

Let $w = u - p_\kappa$. We have:

$$\frac{d}{dr} \frac{1}{r^{n+2\kappa}} \int_{\partial B_r} w^2 dS = \frac{d}{dr} \frac{1}{r^{2\kappa}} \int_{\partial B_1} w(rY)^2 dS$$

$$= \frac{1}{r^{2\kappa+1}} \int_{\partial B_1} [w(rY)[\nabla w(rY) \cdot rY - \kappa w(rY)] dS$$

$$= \frac{2}{r^{n+1+2\kappa}} \int_{\partial B_r} w(X \cdot \nabla w - \kappa w) dS$$

On the other hand, since $\Phi(r; p_\kappa) = \kappa$, it follows that

$$W_\kappa(r; p_\kappa) = 0$$

and we can write:

$$W_\kappa(r; u) = W_\kappa(r; u) - W_\kappa(r; p_\kappa)$$

$$= \frac{2}{r^{n-1+2\kappa}} \int_{B_r} (|\nabla w|^2 + 2\nabla w \cdot \nabla p_\kappa) dX$$

$$- \frac{\kappa}{r^{n+2\kappa}} \int_{\partial B_r} (w^2 + 2w p_\kappa) dS$$

$$= \frac{2}{r^{n-1+2\kappa}} \int_{B_r} |\nabla w|^2 dX - \frac{\kappa}{r^{n+2\kappa}} \int_{\partial B_r} w^2 dS$$

$$+ \int_{\partial B_r} w(X \cdot \nabla p_\kappa - \kappa p_\kappa) dS$$

$$= \frac{2}{r^{n-1+2\kappa}} \int_{B_r} |\nabla w|^2 \, dX - \frac{\kappa}{r^{n+2\kappa}} \int_{\partial B_r} w^2 dS$$

$$= -\frac{2}{r^{n-1+2\kappa}} \int_{B_r} w \Delta w \, dX + \frac{1}{r^{n+2\kappa}} \int_{\partial B_r} w(X \cdot \nabla w - \kappa w) dS$$

$$\leq \frac{1}{r^{n+2\kappa}} \int_{\partial B_r} w(X \cdot \nabla w - \kappa w) dS = \frac{r}{2} \frac{d}{dr} M_\kappa (r; u, p_\kappa)$$

since $w \Delta w = (u - p_\kappa)(\Delta u - \Delta p_\kappa) = -p_\kappa \Delta u \geq 0.$ $\qquad \square$

2.9.3   Proofs of Theorems 2.9.2 and 2.9.3

With the above monotonicity formulas at hands we are ready to prove Theorems 2.9.2 and 2.9.3. Recall that, from the frequency formula, we have the estimate (Corollary 2.5.3):

$$|u(X)| \leq c|X|^\kappa \qquad \text{in } B_{2/3} \tag{31}$$

for any solution of our free boundary problem. At a singular point, we have also a control from below.

Lemma 2.9.6. *Let u be a solution of our zero obstacle problem in B_1. Assume $(0,0) \in \Sigma_\kappa(u)$. Then,*

$$\sup_{\partial B_r} |u| \geq cr^\kappa \qquad (0 < r < 2/3). \tag{32}$$

Proof. Suppose that (32) does not hold. Then, for a sequence $r_j \to 0$ we have

$$h_j = \left(\fint_{\partial B_{r_j}} u^2 dS \right)^{1/2} = o\left(r_j^\kappa\right).$$

We may also assume that (see Lemma 2.9.1)

$$v_j(X) = \frac{u(r_j X)}{h_j} \to q_\kappa(X)$$

uniformly on ∂B_1, for some $q_\kappa \in P_\kappa$. Since $\fint_{\partial B_1} q_\kappa^2 dS = 1$, it follows that q_κ is non trivial.

Under our hypotheses, we have

$$M_\kappa(0+; u, q_\kappa) = \lim_{j \to \infty} \frac{1}{r_j^{n+2\kappa}} \int_{\partial B_{r_j}} (u - q_\kappa)^2 dS = \int_{\partial B_1} q_\kappa^2 dS = \frac{1}{r_j^{n+2\kappa}} \int_{\partial B_{r_j}} q_\kappa^2 dS.$$

Hence,

$$\int_{\partial B_{r_j}} (u - q_\kappa)^2 dS \geq \int_{\partial B_{r_j}} q_\kappa^2 dS$$

or

$$\int_{\partial B_{r_j}} (u^2 - 2u q_\kappa) dS \geq 0.$$

Rescaling, we obtain

$$h_j r_j^\kappa \int_{\partial B_{r_j}} \left(\frac{h_j}{r_j^\kappa} v_j^2 - 2v_j q_\kappa\right) dS \geq 0.$$

Dividing by $h_j r_j^\kappa$ and letting $j \to \infty$ we get

$$-\int_{\partial B_{r_j}} q_\kappa^2 dS \geq 0$$

which gives a contradiction, since q_κ is non trivial. □

Given the estimates (31) and (32) around a point $X_0 \in \Sigma_\kappa(u)$, it is natural to introduce the family of *homogeneous rescalings* given by

$$u_r^{(\kappa)}(X) = \frac{u(rX + X_0)}{r^\kappa}.$$

From the estimate (31) we have that, along a sequence $r = r_j$, $u_r^\kappa \to u_0$ in $C_{loc}^{1,\alpha}(\mathbb{R}^n)$. We call u_0 *homogeneous blow-up*. Lemma 2.9.6 assures that u_0 is non trivial. The next results establishes the uniqueness of these asymptotic profiles and proves the first part of Theorem 2.9.2.

Theorem 2.9.7 (Uniqueness of homogeneous blow-up at singular points). *Assume $(0,0) \in \Sigma_\kappa(u)$. Then there exists a unique non trivial $p_\kappa \in P_\kappa$ such that*

$$u_r^{(\kappa)}(X) = \frac{u(rX)}{r^\kappa} \to p_\kappa(X).$$

As a consequence, (28) holds.

Proof. Consider a homogeneous blow-up u_0. For any $r > 0$ we have:

$$W_\kappa(r; u_0) = \lim_{r_j \to 0} W_\kappa(r; u_j^{(\kappa)}) = \lim_{r_j \to 0} W_\kappa(rr_j; u) = \lim_{r_j \to 0} W_\kappa(0+; u).$$

From Theorem 2.9.2 we infer that u_0 is homogeneous of degree κ. The same arguments in the proof of Lemma 2.9.1 give that u_0 must be a polynomial $p_\kappa \in P_\kappa$.

To prove the uniqueness of u_0, apply the M-monotonicity formula to u and u_0. We have:

$$M_\kappa\,(0+;u,u_0) = c_n \lim_{j\to\infty} \int_{\partial B_1} (u_j^{(\kappa)} - u_0)^2 dS = 0.$$

In particular, by monotonicity, we obtain also that

$$-c_n \int_{\partial B_1} (u_r^{(\kappa)} - u_0)^2 dS = M_\kappa\,(r;u,u_0) \to 0$$

as $r \to 0$, and not just over a subsequence r_j. Thus, if u_0' is a homogeneous blow-up, obtained over another sequence $r_j' \to 0$, we deduce that

$$\int_{\partial B_1} (u_0' - u_0)^2 dS = 0.$$

Since u_0 and u_0' are both homogeneous of degree κ, they must coincide in $\mathbb{R}^{n+1}$.   □

The next lemma gives the second part of Theorem 2.9.2.

Lemma 2.9.8 (Continuous dependence of the blow-ups). *For $X_0 \in \Sigma_\kappa\,(u)$ denote by $p_\kappa^{X_0}$ the blow-up of u obtained in Theorem 2.9.7 so that:*

$$u\,(X) = p_\kappa^{X_0}\,(X - X_0) + o\,(|X - X_0|^\kappa).$$

Then, the mapping $X_0 \longmapsto p_\kappa^{X_0}$ from $\Sigma_\kappa\,(u)$ to P_κ is continuous. Moreover, for any compact $K \subset \Sigma_\kappa\,(u) \cap B_1$, there exists a modulus of continuity σ_K, $\sigma_K\,(0+) = 0$, such that

$$\left| u\,(X) - p_\kappa^{X_0}\,(X - X_0) \right| \le \sigma_K\,(|X - X_0|)\,|X - X_0|^\kappa$$

for every $X_0 \in K$.

Proof. As we have already observed, we endow P_κ with the $L^2\,(\partial B_1)$ norm. The first part of the lemma follows as in Theorem 2.9.7. Indeed, fix $\varepsilon > 0$ and r_ε such that

$$M_\kappa^{X_0}(r_\varepsilon;u,p_\kappa^{X_0}) = \frac{1}{r_\varepsilon^{n+2\kappa}} \int_{\partial B_{r_\varepsilon}} (u\,(X + X_0) - p_\kappa^{X_0})^2 dS < \varepsilon.$$

There exists δ_ε such that if $X_0' \in \Sigma_\kappa\,(u)$ and $|X_0 - X_0'| < \delta_\varepsilon$, then

$$M_\kappa^{X_0'}(r_\varepsilon;u,p_\kappa^{X_0}) = \frac{1}{r_\varepsilon^{n+2\kappa}} \int_{\partial B_{r_\varepsilon}} (u\,(X + X_0') - p_\kappa^{X_0})^2 dS < 2\varepsilon.$$

By monotonicity, we deduce that, for $0 < r < r_\varepsilon$,

$$M_\kappa^{X_0'}(r;u,p_\kappa^{X_0}) = \frac{1}{r^{n+2\kappa}} \int_{\partial B_r} (u\,(X + X_0') - p_\kappa^{X_0})^2 dS < 2\varepsilon.$$

Letting $r \to 0$ we obtain

$$M_\kappa^{X_0'}(0+; u, p_\kappa^{X_0}) = c_n \int_{\partial B_1} (p_\kappa^{X_0'} - p_\kappa^{X_0})^2 dS < 2\varepsilon$$

and therefore the first part of the lemma is proved.

To show the second part, note that if $|X_0 - X_0'| < \delta_\varepsilon$ and $0 < r < r_\varepsilon$, we have:

$$\left\| u\left(\cdot + X_0'\right) - p_\kappa^{X_0'} \right\|_{L^2(\partial B_r)} \leq \left\| u\left(\cdot + X_0'\right) - p_\kappa^{X_0} \right\|_{L^2(\partial B_r)} + \left\| p_\kappa^{X_0} - p_\kappa^{X_0'} \right\|_{L^2(\partial B_r)}$$

$$\leq 2\left(2\varepsilon\right)^{1/2} r^{\kappa + (n-1)/2}.$$

This is equivalent to

$$\left\| w_r^{X_0'} - p_\kappa^{X_0'} \right\|_{L^2(\partial B_1)} \leq 2\left(2\varepsilon\right)^{1/2} \tag{33}$$

where

$$w_r^{X_0'}(X) = \frac{u\left(rX + X_0'\right)}{r^\kappa}.$$

Covering K with a finite number of balls $B_{\delta_\varepsilon(X_0')}(X_0')$ for some $X_0' \in K$, we obtain that (33) holds for all $X_0' \in K$ with $r \leq r_\varepsilon^K$.

We claim that, if $X_0' \in K$ and $0 < r < r_\varepsilon^K$ then

$$\left\| w_r^{X_0'} - p_\kappa^{X_0'} \right\|_{L^\infty(B_{1/2})} \leq C_\varepsilon \quad \text{with } C_\varepsilon \to 0 \text{ as } \varepsilon \to 0. \tag{34}$$

To prove the claim, observe that the two functions $w_r^{X_0'}$ and $p_\kappa^{X_0'}$ are both solution of our zero thin obstacle problem, uniformly bounded in $C^{1,\alpha}(\overline{B}_1^\pm)$. If (34) were not true, by compactness, we can construct a sequence of solutions converging to a non trivial zero trace solutions (from (33)). The uniqueness of the solution of the thin obstacle problem with Dirichlet data implies a contradiction.

It is easy to check that the claim implies the second part of the lemma. $\square$

We are now in position to prove Theorem 2.9.3. The proof uses the Whitney's extension theorem (see [25] or [27]) and the implicit function theorem. We recall that the extension theorem prescribes the compatibility conditions under which there exists a C^k function f in $\mathbb{R}^N$ having prescribed derivatives up to the order k on a given closed set.

Since our reference set is $\Sigma_\kappa(u)$, we first need to show that $\Sigma_\kappa(u)$ is a countable union of closed sets (an F_σ set). This is done in the next Lemma.

Lemma 2.9.9 (Topological structure of $\Sigma_\kappa(u)$). $\Sigma_\kappa(u)$ is a F_σ set.

Proof. Let E_j be the set of points $X_0 \in \Sigma_\kappa(u) \cap \overline{B_{1-1/j}}$ such that

$$\frac{1}{j} \rho^\kappa \leq \sup_{|X-X_0|=\rho} |u(X)| < j\rho^\kappa \tag{35}$$

for $0 < \rho < 1 - |X_0|$. By non degeneracy and (31) we know that

$$\Sigma_\kappa(u) \subset \cup_{j \geq 1} E_j.$$

We want to show that E_j is a closed set. Indeed, if $X_0 \in \overline{E}_j$ then X_0 satisfies (35) and we only need to show that $X_0 \in \Sigma_\kappa(u)$ i.e., from Theorem 2.9.1, that $\Phi^{X_0}(0+; u) = \kappa$.

Since the function $X \mapsto \Phi^X(0+; u)$ is upper-semicontinuous we deduce that $\Phi^{X_0}(0+; u) = \kappa' \geq \kappa$. If we had $\kappa' > \kappa$, we would have

$$|u(X)| \leq |X - X_0|^{\kappa'} \quad \text{in } B_{1-|X_0|}(X_0)$$

which contradicts the estimate from below in (35). Thus $\kappa' = \kappa$ and $X_0 \in \Sigma_\kappa(u)$. $\square$

We are now in position for the proof of Theorem 2.9.3.

Proof of Theorem 2.9.3. We divide the proof into two steps. Recall that $\Sigma_\kappa(u) = F_\kappa(u)$ if $\kappa = 2m$.

Step 1. Whitney's extension. For simplicity it is better to make a slight change of notations, letting $y = x_{n+1}$ and $X = (x_1, ..., x_n, x_{n+1})$. Let $K = E_j$ be one of the compact subset of $\Sigma_\kappa(u)$ constructed in Lemma 2.9.9. We can write

$$p_\kappa^{X_0}(X) = \sum_{|\alpha|=\kappa} \frac{a_\alpha(X_0)}{\alpha!} X^\alpha.$$

The coefficients $a_\alpha(X)$ are continuous on $\Sigma_\kappa(u)$ by Theorem 2.9.2. Since $u = 0$ on $\Sigma_\kappa(u)$ we have

$$\left| p_\kappa^{X_0}(X - X_0) \right| \leq \sigma(|X - X_0|) |X - X_0|^\kappa \qquad X \in K.$$

For every multi-index α, $0 \leq |\alpha| \leq \kappa$, define:

$$f_\alpha(X) = \begin{cases} 0 & \text{if } 0 < |\alpha| < \kappa \\ a_\alpha(X) & \text{if } |\alpha| = \kappa \end{cases} \qquad X \in \Sigma_\kappa(u).$$

We want to construct a function $f \in C^\kappa(\mathbb{R}^{n+1})$, whose derivatives $\partial^\alpha f$ up to the order κ are prescribed and equal to f_α on K. The Whitney extension theorem states that this is possible if, for all $X, X_0 \in K$, the following coherence conditions hold for every multi-index α, $0 \leq |\alpha| \leq \kappa$:

$$f_\alpha(X) = \sum_{|\beta| \leq \kappa - |\alpha|} \frac{f_{\alpha+\beta}(X_0)}{\beta!}(X - X_0)^\beta + R_\alpha(X, X_0) \qquad (36)$$

with

$$|R_\alpha(X, X_0)| \leq \sigma_\alpha^K(|X - X_0|)|X - X_0|^{\kappa - |\alpha|} \qquad (37)$$

where σ_α^K is a modulus of continuity.

Claim: Equations (36) and (37) *hold in our case*.

Proof. Case $|\alpha| = \kappa$. Then we have

$$R_\alpha(X, X_0) = a_\alpha(X) - a_\alpha(X_0)$$

and therefore $|R_\alpha(X, X_0)| \leq \sigma_\alpha(|X - X_0|)$ by the continuity on K of the map $X \mapsto p_\kappa^X$.

Case $0 \leq |\alpha| < \kappa$. We have

$$R_0(X, X_0) = -\sum_{\gamma > \alpha, |\gamma| = \kappa} \frac{a_\gamma(X_0)}{(\gamma - \alpha)!}(X - X_0)^{\gamma - \alpha} = -\partial^\alpha p_\kappa^{X_0}(X - X_0).$$

Now, suppose that there exists no modulus of continuity σ_α such that (37) holds for all $X, X_0 \in K$. Then, there is $\delta > 0$ and sequences $X^i, X_0^i \in K$ with

$$|X^i - X_0^i| = \rho_i \to 0$$

and such that

$$|\partial^\alpha p_\kappa^{X_0}(X - X_0)| \geq \delta|X^i - X_0^i|^{\kappa - |\alpha|}. \qquad (38)$$

Consider the rescalings

$$w^i(X) = \frac{u(X_0^i + \rho_i X)}{\rho_i^\kappa} \qquad \xi^i = \frac{X^i - X_0^i}{\rho_i}.$$

We may assume that $X_0^i \to X_0 \in K$ and $\xi^i \to \xi_0 \in \partial B_1$. From Theorem 2.9.8 we have that

$$\left|w^i(X) - p_\kappa^{X_0^i}(X)\right| \leq \sigma(\rho_i|X|)|X|^\kappa$$

and therefore $w^i(X)$ converges to $p_\kappa^{X_0^i}(X)$, uniformly in every compact subset of $\mathbb{R}^{n+1}$.

Note that, since $X^i, X_0^i \in K$, the inequalities (35) are satisfied there. Moreover, we also have that similar inequalities are satisfied for the rescaled function w^i at 0 and ξ^i.

Thus, passing to the limit, we deduce that

$$\frac{1}{j}\rho^\kappa \le \sup_{|X-X_0|=\rho} \left|p_\kappa^{X_0}(X)\right| < j\rho^\kappa \qquad 0 < \rho < +\infty.$$

This implies that ξ_0 is a point of frequency $\kappa = 2m$ for the polynomial $p_\kappa^{X_0}$ so that, from Theorem 2.9.1, we infer that $\xi_0 \in \Sigma_\kappa(p_\kappa^{\xi_0})$. In particular,

$$\partial^\alpha p_\kappa^{\xi_0} = 0 \qquad \text{for } |\alpha| < \kappa.$$

However, dividing both sides of (38) by $\rho_i^{\kappa-|\alpha|}$ and passing to the limit, we obtain

$$\left|\partial^\alpha p_\kappa^{\xi_0}\right| \ge \delta,$$

a contradiction.

This ends the proof of the claim.

Step 2. Implicit function theorem. Applying Whitney's Theorem we deduce the existence of a function $f \in C^\kappa(\mathbb{R}^{n+1})$ such that

$$\partial^\alpha f = f_\alpha \qquad \text{on } E_j$$

for every $|\alpha| \le \kappa$. Suppose now $X_0 \in \Sigma_\kappa^d(u)$. This means that

$$d = \dim\left\{\xi \in \mathbb{R}^n : \xi \cdot \nabla_x p_\kappa^{X_0}(x,0) \equiv 0\right\}.$$

Then there are $n-d$ linearly independent unit vectors $\nu_i \in \mathbb{R}^n$, such that

$$\nu_i \cdot \nabla_x p_\kappa^{X_0}(x,0) \text{ is not identically zero.}$$

This implies that there exist multi-indices β^i of order $|\beta^i| = \kappa - 1$ such that

$$\partial_{\nu_i}(\partial^{\beta^i} p_\kappa^{X_0}(X_0)) \neq 0.$$

This can be written as

$$\partial_{\nu_i}(\partial^{\beta^i} f(X_0)) \neq 0 \qquad i = 1, ..., n-d. \tag{39}$$

On the other hand, we have

$$\Sigma_\kappa^d(u) \cap E_j \subset \cup_{i=1,...,n-d}\left\{\partial^{\beta^i} f = 0\right\}.$$

From (39) and the implicit function theorem, we deduce that $\Sigma_\kappa^d(u) \cap E_j$ is contained in a d-dimensional C^1 manifold in a neighborhood of X_0. Since $\Sigma_\kappa(u) = \cup E_j$ we conclude the proof. $\qquad\square$

Comments and Further Reading

Sections 2.2 and 2.3 follow closely [8]. The results of these subsections have been generalized in [23] to a Signorini problem for convex fully nonlinear uniformly elliptic operators .

The optimal regularity for global solutions in Sect. 2.4 is taken from [1]. The optimal regularity of a local solution follows a different strategy with respect to this paper, and makes use of the Almgren's formula of Sect. 2.5.

The optimal regularity for the Signorini problem for a non flat smooth manifold $\mathcal{M}$, a smooth non zero obstacle and a divergence form uniformly elliptic operators with smooth coefficients has been proved in [19].

Sections 2.5–2.8 follow [3]. Lemma 2.7.2, crucial in the deduction of the monotonicity of tangential derivatives is a generalization of a corresponding lemma in [7]. Starting from a Lipschitz free boundary, the achievement of further regularity by means of Boundary Harnack Principles is by now a classical technique in both elliptic and parabolic free boundary problems (see once more [7] and [9]).

Section 2.9 is taken from part I of [18]. In part II the case of non zero obstacle is considered. Also, some partial results on the structure of the free boundary around non regular, non singular points are conjectured.

A quadratic Taylor expansion to classify singular points of the free boundary is used in [7] for the classical obstacle problem. The quadratic polynomial (which must be nonnegative!) is constructed through a sequence of approximating polynomials Q_j. In order to control the oscillation of the Q_j and assure the uniqueness of the expansion, the deep monotonicity formula in [4] plays a crucial role.

In [25] a different proof, based on a simpler monotonicity formula, based on the functional $W_2(r; u)$, is given. More recently, another monotonicity formula based on the functional $M_2(r; u, p_2)$ has been proved in [22], which is well designed for studying singular points of the classical obstacle problem.

3 Obstacle Problem for the Fractional Laplacian

This section is devoted to the study of the fractional Laplacian obstacle problem that we recall below.

Given a smooth function $\varphi : \mathbb{R}^n \to \mathbb{R}$, with bounded support (or rapidly vanishing at infinity), we look for a continuous function u satisfying the following conditions:

- $u \geq \varphi$ in $\mathbb{R}^n$
- $(-\Delta)^s u \geq 0$ in $\mathbb{R}^n$
- $(-\Delta)^s u = 0$ when $u > \varphi$
- $u(x) \to 0$ as $|x| \to +\infty$.

The analysis of the problem follows the strategy used for the thin obstacle problem in Sect. 2, with the exception of the preliminary properties in Sects. 1–3, taken from [24]. In particular we borrow the $C^{1,\alpha}$ estimates without proof, for brevity. The main steps are the following ones:

1. Construction of the solution and basic properties.
2. Lipschitz continuity, semiconvexity and $C^{1,\alpha}$ estimates.
3. Reduction to the thin obstacle for the operator L_a.
4. Optimal regularity for tangentially convex global solution.
5. Classification of asymptotic blow-up profiles around a free boundary point.
6. Optimal regularity of the solution.
7. Analysis of the free boundary at stable points: Lipschitz continuity.
8. Boundary Harnack Principles and $C^{1,\alpha}$ regularity of the free boundary at stable points.

3.1 Construction of the Solution and Basic Properties

We start by proving the existence of a solution. Observe that the proof fails for $n = 1$ and $s > 1/2$, because in this case it is impossible to have $(-\Delta)^s u \geq 0$ in $\mathbb{R}$ with u vanishing at infinity.

Let S be the set of rapidly decreasing C^∞ functions in $\mathbb{R}^n$. We denote by $\dot{H}^s$ the completion of S in the norm

$$\|f\|_{\dot{H}^s}^2 = \int_{\mathbb{R}^n} \int_{\mathbb{R}^n} \frac{|f(x) - f(y)|^2}{|x - y|^{n+2s}} dx dy \sim \int_{\mathbb{R}^n} |\xi|^{2s} \left| \hat{f}(\xi) \right|^2 d\xi.$$

With the inner product

$$\langle f, g \rangle_{\dot{H}^s} = \int_{\mathbb{R}^n} \int_{\mathbb{R}^n} \frac{(f(x) - f(y))(g(x) - g(y))}{|x - y|^{n+2s}} dx dy$$

$$= 2 \int_{\mathbb{R}^n} f(x)(-\Delta)^s g(x) dx \sim \int_{\mathbb{R}^n} |\xi|^{2s} \hat{f}(\xi) \overline{\hat{g}(\xi)} d\xi,$$

$\dot{H}^s$ is a Hilbert space. Since we are considering $n \geq 2$ and $s < n/2$, it turns out that $\dot{H}^s$ coincides with the set of functions in $L^{2n/(n-2s)}$, for which the $\dot{H}^s$-norm is finite.

The solution u_0 of the obstacle problem is constructed as the unique minimizer of the strictly convex functional

$$J(v) = \|v\|_{\dot{H}^s}^2$$

over the closed, convex set $\mathbb{K}_s = \left\{ v \in \dot{H}^s : v \geq \varphi \right\}$.

In the following proposition we gather some basic properties of u.

Proposition 3.1.1. *Let u_0 be the minimizer of the functional J over $\mathbb{K}_s$. Then:*

(a) *The function u_0 is a supersolution, that is $(-\Delta)^s u_0 \geq 0$ in $\mathbb{R}^n$ in the sense of measures. Thus, is lower semicontinuous and in particular, the set $\{u_0 > \varphi\}$ is open.*

(b) *u_0 is actually continuous in $\mathbb{R}^n$.*

(c) *If $u_0(x) > \varphi(x)$ in some ball B then $(-\Delta)^s u_0 = 0$ in B.*

Proof. (a) Let $h \geq 0$ be any smooth function with compact support. If $t > 0$, $u_0 + th \geq \varphi$ so that

$$\langle u_0, u_0 \rangle_{\dot{H}^s} \leq \langle u_0 + th, u_0 + th \rangle_{\dot{H}^s}$$

or

$$0 \leq 2t \langle u_0, h \rangle_{\dot{H}^s} + t^2 \langle h, h \rangle_{\dot{H}^s} = 2t \langle u_0, (-\Delta)^s h \rangle_{L^2} + t^2 \langle h, h \rangle_{\dot{H}^s}$$

from which

$$\langle u_0, (-\Delta)^s h \rangle_{L^2} = \langle (-\Delta)^s u_0, h \rangle_{L^2} \geq 0.$$

Therefore $(-\Delta)^s u_0$ is a nonnegative measure and therefore is lower semicontinuous by Propositions A2.

(b) The continuity follows Proposition A3.

(c) For any test function $h \geq 0$, supported in B the proof in (a) holds also for $t < 0$. Therefore $(-\Delta)^s u_0 = 0$ in B. $\qquad\square$

Corollary 3.1.2. *The minimizer u_0 of the functional J over $\mathbb{K}_s$ is a solution of the obstacle problem.* $\qquad\square$

3.2   Lipschitz Continuity, Semiconvexity and $C^{1,\alpha}$ Estimates

Following our strategy, we first show that, if φ is smooth enough then the solution of our obstacle problem is Lipschitz and semiconvex. We are mostly interested in the case $\varphi \in C^{1,1}$. When φ has weaker regularity, u_0 inherits corresponding weaker regularity (see [24]). We emphasize that the proof in this section depends only on maximum principle and translation invariance.

Lemma 3.2.1. *The function u_0 is the least supersolution of $(-\Delta)^s$ such that $u_0 > \varphi$ and $\liminf_{|x| \to \infty} u_0(x) \geq 0$. Proof.* Let v such that $(-\Delta)^s v \geq 0$, $v > \varphi$ and $\liminf_{|x| \to \infty} v(x) \geq 0$. Let $w = \min\{u_0, v\}$. Then w is lower-semicontinuous in $\mathbb{R}^n$ (by Proposition A.4) and is another supersolution above φ. We show that $w \geq u_0$.

Since $\varphi \leq w \leq u_0$, we have $w(x) = u_0(x)$ for every x in the contact set $\Lambda(u_0) = \{u_0 = \varphi\}$. In $\Omega = \{u_0 > \varphi\}$, u_0 solves $(-\Delta)^s u_0 = 0$ and w is a supersolution. By Proposition 3.1.1 (b), u_0 is continuous. Then $w - u_0$ is lower-semicontinuous and $w \geq u_0$ from comparison. $\qquad\square$

Corollary 3.2.2. *The function u_0 is bounded and* $\sup u_0 \leq \sup \varphi$. *If the obstacle φ has a modulus of continuity ω, then u_0 has the same modulus of continuity.*

In particular, if φ is Lipschitz, then u_0 is Lipschitz and $\mathrm{Lip}(u_0) \leq \mathrm{Lip}(\varphi)$.

Proof. By hypothesis $u_0 \geq 0$. The constant function $v(x) = \sup \varphi$ is a supersolution that is above φ. By Lemma 3.2.1, $u_0 \leq v$ in $\mathbb{R}^n$.

Moreover, since ω is a modulus of continuity for φ, for any $h \in \mathbb{R}^n$,

$$\varphi(x + h) + \omega(|h|) \geq \varphi(x)$$

for all $x \in \mathbb{R}^n$. Then, the function $u_0(x + h) + \omega(|h|)$ is a supersolution above $\varphi(x)$. Thus $u_0(x + h) + \omega(|h|) \geq u_0(x)$ for all $x, h \in \mathbb{R}^n$. Therefore u_0 has a modulus of continuity not larger than ω. $\qquad\square$

Lemma 3.2.3. *Let $\varphi \in C^{1,1}$ and assume that* $\inf \partial_{\tau\tau} \varphi \geq -C$, *for any unit vector τ. Then $\partial_{\tau\tau} u_0 \geq -C$ too. Thus u_0 is semiconvex.*

Proof. Since $\partial_{\tau\tau} \varphi \geq -C$, we have

$$\frac{\varphi(x + h\tau) + \varphi(x - h\tau)}{2} + Ch^2 \geq \varphi(x)$$

for every $x \in \mathbb{R}^n$ and $h > 0$. Therefore:

$$V(x) \equiv \frac{u_0(x + h\tau) + u_0(x - h\tau)}{2} + Ch^2 \geq \varphi(x)$$

and V is also a supersolution: $(-\Delta)^s V \geq 0$. Thus, by Lemma 3.2.1, $V \geq u_0$ so that:

$$\frac{u_0(x + h\tau) + u_0(x - h\tau)}{2} + Ch^2 \geq u_0(x)$$

for every $x \in \mathbb{R}^n$ and $h > 0$. This implies $\partial_{\tau\tau} u_0 \geq -C$. $\qquad\square$

From the results in [24] we can prove a partial regularity result, under the hypothesis that φ is smooth. The proof is long and non elementary, so we refer to the original paper [24].

Theorem 3.2.4. *Let $\varphi \in C^2$. Then $u_0 \in C^{1,\alpha}$ for every $\alpha < s$ and* $(-\Delta)^s u_0 \in C^\beta$ *for every $\beta < 1 - s$.*

3.3 Thin Obstacle for the Operator L_a: Local $C^{1,\alpha}$ Estimates

To achieve optimal regularity we now switch to the equivalent thin obstacle problem for the operator L_a as mentioned in the introduction and that we restate here:

$$\begin{cases} u\left(x,0\right) \geq \varphi\left(x\right) & \text{in } B_1' \\ u\left(x,-y\right) = u\left(x,y\right) & \text{in } B_1 \\ L_a u = 0 & \text{in } B_1 \backslash \varLambda\left(u\right) \\ L_a u \leq 0 & \text{in } B_1, \text{ in the sense of distributions.} \end{cases} \tag{40}$$

In the global setting (i.e. with B_1 replaced by $\mathbb{R}^n$), $u_0\left(x\right) = u\left(x,0\right)$ is the solution of the global obstacle problem for $\left(-\varDelta\right)^s$ and, from Appendix C,

$$\left(-\varDelta\right)^s u_0 = \kappa_a \lim_{y \to 0+} y^a u_y\left(x,y\right).$$

The estimates in Corollary 3.2.2 and Lemma 3.2.3 translate, after an appropriate localization argument and the use of boundary estimates for the operator L_a, into corresponding estimates for the solution of u. Namely:

Lemma 3.3.1. *Let $\varphi \in C^{2,1}\left(B_1'\right)$ and u be the solution of (40). Then*

1. $\nabla_x u\left(X\right) \in C^\alpha\left(B_{1/2}\right)$ *for every $\alpha < s$;*
2. $|y|^a u_y\left(X\right) \in C^\alpha\left(B_{1/2}\right)$ *for every $\alpha < 1 - s$;*
3. $u_{\tau\tau}\left(X\right) \geq -C$ *in $B_{1/2}$.*

Proof. From Corollary 3.2.2 and Lemma 3.2.3 we have that the above estimates holds on $y = 0$. Since $\partial_{x_j} u$ and $u_{\tau\tau}$ also solve the equation $L_a w = 0$ in $B_1 \backslash \varLambda\left(u\right)$, the estimates 1 and 3 extends to the interior. On the other hand $w\left(x,y\right) = |y|^a u_y\left(X\right)$ solves the conjugate equation $\text{div}(|y|^{-a} \nabla w\left(X\right))$ and we obtain 2. $\qquad\square$

Remark 3.3.1. Observe that u can only be $C^{1,\alpha}$ in both variables up to $y = 0$ only if $a \leq 0$. If $a > 0$, since $y^a u_y\left(X\right)$ has a non-zero limit for some x in the contact set, it follows that u_y cannot be bounded.

We close this section with a compactness result, useful in dealing with blow-up sequences.

Lemma 3.3.2. *Let $\{v_j\}$ be a bounded sequence of functions in $W^{1,2}\left(B_1, |y|^a\right)$. Assume that there exists a constant C such that, in B_1:*

$$|\nabla_x v_j\left(X\right)| \leq C \quad \text{and} \quad |\partial_y v_j\left(X\right)| \leq C |y|^{-a} \tag{41}$$

and that, for each small $\delta > 0$, v_j is uniformly $C^{1,\alpha}$ in $B_{1-\delta} \cap \{|y| > \delta\}$.

Then, there exists a subsequence $\{v_{j_k}\}$ strongly convergent in $W^{1,2}\left(B_{1/2}, |y|^a\right)$.

Proof. From the results in [20], there is a subsequence, that we still call $\{v_j\}$, that converges strongly in $L^2\left(B_{1/2}, |y|^a\right)$. Since for each $\delta > 0$, v_j is uniformly bounded in $C^{1,\alpha}$ in the set $B_{1-\delta} \cap \{|y| > \delta\}$, we can extract a subsequence so that ∇v_j converges uniformly in $B_{1-\delta} \cap \{|y| > \delta\}$. Thus, ∇v_j converges pointwise in $B_1 \backslash \{y = 0\}$.

Now, from (41) and the fact that C and $|y|^{-a}$ both belongs to $L^2\left(B_{1/2}, |y|^a\right)$, the convergence of each partial derivative of v_j in $L^2\left(B_{1/2}, |y|^a\right)$ follows from the dominated convergence theorem. $\qquad\square$

3.4   Minimizers of the Weighted Rayleigh Quotient and a Monotonicity Formula

The next step towards optimal regularity is to consider tangentially convex global solutions. The following lemma is the analogue to Lemma 2.4.1.

Lemma 3.4.1. *Let ∇_θ denote the surface gradient on the unit sphere ∂B_1. Set, for $-1 < a < 1$,*

$$\lambda_{0,a} = \inf \left\{ \frac{\int_{\partial B_1^+} |\nabla_\theta w|^2 \, y^a dS}{\int_{\partial B_1^+} w^2 y^a dS} : w \in W^{1/2}\left(\partial B_1^+, y^a dS\right) : w = 0 \text{ on } (\partial B_1')^+ \right\}$$

where $(\partial B_1')^+ = \{(x', x_n) \in \partial B_1', \ x_n > 0\}$. Then the first eigenfunction, up to a multiplicative factor, is given by

$$w(x, y) = \left(\sqrt{x_n^2 + y^2} - x_n\right)^s \qquad s = (1 - a)/2$$

and[5]

$$\lambda_{0,a} = \frac{1-a}{4}(2n + a - 1).$$

Lemma 3.4.2 *Let w be continuous in B_1, $w(0) = 0$, $w(x, 0) \leq 0$, $w(x, 0) = 0$ on $\Lambda \subset \{y = 0\}$, $L_a w = 0$ in $B_1 \backslash \Lambda,$. Assume that the set*

$$\{x \in B_r' : w(x, 0) < 0\}$$

is non empty and convex. Set

$$\beta(r) = \beta(r; w) = \frac{1}{r^{1-a}} \int_{B_r^+} \frac{y^a |\nabla w(X)|^2}{|X|^{n+a-1}} dX.$$

Then, $\beta(r)$ is bounded and increasing for $r \in (0, 1/2]$.

Proof. We have $L_a w^2 = 2w L_a w + 2y^a |\nabla w|^2 = 2y^a |\nabla w|^2$, so that

$$\beta(r) = \frac{1}{r^{1-a}} \int_{B_r^+} \frac{y^a |\nabla w(X)|^2}{|X|^{n+a-1}} dX = \frac{1}{2r^{1-a}} \int_{B_r^+} \frac{L_a(w^2)}{|X|^{n+a-1}} dX.$$

Now:

$$\beta'(r) = \frac{a-1}{2r^{2-a}} \int_{B_r^+} \frac{L_a(w^2)}{|X|^{n+a-1}} dX + \frac{1}{r^n} \int_{\partial B_r^+}^2 y^a |\nabla w|^2 dS.$$

---

[5] Formally, the first eigenvalue can be obtained plugging $\alpha = s = (1-a)/2$ and $n + a$ instead of n into the formula $\alpha(\alpha - 1) + n\alpha$.

Since $w(0,0) = 0$ and $y^a w_y(x,y) w(x,y) \to 0$ as $y \to 0^+$, we can write

$$\frac{1}{2r^{2-a}} \int_{B_r^+} \frac{L_a(w^2)}{|X|^{n+a-1}} dX$$

$$= \frac{1}{r^{n+1}} \int_{\partial B_r^+} y^a w w_\nu dS - \frac{1}{2r^{2-a}} \int_{B_r^+} y^a \nabla w^2 \cdot \nabla \left(\frac{1}{|X|^{n+a-1}}\right) dX$$

$$= \frac{1}{r^{n+1}} \int_{\partial B_r^+} y^a w w_\nu dS + \frac{(n+a-1)}{2r^{n+2}} \int_{\partial B_r^+} y^a w^2 dS$$

$$\leq \left(\frac{1-a}{2r^{n+2}} \int_{\partial B_r^+} y^a w^2 dS\right)^{1/2} \left(\frac{2}{(1-a)r^n} \int_{\partial B_r^+} y^a w_\nu^2 dS\right)^{1/2}$$

$$+ \frac{(n+a-1)}{2r^{n+2}} \int_{\partial B_r^+} y^a w^2 dS$$

$$\leq \frac{(2n+a-1)}{2r^{n+2}} \int_{\partial B_r^+} y^a w^2 dS + \frac{1}{(1-a)r^n} \int_{\partial B_r^+} y^a w_\nu^2 dS.$$

On the other hand,

$$\int_{\partial B_r^+} y^a |\nabla w|^2 dS = \int_{\partial B_r^+} y^a |\nabla_\theta w|^2 dS + \int_{\partial B_r^+} y^a w_\nu^2 dS$$

so that we obtain:

$$\beta'(r) \geq -(1-a)\frac{(2n+a-1)}{4r^{n+2}} \int_{\partial B_r^+} y^a w^2 dS + \frac{1}{r^n} \int_{\partial B_r^+} y^a |\nabla_\theta w|^2 dS.$$

The convexity of $\{x \in B_r' : w(x,0) < 0\}$ implies that the Rayleigh quotient must be greater than $\lambda_{0,a}$ and therefore we conclude $\beta'(r) \geq 0$ and in particular, $\beta(r) \leq \varphi(1/2)$. $\qquad\square$

3.5 Optimal Regularity for Tangentially Convex Global Solutions

In this section we consider global solutions that represent possible asymptotic profiles, obtained by a suitable blow-up of the solution at a free boundary point.

First of all we consider functions $u : \mathbb{R}^n \to \mathbb{R}$, *homogeneous of degree k*, solutions of the following problem:

$$\begin{cases} u\,(x,0) \geq 0 & \text{in } \mathbb{R}^n \\ u\,(x,-y) = u\,(x,y) & \text{in } \mathbb{R}^n \times \mathbb{R} \\ L_a u = 0 & \text{in } (\mathbb{R}^n \times \mathbb{R}) \setminus \Lambda \\ L_a u \leq 0 & \text{in the sense of distributions in } \mathbb{R}^n \times \mathbb{R} \\ u_{\tau\tau} \geq 0 & \text{in } \mathbb{R}^n \times \mathbb{R}, \text{ for every tangential unit vector } \tau \end{cases} \tag{42}$$

where $\Lambda = \Lambda\,(u) = \{(x,0) : u\,(x,0) = 0\}$. The following proposition gives a lower bound for the degree k, which implies the optimal regularity of the solution.

Lemma 3.5.1. *If there exists a solution u of problem (42), then $k \geq 1 + s = (3-a)/2$.*

Proof. Apply the monotonocity formula in Lemma 3.4.2 to $w = u_\tau$. Then, $L_a w = 0$ in $(\mathbb{R}^n \times \mathbb{R}) \setminus \Lambda$ and, by symmetry, $w\,(x,0)\,w_y\,(x,0) = 0$. Moreover, the contact set where $w = 0$ is convex, since $u_{\tau\tau} \geq 0$. Therefore w satisfies all the hypotheses of that lemma. Recall that we always assume that $(0,0) \in F\,(u)$ so that $w\,(0,0) = 0$. Thus

$$\beta\,(r;w) = \frac{1}{r^{1-a}} \int_{B_r^+} \frac{y^a\,|\nabla w\,(X)|^2}{|X|^{n+a-1}}\,dX \leq \beta\,(1,w)\,.$$

On the other hand, since w is homogeneous of degree $k-1$, we have

$$\beta\,(r;w) = \frac{r^{2k-2}}{r^{1-a}} \int_{B_1^+} \frac{y^a\,|\nabla w\,(X)|^2}{|X|^{n+a-1}}\,dX = \frac{r^{2k-2}}{r^{1-a}}\,\beta\,(1,w)\,.$$

This implies $r^{2k-2} \leq r^{1-a}$ or $k \geq 1 + s$. $\qquad\square$

From Lemma 3.5.1 it would be possible to deduce the optimal regularity of the solution u. However, to study the free boundary regularity we need to classify precisely the solution to problem (42).

Let

$$\Lambda_* = \left\{ (x,0) \in \mathbb{R}^n : \lim_{y \to 0+} y^a u_y\,(x,y) < 0 \right\}.$$

Notice that $\overline{\Lambda_*}$ is the support of $L_a u$ and since $L_a u = 0$ in $(\mathbb{R}^n \times \mathbb{R}) \setminus \Lambda$, we have $\overline{\Lambda_*} \subset \Lambda$ (Λ is closed).

Lemma 3.5.2. *Let u be a solution of problem (42). If $\mathcal{H}^n(\Lambda_*) = 0$ then u is a polynomial of degree k.*

Proof. We know from Lemma 3.3.1 that $|y|^a\,u_y\,(x,y)$ is locally bounded. If $\mathcal{H}^n(\Lambda_*) = 0$, then

$$\lim_{y \to 0} |y|^a\,u_y\,(x,y) = 0 \text{ a.e. } x \in \mathbb{R}^n.$$

Thus $\lim_{y \to 0} |y|^a u_y(x, y) = 0$ weak* in L^∞ and from Lemma B1 we infer that u is a global solution of $L_a u = 0$ in $\mathbb{R}^n \times \mathbb{R}$. Using Lemma B5 we conclude the proof.   □

Lemma 3.5.3. *Let u be a solution of problem (42). If $\mathcal{H}^n(\Lambda) \neq 0$ then, either $u \equiv 0$ or $k = 1 + s$ and Λ is a half n-dimensional space.*

Proof. First observe that if $\mathcal{H}^n(\Lambda_*) = 0$, then $u \equiv 0$, otherwise, from Lemma 3.5.2, $u(x, 0)$ would be a polynomial vanishing on a set of positive measure and therefore identically zero. Thus, the polynomial u must have the form

$$u(x, y) = p_1(x) y^2 + \dots + p_j(x) y^{2j}.$$

We have

$$L_a u = \Delta p_1(x) y^2 + \dots + \Delta p_j(x) y^{2j} + 2p_1(x) + \dots$$
$$+ (2j - 2)p_j(x) y^{2j-2} + ap_1(x) y + \dots + p_j(x) y^{2j-1}$$
$$= 0$$

which implies $p_1 = 0$. Iterating the computation of L_a we get $p_2 = \dots = p_j = 0$.

Consider now the case $\mathcal{H}^n(\Lambda_*) \neq 0$. Then Λ_* is a *thick convex* cone. Assume that e_n is a direction inside Λ_* such that a neighborhood of e_n is contained in Λ_*. Therefore, for any $x \in \mathbb{R}^n$, $x + he_n \in \Lambda_*$ for $h > 0$, large enough.

Let $x/|x|$ be close to e_n so that $x \in \Lambda_*$ and

$$\lim_{y \to 0} \frac{u(x, y) - u(x, 0)}{|y|^{-a} y} = \lim_{y \to 0} |y|^a u_y(x, y) < 0.$$

Then, $u \leq 0$ near the direction e_n. On the other hand, if $y > 0$, then

$$\partial_y (y^a u_y(x, y)) = -\Delta_x u \leq 0,$$

and therefore, for any $X \in \mathbb{R}^{n+1}$ we have that $u(X + he_n) \leq 0$ for h large enough.

Since $u_{\tau\tau} \geq 0$, u is convex in the e_n direction. Thus $w = u_{x_n}$ is increasing along the direction e_n and cannot be positive at any point X, because, otherwise, $\lim_{h \to +\infty} u(X + he_n) = +\infty$, a contradiction.

Moreover, $w = 0$ on Λ and

$$L_a w(X) = 0 \text{ in } (\mathbb{R}^n \times \mathbb{R}) \setminus \Lambda_* \supseteq (\mathbb{R}^n \times \mathbb{R}) \setminus \Lambda.$$

Thus w must coincide with the first eigenfunction of the weighted spherical Laplacian, minimizer of $\int_{S_1} |\nabla_\theta v|^2 |y|^a \, dS$ over all v vanishing on Λ and such that $\int_{S_1} v^2 |y|^a \, dS = 1$.

Since Λ is convex, $\Lambda \cap B_1$ is contained in half of the sphere $B_1 \cap \{y = 0\}$. If it were exactly half of the sphere then it would be given by the first eigenfunction defined in Lemma 3.4.1, up to a multiplicative constant, by he explicit expression

$$w(x,y) = \left(\sqrt{x_n^2 + y^2} - x_n\right)^s \qquad s = (1-a)/2.$$

On the other hand, the above function is not a solution across $\{y = 0, x_n \geq 0\}$. Therefore, if $\Lambda \cap B_1$ is *strictly* contained in half of the sphere $B_1 \cap \{y = 0\}$, there must be another eigenfunction corresponding to a smaller eigenvalue and consequently to a degree of homogeneity *smaller than* s. This would imply $k < 1 + s$, contradicting Lemma 3.5.1. The only possibility is therefore $k = 1 + s$, with $\Lambda = \{y = 0, x_n \geq 0\}$.   □

The next theorem gives the classification of asymptotic profiles.

Theorem 3.5.4. *Let u be a non trivial solution of problem (42). There are only two possibilities:*

(1) $k = 1 + s$, Λ *is a half n-dimensional space and u depends only on two variables. Up to rotations and multiplicative constants u is unique and there is a unit vector τ such that $\Lambda = \{(x,0) : x \cdot \tau \geq 0\}$ and*

$$u_\tau(x,y) = \left(\sqrt{(x \cdot \tau)^2 + y^2} - (x \cdot \tau)\right)^s$$

(2) k *is an integer greater than equal to 2, u is a polynomial and $\mathcal{H}^n(\Lambda) = 0$.*

Proof. If $\mathcal{H}^n(\Lambda) \neq 0$, from Lemma 3.5.3 we deduce that, up to rotations and multiplicative constants, there is a unique solution of problem (42), homogeneous of degree $k = 1 + s$. Moreover, for this solution the free boundary $F(u)$ is flat, that is there is a unit vector (say) e_n such that

$$\Lambda = \{(x,0) : x_n \geq 0\}$$

and

$$u_{x_n}(x,y) = \left(\sqrt{x_n^2 + y^2} - x_n\right)^s.$$

Integrating u_{x_n} from $F(u)$ along segments parallel to e_n we uniquely determine $u(x,0) = u(x_n,0)$. If we had another solution v, homogeneous of degree $1 + s$, with $v(x,0) = u(x,0)$, then necessarily (see the proof of the Liouville-type Lemma B5), for some constant c and $y \neq 0$, we have

$$v(x,y) - u(x,y) = c|y|^s y.$$

But the constant c must be zero, otherwise $v - u$ cannot be solution across $\{y = 0\} \backslash \Lambda$.

As a consequence, if u is a solution homogeneous of degree $1 + s$, with e_n normal to $F(u)$, then $u = u(x_n, y)$. Indeed, translating in any direction orthogonal to x_n and y we get another global solution with the same free boundary. By uniqueness, u must be invariant in those directions.

If $\mathcal{H}^n(\Lambda) = 0$, then $\mathcal{H}^n(\Lambda_*) = 0$ and from Lemma 3.5.2 we conclude that u is a polynomial and $k \geq 2$. $\qquad\square$

3.6 Frequency Formula

As in the zero thin obstacle case, a crucial tool in order to achieve optimal regularity is given by a frequency formula of Almgren type.

If the obstacle were zero, then the frequency formula states that the quantity

$$D_a(r; u) = r \frac{\int_{B_r} |y|^a |\nabla u|^2 \, dX}{\int_{\partial B_r} |y|^a u^2 dS}$$

is bounded and monotonically increasing. The proof mimics the case $a = 0$ and the conclusion is the following.

Theorem 3.6.1. *Let u be a solution of the zero thin obstacle for the operator L_a in B_1. Then $D_a(r; u)$ is monotone nondecreasing in r for $r < 1$. Moreover, $D_a(r; u)$ is constant if and only if u is homogeneous.*

When the obstacle φ is non zero we cannot reduce to that case. Instead, assuming that $\varphi \in C^{2,1}$, we let

$$\tilde{u}(x, y) = u(x, y) - \varphi(x) + \frac{\Delta \varphi(0)}{2(1 + a)} y^2$$

so that $L_a \tilde{u}(0) = 0$. Moreover $\Lambda = \Lambda(u) = \{\tilde{u} = 0\}$. The function $\tilde{u}$ is a solution of the following system:

$$(43) \quad \begin{cases} \tilde{u}(x, 0) \geq 0 & \text{in } B_1' \\ u(x, -y) = u(x, y) & \text{in } B_1 \\ L_a \tilde{u}(x, y) = |y|^a g(x) & \text{in } B_1 \backslash \Lambda \\ L_a(x, y) \leq |y|^a g(x) & \text{in } B_1, \text{ in the sense of distributions} \end{cases}$$

where

$$g(x) = \Delta \varphi(x) - \Delta \varphi(0)$$

is Lipschitz.

Notice that $|y|^a g(x) \to 0$ as $x \to 0$ and

$$|L_a \tilde{u}(x, y)| \leq C |y|^a g(x).$$

What we expect is a small variation of Almgren's formula. Since $u - \tilde{u}$ is a $C^{2,1}$ function, it is enough to prove any regularity result for $\tilde{u}$ instead of u. *In order to simplify the notation we will still write u for $\tilde{u}$.*

Define

$$F(r) = F(r; u) = \int_{\partial B_r} u^2 |y|^a \, d\sigma = r^{n+a} \int_{\partial B_1} (u(rX))^2 |y|^a \, dS.$$

We have:

$$F'(r) = (n+a) r^{n+a-1} \int_{\partial B_1} (u(rX))^2 |y|^a \, dS$$

$$+ r^{n+a} \int_{\partial B_1} 2u(rX) \nabla u(rX) \cdot X |y|^a \, dS$$

$$= (n+a) r^{-1} \int_{\partial B_r} (u(X))^2 |y|^a \, dS + \int_{\partial B_r} 2u(X) u_\nu(X) |y|^a \, dS$$

Thus have that $\log F(r)$ is differentiable for $r > 0$ and:

$$\frac{d}{dr} \log F(r) = \frac{F'(r)}{F(r)} = \frac{n+a}{r} + \frac{\int_{\partial B_r} 2u u_\nu |y|^a \, dS}{\int_{\partial B_r} u^2 |y|^a \, dS}.$$

Note that the monotonicity of $D_a(r; u)$ when $\varphi = 0$ amounts to say that the function

$$r \longmapsto r \frac{d}{dr} \log F(r)$$

is increasing, since in this case

$$\int_{\partial B_r} 2u u_\nu |y|^a \, dS = \int_{B_r} L_a(u^2) = \int_{B_r} (|y|^a |\nabla u|^2 + 2u L_a u) dX$$

$$= \int_{B_r} 2 |y|^a |\nabla u|^2 \, dX.$$

We will use the following modification:

$$\Phi(r) = \Phi(r; u) = (r + c_0 r^2) \frac{d}{dr} \log \max \left[F(r), r^{n+a+4} \right]. \tag{44}$$

Then:

Theorem 3.6.2 (Monotonicity formula). *Let u be a solution of problem (43). Then, there exists a small r_0 and a large c_0, both depending only on $a, n, \|\varphi\|_{C^{2,1}}$, such that $\Phi(r; u)$ is monotone nondecreasing for $r < r_0$.*

We first need a Poincaré type estimate. Recall that $u(0,0) = 0$ since the origin belongs to the free boundary.

To prove the theorem we need two lemmas.

Lemma 3.6.3. *Let u be a solution of problem (43), $u(0,0) = 0$. Then*

$$\int_{\partial B_r} (u(X))^2 |y|^a \, dS \leq Cr \int_{B_r} |\nabla u(X)|^2 |y|^a \, dX + c(a,n) \, r^{6+a+n}$$

and

$$\int_{B_r} (u(X))^2 |y|^a \, dX \leq Cr^2 \int_{B_r} |\nabla u(X)|^2 |y|^a \, dX + c(a,n) \, r^{7+a+n}$$

where c, C depend only on a, n and $\|\varphi\|_{C^{2,1}}$.

Proof. By Lemma B7 we have

$$\int_{\partial B_r} (u(X))^2 |y|^a \, dS \leq Cr \int_{B_r} |\nabla u(X)| |y|^a \, dX + \bar{u} \int_{\partial B_r} u(X) |y|^a \, dS$$

so that we have to prove that

$$\int_{\partial B_r} u(X) |y|^a \, dS \leq c(a,n) \, r^{6+a+n}.$$

From Lemma B6 we have (here $k = 1$):

$$0 = u(0) \geq \frac{1}{\omega_{n+a} r^{n+a}} \int_{\partial B_r} u(X) |y|^a \, dS - Cr^3$$

and therefore

$$\int_{\partial B_r} u^+(X) |y|^a \, dS \leq \int_{\partial B_r} u^-(X) |y|^a \, dS + Cr^{3+a+n}. \tag{45}$$

We estimate u^- by integrating along the straight line from $(x,0)$ to (x,y) and applying Cauchy-Schwarz inequality. Since $u(x,0) \geq 0$, we can write

$$u^-(x,y) = u^-(x,y) - u^-(x,0) \leq \int_0^y |\nabla u(x,t)| \, dt$$

$$\leq \left(\int_0^y |\nabla u(x,t)|^2 \, t^a dt \right)^{1/2} \left(\int_0^y t^{-a} dt \right)^{1/2}$$

$$\leq Cy^{\frac{1-a}{2}} \left(\int_0^y |\nabla u(x,t)|^2 \, t^a dt \right)^{1/2}.$$

Integrating on ∂B_r and using Cauchy-Schwarz inequality again, we get, observing that on ∂B_r we have $dS = \frac{r}{|y|}dx$:

$$\int_{\partial B_r} u^-(X)\,|y|^a\,dS = \int_{\partial B_r} u^-(X)\,|y|^{a-1}\,r\,dx$$

$$\leq Cr \int_{\partial B_r} |y|^{\frac{a-1}{2}} \left(\int_0^y |\nabla u(x,t)|^2\,t^a\,dt \right)^{1/2} dx$$

$$\leq Cr \left(\int_{B_r'} \int_0^{\sqrt{r^2-|x|^2}} |\nabla u(x,t)|^2\,t^a\,dt\,dx \right)^{1/2}$$

$$\times \left(\int_{\partial B_r} |y|^{a-1}\,dx \right)^{1/2}$$

$$\leq Cr^{(n+a+1)/2} \left(\int_{B_r} |\nabla u(X)|^2\,|y|^a\,dX \right)^{1/2}.$$

Combining with (45) we obtain:

$$\int_{\partial B_r} u(X)\,|y|^a\,d\sigma \leq Cr^{(n+a+1)/2} \left(\int_{B_r} |\nabla u(X)|^2\,|y|^a\,dX \right)^{1/2} + Cr^{3+a+n}.$$

Therefore,

$$\int_{\partial B_r} |u(X)|^2\,|y|^a\,dS \leq Cr \int_{B_r} |\nabla u(X)|\,|y|^a\,dX + \frac{1}{\omega_{n+a}r^{n+a}} \left(\int_{\partial B_r} u(X)\,|y|^a\,dS \right)^2$$

$$\leq Cr \int_{B_r} |\nabla u(X)|\,|y|^a\,dX + Cr^{6+a+n}.$$

This gives the first inequality.

The second one can be obtained by integrating in r. $\qquad\square$

Lemma 3.6.4. *The following identity holds for any $r \leq 1$.*

$$r\int_{\partial B_r} \left(|\nabla_\theta u|^2 - u_\nu^2 \right) |y|^a\,dS = \int_{B_r} [(n+a-1)\,|\nabla u(X)|^2 - 2\,\langle X, \nabla u\rangle\,g(X)]\,|y|^a\,dX \tag{46}$$

where $\nabla_\theta u$ denotes the tangential gradient.

Proof. Consider the vector field

$$\mathbf{F} = \frac{1}{2} y^a\,|\nabla u|^2\,X - y^a\,\langle X, \nabla u\rangle\,\nabla u \qquad (y > 0).$$

We have:

$$\mathrm{div}\mathbf{F} = \frac{1}{2}(n+a-1)y^a |\nabla u|^2 - \langle X, \nabla u \rangle L_a u.$$

Since $\langle X, \nabla u \rangle$ is a continuous function on B'_r that vanishes on $\Lambda = \{u = 0\}$ we have that $\langle X, \nabla u \rangle L_a u$ has no singular part and coincides with $\langle X, \nabla u \rangle |y|^a g(x)$.

An application of the divergence theorem gives (46). $\qquad\square$

Proof of Theorem 3.6.2. First we observe that by taking the maximum in (44) it may happen that we get a non differentiable functions. However, $\max[F(r), r^{n+a+4}]$ is absolutely continuous (it belongs to $W^{1,1}_{loc}(0,1)$) and in any case, the jump in the derivative will be in the positive direction.

When $F(r) \le r^{n+a+4}$ we have

$$\Phi(r) = (r + c_0 r^2) \frac{d}{dr} \log r^{n+a+4}$$

and $\Phi'(r) = (n+a+4)c_0 > 0$.

Thus we can concentrate on the case $F(r) > r^{n+a+4}$ where

$$\Phi(r) = (r + c_0 r^2) \frac{d}{dr} \log F(r).$$

We have:

$$\Phi(r) = (r + c_0 r^2) \frac{\int_{\partial B_r} 2u u_\nu |y|^a \, dS}{\int_{\partial B_r} u^2 |y|^a \, dS} + (1 + c_0 r)(n+a)$$

$$\equiv 2\Psi(r) + (1 + c_0 r)(n+a).$$

We show that the first term is increasing, by computing its logarithmic derivative. We find:

$$\frac{d}{dr} \log \Psi(r) = \frac{d}{dr} \left[\log(r + c_0 r^2) + \log \int_{\partial B_r} u u_\nu |y|^a \, dS - \log \int_{\partial B_r} u^2 |y|^a \, dS \right]$$

$$= \frac{1}{r} + \frac{c_0}{1+c_0 r} + \frac{\frac{d}{dr} \int_{\partial B_r} u u_\nu |y|^a \, dS}{\int_{\partial B_r} u u_\nu |y|^a \, dS} - \frac{\int_{\partial B_r} 2u u_\nu |y|^a \, dS}{\int_{\partial B_r} u^2 |y|^a \, dS} - \frac{n+a}{r}.$$

We estimate $\frac{d}{dr} \int_{\partial B_r} u u_\nu |y|^a \, dS$ from below. Since

$$\int_{\partial B_r} u u_\nu |y|^a \, dS = \int_{B_r} (|y|^a |\nabla u|^2 + u L_a u) dX$$

we can write, recalling that $|L_a u| \le c|y|^a |x|$,

$$\frac{d}{dr} \int_{\partial B_r} u u_\nu |y|^a \, dS = \int_{\partial B_r} (|y|^a |\nabla u|^2 + u L_a u) dS$$

$$\geq \int_{\partial B_r} |y|^a |\nabla u|^2 \, dS - c r^{(n+a+2)/2} \left[\int_{\partial B_r} u^2 |y|^a \, dS \right]^{1/2}$$

$$= \int_{\partial B_r} |y|^a |\nabla u|^2 \, dS - c r^{(n+a+2)/2} \left[F(r) \right]^{1/2}.$$

We now use Lemma 3.6.4 to estimate $\int_{\partial B_r} |y|^a |\nabla u|^2 \, d\sigma$ from below.

$$\int_{\partial B_r} |y|^a |\nabla u|^2 \, d\sigma = \int_{\partial B_r} \left(|u_\tau|^2 + u_\nu^2 \right) |y|^a \, d\sigma$$

$$= 2 \int_{\partial B_r} u_\nu^2 |y|^a \, dS + \frac{1}{r} \int_{B_r} [(n+a-1) |\nabla u(X)|^2$$

$$-2 \langle X, \nabla u \rangle g(X)] |y|^a \, dX$$

$$= 2 \int_{\partial B_r} u_\nu^2 |y|^a \, dS + \frac{n+a-1}{r} \int_{\partial B_r} u u_\nu |y|^a \, dS$$

$$-\frac{1}{r} \int_{B_r} [(n+a-1)u - 2 \langle X, \nabla u \rangle] g(X) |y|^a \, dX.$$

Therefore

$$\frac{d}{dr} \int_{\partial B_r} u u_\nu |y|^a \, dS \geq 2 \int_{\partial B_r} u_\nu^2 |y|^a \, dS + \frac{n+a-1}{r} \int_{\partial B_r} u u_\nu |y|^a \, dS$$

$$-c r^{n+a+1} \left[\sqrt{G(r)} + r \sqrt{H(r)} + \sqrt{r F(r)} \right]$$

where

$$G(r) = \int_{B_r} u^2 |y|^a \, dX \quad \text{and} \quad H(r) = \int_{B_r} |\nabla u|^2 |y|^a \, dX.$$

Collecting all the above estimates, we can write:

$$\frac{d}{dr} \log \Psi(r) = P(r) + Q(r)$$

with

$$P(r) = \frac{2 \int_{\partial B_r} u_\nu^2 |y|^a \, dS}{\int_{\partial B_r} u u_\nu |y|^a \, dS} - \frac{\int_{\partial B_r} 2 u u_\nu |y|^a \, dS}{\int_{\partial B_r} u^2 |y|^a \, dS} \geq 0$$

and

$$Q\left(r\right) = \frac{c_0}{1 + c_0 r} - cr^{(n+a+1)/2} \frac{\sqrt{G\left(r\right)} + \sqrt{rF\left(r\right)} + r\sqrt{H\left(r\right)}}{\int_{\partial B_r} uu_\nu \left|y\right|^a d\sigma}$$

$$\geq \frac{c_0}{1 + c_0 r} - cr^{(n+a+1)/2} \frac{\sqrt{G\left(r\right)} + \sqrt{rF\left(r\right)} + r\sqrt{H\left(r\right)}}{H\left(r\right) - r^{(n+a+2)/2}\sqrt{G\left(r\right)}}.$$

First we estimate F, G, H. Since $F\left(r\right) > r^{n+4+a}$, from the Poincaré Lemma 3.6.3 we have:

$$r^{n+4+a} < F\left(r\right) \leq Cr \int_{B_r} \left|\nabla u\left(X\right)\right|^2 \left|y\right|^a dX + c\left(a, n\right) r^{6+a+n}$$

$$= CrH\left(r\right) + c\left(a, n\right) r^{6+a+n}.$$

Integrating the above inequalities in r, we get:

$$G\left(r\right) = \int_0^r F\left(s\right) ds \leq Cr^2 H\left(r\right) + c\left(a, n\right) r^{7+a+n}.$$

This means that, for small enough r_0 and $r < r_0$:

$$F\left(r\right) \leq crH\left(r\right) \quad \text{and} \quad G\left(r\right) \leq Cr^2 H\left(r\right)$$

so that:

$$Q\left(r\right) \geq \frac{c_0}{1 + cr} - cr^{(n+a+1)/2} \frac{r\sqrt{H\left(r\right)}}{H\left(r\right) - r^{(n+a+4)/2}\sqrt{H_r}}.$$

Since $r^{n+4+a} < F\left(r\right) \leq CrH\left(r\right)$, we also have $H\left(r\right) \geq cr^{n+a+3}$ and for r_0 small:

$$Q\left(r\right) \geq \frac{c_0}{1 + c_0 r} - cr^{(n+a+1)/2} \frac{r}{\sqrt{H\left(r\right)} - r^{(n+a+4)/2}}$$

$$\geq \frac{c_0}{1 + c_0 r} - cr^{\frac{n+a+1}{2} + 1 - \frac{n+a+3}{2}} = \frac{c_0}{1 + cr} - c$$

which is positive if c_0 is large and r_0 small. $\qquad\qquad\square$

3.7 Blow-up Sequences and Optimal Regularity

The optimal regularity of the solution can be obtained by a careful analysis of the possible values of $\Phi\left(0+\right)$. When Φ is constant and the obstacle is zero, $\Phi\left(0+\right) - n - a$ represents the degree of homogeneity at the origin.

Thus, by a suitable blow-up of the solution, we will be able to classify the possible asymptotic behaviors at the origin, using the results of Sect. 5. The first result is the following (compare with Theorem 2.6.2).

Theorem 3.7.1. *Let u be a solution of problem (43), $u(0,0) = 0$. Then either $\Phi(0+, u) + a + 2(1 + s)$ or $\Phi(0+, u) \geq n + a + 4$.*

To prove the lemma, we introduce a rescaling similar to (16) in the case of the zero obstacle problem. We set:

$$u_r(X) = \frac{u(rX)}{d_r} \tag{47}$$

where

$$d_r = \left(r^{-(n+a)} \int_{\partial B_r} u^2 |y|^a \, d\sigma \right)^{1/2} = \left(r^{-(n+a)} F(r) \right)^{1/2}.$$

Two things can occur:

$$\liminf_{r \to 0} \frac{d_r}{r^2} \begin{cases} = +\infty & \text{first case} \\ < +\infty & \text{second case.} \end{cases} \tag{48}$$

The next lemma takes care of the first case.

Lemma 3.7.2. *Let u be as in Theorem 3.7.1. Assume that*

$$\liminf_{r \to 0} \frac{d_r}{r^2} = +\infty.$$

Then, there is a sequence $r_k \to 0$ and a function $u_0 : \mathbb{R}^{n+1} \to \mathbb{R}$, non identically zero, such that:

1. $u_{r_k} \to u_0$ *in* $W^{1,2}(B_{1/2}, |y^a|)$
2. $u_{r_k} \to u_0$ *uniformly in* $B_{1/2}$.
3. $\nabla_x u_{r_k} \to \nabla_x u_0$ *uniformly in* $B_{1/2}$.
4. $|y|^a \partial_y u_{r_k} \to |y|^a \partial_y u_0$ *uniformly in* $B_{1/2}$.

Moreover, u_0 is a solution of system (42) and its degree of homogeneity is $[\Phi(0+; u) - n - a]/2$.

Proof. First of all, observe that $\|u_r\|_{L^2(\partial B_1, |y|^a)} = 1$ for every r.

Claim: u_r is bounded in $W^{1,2}(B_1, |y^a|)$.

To prove the claim we use the frequency formula. Since $\liminf_{r \to 0} \frac{d_r}{r^2} = +\infty$, then $F(r) > r^{n+a+4}$ for r small enough. In particular we may assume $r < r_0$ where r_0 is defined in Theorem 3.6.2.

Thus we can write ($\Phi(r) = \Phi(r, u)$):

$$\Phi(r_0) \geq \Phi(r) \geq \left(r + c_0 r^2\right) \frac{d}{dr} \log \max\left[F(r), r^{n+a+4}\right]$$

$$\geq \left(r + c_0 r^2\right) \frac{\int_{B_r} \left(uL_a u + |\nabla u|^2 |y|^a\right) dX}{\int_{\partial B_r} u^2 |y|^a dS} + (n+a)(1+c_0 r)$$

$$\geq 2\left(r + c_0 r^2\right) \frac{\int_{\partial B_r} 2uu_\nu |y|^a dS}{\int_{\partial B_r} u^2 |y|^a dS} + (n+a)(1+c_0 r). \tag{49}$$

Recalling that $|L_a u| \leq C |y|^a |x|$ in $\mathbb{R}^{n+1} \backslash \Lambda(u)$ and that $u = 0$ on $\Lambda(u)$, we get:

$$\left|\int_{B_r} uL_a u dX\right| \leq cr^{(n+a+3)/2} \left(\int_{B_r} u^2 |y|^a dX\right)^{1/2}.$$

From Lemma 3.6.4:

$$\int_{B_r} u^2 |y|^a dX \leq Cr^2 \int_{B_r} |\nabla(X)|^2 |y|^a dX + cr^{7+a+n}$$

so that

$$\left|\int_{B_r} uL_a u dX\right| \leq Cr^{(n+a+5)/2} \left(\int_{B_r} |\nabla u(X)|^2 |y|^a dX\right)^{1/2} + cr^{5+a+n}. \tag{50}$$

From the same lemma:

$$\int_{B_r} |\nabla u(X)|^2 |y|^a dX \geq \frac{1}{r} \int_{\partial B_r} u^2 |y|^a d\sigma - cr^{5+a+n}$$

but, since $d_r/r^2 \to +\infty$, we deduce

$$\int_{\partial B_r} u^2 |y|^a d\sigma \geq C_1 r^{n+a+4}$$

with a constant C_1 as large as we wish as $r \searrow 0$. Therefore, for r small:

$$\int_{B_r} |\nabla u(X)|^2 |y|^a dX \geq C_1 r^{n+a+3} - cr^{5+a+n} \geq \frac{C_1}{2} r^{3+a+n} \tag{51}$$

and

$$\left|\int_{B_r} uL_a u dX\right| \leq Cr \int_{B_r} |\nabla u(X)|^2 |y|^a dX. \tag{52}$$

Comparing (50) with (52) we see that, for r small:

$$\int_{B_r} \left(u L_a u + |\nabla u|^2 |y|^a \right) dX \geq \frac{1}{2} \int_{B_r} |\nabla u (X)|^2 |y|^a dX.$$

Continuing from (49) we get:

$$\Phi(r_0) \geq (r + c_0 r^2) \frac{\int_{B_r} |\nabla u (X)|^2 |y|^a dX}{\int_{\partial B_r} u^2 |y|^a d\sigma} + (n + a)(1 + c_0 r)$$

$$\geq \int_{B_1} |\nabla u_r (X)|^2 |y|^a dX.$$

Combining the above inequality with $\int_{\partial B_r} u_r^2 |y|^a dS = 1$ we infer that u_r is bounded in $W^{1,2}(B_1, |y^a|)$.

Now, u_r^+ and u_r^- are subsolutions of the equation

$$L_a u \geq -cr |y|^a |x|$$

and therefore (see [16]) u_r is bounded in $L^\infty (B_{3/4})$.

We also know that u is semiconvex in the x variable, that is $u_{\tau\tau} \geq -C$ for every tangential direction τ. This implies a similar bound for u_r, namely:

$$\partial_{\tau\tau} u_r = \frac{r^2}{d_r} u_{\tau\tau} (rX) \geq -C \frac{r^2}{d_r}. \tag{53}$$

The functions u_r are solutions of a uniformly elliptic equation with smooth coefficients in $B_1 \cap \{|y| > \delta\}$ for every $\delta > 0$. Then the u_r are uniformly bounded in $C^{1,\alpha}$ in $B_{1-\delta} \cap \{|y| > \delta\}$ for any $\alpha, \delta \leq 1$. Therefore, we can apply Lemma 3.3.2 to obtain a subsequence u_{r_k} strongly convergent in $W^{1,2}(B_{1/2}, |y^a|)$ to some function u_0 as $r_k \searrow 0$.

From Theorem 3.6.2, we know that $\Phi(r, u)$ is monotone and converges to $\Phi(0+; u)$ as $r \searrow 0$. We have:

$$\Phi(rs, u) \sim rs \frac{\int_{B_{rs}} |\nabla u (X)|^2 |y|^a dX}{\int_{\partial B_{rs}} u^2 |y|^a dS} + (n + a)$$

$$= r \frac{\int_{B_r} |\nabla u_s (X)|^2 |y|^a dX}{\int_{\partial B_r} u_s^2 |y|^a dS} + (n + a).$$

We want to set $s = r_k \searrow 0$ and pass to the limit in the above expression to obtain:

$$\Phi(0+, u) - (n + a) = r \frac{\int_{B_r} |\nabla u_0 (X)|^2 |y|^a dX}{\int_{\partial B_r} u_0^2 |y|^a dS}. \tag{54}$$

This is possible since $\int_{\partial B_r} u_s^2 |y|^a \, dS \geq c > 0$. In fact, if $\delta > 0$ is small and $r \geq 1 - \delta$, from the Poincaré inequality (64) we have

$$\int_{\partial B_{(1-\delta)r}} u_s^2 |y|^a \, dS \geq c \int_{\partial B_r} u_s^2 |y|^a \, dS$$

where c depends only on δ, n and $\Phi(1; u)$.

Iterating this inequality k times until $(1 - \delta)^k < r$, we obtain the desired uniform bound from below.

From (53), we have that

$$\partial_{\tau\tau} u_{r_k} \geq -C \frac{r_k^2}{d_{r_k}} \to 0$$

and therefore u_0 is tangentially convex. On the other hand, each u_r satisfies the following conditions:

(a)
$$u_r(x, 0) \geq 0 \quad \text{in } B_1'.$$

(b)
$$L_a u_r = \frac{r^{2-a}}{d_r} L_a u(rX) = \frac{r^2}{d_r} |y|^a g(rx) \quad \text{in } B_1 \backslash \Lambda(u_r)$$

(c)
$$|L_a u_r| \leq \frac{r^2}{d_r} |y|^a |g(rx)| \quad \text{in } B_1.$$

Since

$$\frac{r^2}{d_r} |y|^a |g(rx)| \leq c \frac{r^2}{d_r} |y|^a r |x| \to 0 \quad \text{as } r \to 0$$

it follows that u_0 is a solution of the homogeneous problem

$$\left\{ \begin{array}{ll} u_0(x, 0) \geq 0 & \text{in } B_1' \\ u(x, -y) = u(x, y) & \text{in } B_1 \\ L_a u_0 = 0 & \text{in } B_1 \backslash \Lambda \\ L_a(x, y) \leq 0 & \text{in } B_1, \text{ in the sense of distributions} \end{array} \right\}$$

For this problem, the frequency formula holds as in Theorem 3.7.1 without any error correction. Thus we conclude that u_0 is homogeneous in $B_{1/2}$ and its degree of homogeneity is exactly $(\Phi(0+; u_0) - (n + a))/2$. Since it is homogeneous, then it can be extended to $\mathbb{R}^{n+1}$ as a global solution of the homogeneous problem.

Finally, from the a priori estimates in Lemma 3.3.1, it follows that we can choose r_k so that the sequences u_{r_k}, ∇u_{r_k} and $|y|^a \partial u_{r_k}$ converge uniformly in $B_{1/2}$. $\qquad \square$

Proof of Theorem 3.7.1. In the *first case* of (48), we use Lemma 3.7.2 and Theorem 3.5.4 to find the blow-up profile u_0 and to obtain that the degree of homogeneity of u_0 is $1 + s$ or at least 2. Therefore

$$\Phi(0+; u) = \Phi(0+; u_0) + a + 2(1+s) \quad \text{or} \quad \Phi(0+; u) = \Phi(0+; u_0) \geq n + a + 4.$$

Consider now the *second case* of (48).

If $F(r_k; u) < r_k^{n+a+4}$ for some sequence $r_k \to 0$, then, for these values of r_k,

$$\Phi(r_k; u) = (n + a + 4)(1 + c_0 r_k)$$

so that $\Phi(0+; u_0) + a + 4$.

On the other hand, assume that $F(r; u) \geq r^{n+a+4}$ for r small. Since we are in the second case, for some sequence $r_j \searrow 0$ we have $d_{r_j}/r_j^2 \leq C$ so that

$$r_j^{n+a+4} \leq F(r_j; u) \leq C r_j^{n+a+4}.$$

Taking logs in the last inequality, we get

$$(n + a + 4) \log r_j \leq \log F(r_j; u) \leq C + (n + a + 4) \log r_j.$$

We want to show that $\Phi(0+; u) \geq n + a + 4$ in this case. By contradiction, assume that for small r_j we have $\Phi(r_j; u) \leq n + a + 4 - \varepsilon_0$. Take $r_m < r_j \ll 1$ and write:

$$(n + a + 4)(\log r_j - \log r_m) - C \leq \log F(r_j; u) - \log F(r_m; u)$$

$$= \int_{r_m}^{r_j} \frac{d}{dr} \log F(s; u) \, ds \leq \int_{r_m}^{r_j} (r + c_0 r^2)^{-1} \Phi(s; u) \, ds \leq \int_{r_m}^{r_j} r^{-1} \Phi(r_j; u) \, ds$$

$$\leq (n + a + 4 - \varepsilon_0)(\log r_j - \log r_m)$$

which gives a contradiction if we make $(\log r_j - \log r_m) \to +\infty$. $\qquad\square$

From the classification of the homogeneity of a global profile, we may proceed, using Theorem 3.6.4, to identify u_0, modulus rotations. We have:

Lemma 3.7.3 *Let u be as in Lemma 3.7.1. Assume that $\Phi(0+; u) + a + 2(1+s)$ There is a family of rotations A_r, with respect to x, such that $u_r \circ A_r$ converges to the unique profile u_0 of degree $1 + s$, where u_0 is defined in Theorem 3.6.4. More precisely:*

1. $u_r \circ A_r \to u_0$ in $W^{1,2}(B_{1/2}, |y^a|)$
2. $u_r \circ A_r \to u_0$ uniformly in $B_{1/2}$.
3. $\nabla_x(u_r \circ A_r) \to \nabla_x u_0$ uniformly in $B_{1/2}$.
4. $|y|^a \partial_y(u_r \circ A_r) \to |y|^a \partial_y u_0$ uniformly in $B_{1/2}$.

We now control the decay of u at $(0,0)$ in terms of the decay of $F(r; u)$.

Lemma 3.7.4. *If*

$$F(r; u) \leq cr^{n+a+2(1+\alpha)} \tag{55}$$

for every $r < 1$, *then* $u(0,0) = 0$, $|\nabla u(0,0)| = 0$ *and* u *is* $C^{1,\alpha}$ *at the origin in the sense that*

$$|u(X)| \leq C_1 |X|^{1+\alpha}$$

for $|X| \leq 1/2$ *and* $C_1 = C_1(C, n, a)$.

Proof. Consider u^+ and u^-, the positive and negative parts of u. We have

$$L_a u^+ \geq |y|^a g(x) \geq -C |y|^a |x|$$
$$L_a u^+ \geq -|y|^a g(x) \geq -C |y|^a |x|.$$

For some $r > 0$, let U be the L_a-harmonic replacement of u^+ in B_r. Note that

$$0 = L_a U \leq L_a \left(U + C \frac{|X|^2 - r^2}{2(n+a+1)} \right).$$

Hence, by comparison principle

$$U \geq u^+ - C'r^2.$$

From (55) we have:

$$\int_{\partial B_r} U^2 |y|^a \, d\sigma = \int_{\partial B_r} (u^+)^2 |y|^a \, d\sigma \leq cr^{n+a+2(1+\alpha)}.$$

Since $w(X) = |y|^a$ is a A_2 weight, from the local L^∞ estimates (see [16]) we conclude

$$\sup_{B_{1/2}} |U(X)| \leq C_1 |r|^{1+\alpha}.$$

Then $u^+ \leq U + r^2 \leq Cr^{1+\alpha}$ in $B_{1/2}$. A similar estimate holds for u^- and the proof is complete. □

The frequency formula provides the precise control of $F(r)$ from above.

Lemma 3.7.5. *If* $\Phi(0+; u) = \mu$ *then*

$$F(r; u) \leq cr^\mu$$

for any $r < 1$ *and* $c = c(F(1; u), c_0)$.

Proof. Let $f(r) = \max\{F(r), r^{n+a+4}\} \geq F(r)$. Since Φ is nondecreasing:

$$\mu = \Phi(0+) \leq \Phi(r) = (r + c_0 r^2) \frac{d}{dr} \log f(r)$$

and then

$$\frac{d}{dr} \log f(r) \geq \frac{\mu}{(r + c_0 r^2)}.$$

An integration gives:

$$\log f(1) - \log f(r) \geq \int_r^1 \frac{\mu}{(s + c_0 s^2)} ds$$

$$\geq -\mu \log r - \mu \int_r^1 \left[\frac{1}{s} - \frac{\mu}{(s + c_0 s^2)} \right] ds$$

$$\geq -\mu \log r - C_1 \mu$$

so that

$$\log f(r) \leq \log f(1) + \mu \log r + C_1 \mu.$$

Taking the exponential of the two sides we infer

$$F(r) \leq f(r) \leq C r^\mu$$

with $C = f(1) e^{C_1 \mu}$. □

From Lemmas 3.7.4 and 3.7.5 the optimal regularity of a solution u of (43) follows easily. Precisely:

Theorem 3.7.6. Let u be a solution of (43), with $u(0,0) = 0$. Then

$$|u(X)| \leq C |X|^{1+s} \sup_{B_1} |u|$$

where $C = C \left(n, a, \|g\|_{Lip} \right).$

Proof. From Lemma 3.7.1, $\mu = \Phi(0+) \geq n + a + 2(1 + s)$ and the conclusion comes from Lemmas 3.7.5 and 3.7.4. □

Corollary 3.7.7. Let $\varphi \in C^{2,1}$. Then the solution u of the obstacle problem for the operator $(-\Delta)^s$ belongs to $C^{1,s}(\mathbb{R}^n)$.

Proof. Using the equivalence between the obstacle problem for $(-\Delta)^s$ and the thin obstacle for L_a, Theorem 3.7.6 shows that $u - \varphi$ has the right decay at free boundary points. This is enough to prove that $u \in C^{1,s}(\mathbb{R}^n)$. □

Remark 3.7.1. Observe that it is not true that the solution of the thin obstacle problem for L_a is $C^{1,s}$ in both variables x and y. It is quite interesting however, that the optimal decay takes place in both variables at a free boundary point.

In any case, we have:

Corollary 3.7.7. Let u be a solution of one of the systems (40) or (43). Then

$$u(x, y_0) \in C^{1,s}(B'_{1/2}) \quad \text{for every } y_0 \in (0, 1/2).$$

3.8   Nondegenerate Case: Lipschitz Continuity of the Free Boundary

As we have already seen from the examples in Sect. 2.1, we cannot assure any minimal growth of u at a free boundary point. Moreover, the free boundary could be composed entirely by singular points. Indeed, in analogy with what happen in the zero obstacle problem, the regularity of the free boundary can be inferred for points around which u has an asymptotic profile corresponding to the optimal homogeneity degree $\Phi(0+) + a + 2(1+s)$. Accordingly, we say that $X_0 \in F(u)$ is *regular or stable* if

$$\mu(X_0) = \Phi(0+) + a + 4.$$

As always, we refer to the origin $(X_0 = (0,0))$.

Theorem 3.8.1. *Assume* $\mu = \Phi(0+) + a + 4$. *Then, there exists a neighborhood of the origin* B_ρ *and a tangential cone* $\Gamma'(\theta, e_n) \subset \mathbb{R}^n \times \{0\}$ *such that, for every* $\tau \in \Gamma'(\theta, e_n)$, *we have* $\partial_\tau u \geq 0$ *in* B_ρ.

In particular, the free boundary is the graph $x_n = f(x_1, ..., x_{n-1})$ *of a Lipschitz function* f.

The theorem follows by applying the following lemma (compare with Lemma 2.7.2) to a tangential derivative $h = \partial_\tau u_r$, where u_r is the blow-up family (47) that defines the limiting profile u_0, for r small.

Lemma 3.8.2. *Let* Λ *be a subset of* $\mathbb{R}^n \times \{0\}$. *Assume* h *is a continuous function with the following properties:*

1. $L_a h \leq \gamma |y|^a$ *in* $B_1 \backslash \Lambda$.
2. $h \geq 0$ *for* $|y| \geq \sigma > 0$, $h = 0$ *on* Λ.
3. $h \geq c_0$ *for* $|y| \geq \sqrt{1+a}/8n$.
4. $h \geq -\omega(\sigma)$ *for* $|y| < \sigma$, *where* ω *is the modulus of continuity of* h.

There exist $\sigma_0 = \sigma_0(n, a, c_0, \omega)$ *and* $\gamma_0 = \gamma_0(n, a, c_0, \omega)$ *such that, if* $\sigma < \sigma_0$ *and* $\gamma < \gamma_0$, *then* $h \geq 0$ *in* $B_{1/2}$.

Proof. Suppose $X_0 = (x_0, y_0) \in B_{1/2}$ and $h(X_0) < 0$. Let

$$\mathcal{Q} = \left\{ (x, y) : |x - x_0| < \frac{1}{3}, |y| \geq \frac{\sqrt{1+a}}{4n} \right\}$$

and

$$P(x, y) = |x - x_0|^2 - \frac{n}{a+1} y^2.$$

Observe that $L_a P = 0$. Define

$$v(X) = h(X) + \delta P(X) - \frac{\gamma}{2(a+1)} y^2.$$

Then:

- $v(X_0) = h(X_0) + \delta P(X_0) - \frac{\gamma}{2(a+1)} y_0^2 < 0$
- $v(X) \geq 0$ on Λ
- $L_a v = L_a h + \delta L_a P - \gamma |y|^a \leq 0$ outside Λ.

Thus, v must have a negative minimum on ∂Q. However, on

$$\partial Q \cap \left\{ |y| \geq \frac{\sqrt{1+a}}{8n} \right\}$$

we have $v \geq c_0 - \frac{\delta}{16n} - \frac{\gamma}{32n} > 0$ if δ, γ are small depending on c_0 and n.
    On

$$\left\{ |x - x_0| = \frac{1}{3}, \ \sigma \leq |y| \leq \frac{\sqrt{1+a}}{8n} \right\}$$

we have

$$v \geq \delta \left(\frac{1}{9} - \frac{1}{64n} \right) - \frac{\gamma}{128n^2} > 0$$

if γ is small compared to δ.
    Finally, on

$$\left\{ |x - x_0| = \frac{1}{3}, \ |y| \leq \sigma \right\}$$

we have

$$v \geq -\omega(\sigma) + \left(\frac{1}{9} - \frac{n^2}{1+a} \sigma^2 \right) - \frac{\gamma \sigma^2}{2+a} > 0$$

if σ is small, depending on δ, γ and ω.
    Hence $v \geq 0$ on ∂Q and we have reached a contradiction. Therefore $h \geq 0$ in $B_{1/2}$. $\qquad\square$

Proof of Theorem 3.8.1. Since $\mu = \Phi(0+) < n + a + 4$, Lemma 3.7.1 gives $\mu + a + 2(1 + s)$. Moreover, the blow-up sequence u_r converges (modulus subsequences) to the global profile u_0, whose homogeneity degree is $1 + s$ and whose free boundary is flat.

Let us assume that e_n is the normal to the free boundary of u_0. Then

$$\partial_n u_0(x, y) = c \left(\sqrt{x_n^2 + y^2} - x_n \right)^s.$$

For some $\theta_0 > 0$, let σ any vector orthogonal to y and x_n such that $|\sigma| < \theta_0$. From Theorem 3.6.4 we know that u_0 is constant in the direction of σ and therefore, if $\tau = e_n + \sigma$, $\partial_\tau u_0 = \partial_n u_0$.

On the other hand, $\nabla_x u_r \to \nabla_x u_0$ uniformly in every compact subset of $\mathbb{R}^{n+1}$. Thus, for every δ_0, there is an r for which

$$|\partial_\tau u_0 - \partial_\tau u_r| \leq \delta_0$$

where $\tau = e_n + \sigma$. If we differentiate the equation

$$L_a u_r (X) = \frac{r^2}{d_r} |y|^a g(rX)$$

we get

$$L_a [\partial_\tau u_r (X)] = \frac{r^2}{d_r} |y|^a r \partial_\tau g(rX) \le Cr |y|^a \qquad \text{in } B_1 \backslash \Lambda(u_r) \qquad (56)$$

and the right hand side tends to zero as $r \to 0$.

Thus, for r small enough, $\partial_\tau u_r$ satisfies all the hypotheses of Lemma 3.8.2 and therefore is nonnegative in $B_{1/2}$. This implies that near the origin, the free boundary is a Lipschitz graph. $\qquad \square$

3.9 Boundary Harnack Principles and $C^{1,\alpha}$ Regularity of the Free Boundary

3.9.1 Growth Control for Tangential Derivatives

We continue to examine the regularity of $F(u)$ at stable points. As we have seen, at these point we have an exact asymptotic picture and this fact allows us to get a minimal growth for any tangential derivative when u_r is close to the blow-up limit u_0. This is needed in extending the Carleson estimate and the Boundary Harnack principle in our non-homogeneous setting.

First, we need to refine Lemma 3.8.2.

Lemma 3.9.1. *Let* $\delta_0 = (12n)^{-1/2s}$. *Let* v *be a function satisfying the following properties:*

1. $L_a v(X) \le \varepsilon_0$ *for* $X \in B_1' \times (0, \delta_0)$
2. $v(X) \ge 0$ *for* $X \in B_1' \times (0, \delta_0)$
3. $v(x, \delta_0) \ge \frac{1}{4n}$ *for* $x \in B_1'$.

Then, there exists $\varepsilon_0 = \varepsilon_0(n, a) > 0$ *such that*

$$v(x, y) \ge C |y|^{2s} \qquad \text{in } B_{1/2}' \times [0, \delta_0].$$

Proof. Let $X_0 = (x_0, y_0) \in B_{1/2}' \times [0, \delta_0]$. We compare v with

$$w(x, y) = \left(1 + \frac{\varepsilon_0}{2}\right) y^2 - \frac{|x - x_0|^2}{n} + y^{2s}.$$

Inside $B_1' \times (0, \delta_0)$, we have $L_a w(X) = \varepsilon_0 \ge L_a v$.

For $y = 0$, $w(x,0) \leq 0 \leq v(x,0)$. For $y = \delta_0$,

$$w(x, \delta_0) \leq \left(1 + \frac{\varepsilon_0}{2}\right)\delta_0^2 + \delta_0^{2s} \leq 3\delta_0^{2s} = \frac{1}{4n} \leq v(x, \delta_0).$$

Finally, if $|x - x_0| = 1/2$, we have

$$w(x, y) \leq 3\delta_0^{2s} - \frac{1}{4n} \leq 0 \leq v(x, y).$$

Therefore, by comparison, we deduce that $w \leq v$ in $B_1' \times (0, \delta_0)$. In particular,

$$v(x_0, y_0) \geq w(x_0, y_0) \geq y^{2s}. \qquad \square$$

Corollary 3.9.2. *Let u be a solution of (43) with $|g|, |\nabla g| \leq \varepsilon_0$. Let u_0 the usual asymptotic nondegenerate profile and assume that $|\nabla_x u - \nabla u_0| \leq \varepsilon_0$. Then, if ε_0 is small enough, there exists $c = c(n, a)$ such that*

$$u_\tau(X) \geq c \operatorname{dist}(X, \Lambda)^{2s}$$

for every $X \in B_{1/2}$ and every tangential direction τ such that $|\tau - e_n| < 1/2$.

Proof. From (56) and Theorem 3.8.1, we know that u_τ is positive in $B_{1/2}$. Applying Lemma 3.9.1 we get

$$u_\tau \geq c|y|^{2s} \quad \text{in } B_{1/4}.$$

Let now $X = (x, y) \in B_{1/8}$ and $d = \operatorname{dist}(X, \Lambda)$. Consider the ball $B_{d/2}(X)$. At the top point of this ball, say (x_T, y_T) we have $y_T \geq d/2$. Therefore

$$u_\tau(x_T, y_T) \geq cd^{2s}.$$

By Harnack inequality,

$$u_\tau(x, y) \geq cu_\tau(x_T, y_T) \geq cd^{2s}. \qquad \square$$

3.9.2 Boundary Harnack Principles

Using the growth control from below provided by Lemma 3.9.2 it is easy to extend the Carleson estimate to our nonhomogeneous setting.

Definition 2.8.1 of $\Delta - NTA$ domain can be generalized to $L_a - NTA$ domains replacing Δ-capacity by L_a-capacity. Again a flat n-dimensional manifold with Lipschitz free boundary is a simple example of $L_a - NTA$ domain.

We keep the same notation of Sect. 2.8:

- $D \subset B_1$ is a $L_a - NTA$ domain
- $\partial^* D = \partial D \cap B_1$.
- For $Q \in \partial^* D$, $A_r(Q)$ is a point such that $B_{\eta r}(A_r(Q)) \subset B_r(Q) \cap D$. In particular, $|A_r(Q) - Q| \sim r$.

Lemma 3.9.3 (Carleson estimate). *Let $D \subset B_1$ be a $L_a - NTA$ domain. Suppose w is a positive function in D vanishing on $\partial^* D$. Assume that:*

1. $|L_a w| \leq c |y|^a$
2. *nondegeneracy:*

$$w(X) \geq C d_X^\beta$$

for some $\beta \in (0,2)$, where $d_X = \mathrm{dist}(X, \partial^* D)$.

Then, for every $Q \in \partial^* D \cap B_{1/2}$ and r small:

$$\sup_{B_r(Q) \cap D} w \leq C(n, a, D) w(A_r(Q)).$$

Proof. Let w^* be the harmonic replacement of w in $B_{2r}(Q) \cap D$, r small. From Theorem 2.8.2 adapted to the operator L_a, with identical proof, we get

$$w^*(X) \leq C w^*(A_r(Q)) \qquad \text{in } B_r(Q) \cap D.$$

On the other hand, we claim that

$$|w^* - w| \leq cr^2 \quad \text{in } B_{2r}(Q) \cap D.$$

In fact, compare w with the function $w^*(X) + C\left(|X - Q|^2 - r^2\right)$. On $\partial[B_{2r}(Q) \cap D]$ we have:

$$w^*(X) + C\left(|X - Q|^2 - r^2\right) \leq w(X).$$

Also:

$$L_a\left[w^*(X) + C\left(|X - Q|^2 - r^2\right)\right] = C|y|^a \geq L_a w.$$

Thus $w - w^* \leq Cr^2$. Similarly we obtain the other inequality. The claim follows.

Collecting our estimate we can write

$$w(X) \leq C\left[w(A_r(Q)) + cr^2\right] \quad \text{in } B_r(Q) \cap D.$$

From the nondegeneracy condition $w(A_r(Q)) \geq cr^\beta$ and since $\beta < 2$, the theorem follows. $\qquad\square$

The following theorem expresses a boundary Harnack principle valid in our nonhomogeneous setting. Here $D = B_1 \backslash \Lambda$ where Λ is a piece of the hyperplane $y = 0$. We require that at each point of Λ, the L_a capacity of $B_r(Q) \cap \Lambda$ has a natural minimal growth of order r^{n+a-1}. This assures that D is a $L_a - NTA$ domain.

Theorem 3.9.3 (Boundary Harnack principle). *Let $D = B_1 \backslash \Lambda$ where $\Lambda \subset \{y = 0\}$. Assume that for every $X \in \Lambda$ and $r < 1/2$,*

$$\mathrm{cap}_{L_a}(B_r(X) \cap \Lambda) \geq \kappa r^{n+a-1}.$$

Let v, w positive functions in D satisfying the hypotheses 1 and 2 of Lemma 3.9.3 and symmetric in y.

Then there is a constant $c = c(n, a, \kappa)$ such that

$$\frac{v(X)}{w(X)} \leq c \frac{v\left(0, \frac{1}{2}\right)}{w\left(0, \frac{1}{2}\right)} \quad \text{in } B_{1/2}.$$

Moreover, the ratio v/w is Hölder continuous in $B_{1/2}$, uniformly up to Λ.

Proof. Let us normalize v, w setting $v\left(0, \frac{1}{2}\right) = w\left(0, \frac{1}{2}\right) = 1$. From Carleson estimate and Harnack inequality, for any $\delta > 0$ we get:

$$v(X) \leq C \quad \text{in } B_{3/4}$$

and

$$w(X) \geq c \quad \text{in } B_{3/4} \cap \{|y| > \delta\}.$$

This implies that, for a constant s small enough, $v - sw$ fulfills the conditions of Lemma 3.9.2. Therefore $v - sw \geq 0$ in $B_{1/2}$ or, in other words:

$$\frac{v(X)}{w(X)} \leq s \quad \text{in } B_{1/2}.$$

At this point, the rest of proof of (c) follows again, for instance, as in [9, Sect. 11.2]. $\qquad \square$

3.9.3 $C^{1,\alpha}$ Regularity of the Free Boundary

As in the case of the thin-obstacle from the Laplace operator, the $C^{1,\alpha}$ regularity of the free boundary follows by applying Theorem 3.9.3 to the quotient of two positive tangential derivatives. Precisely we have:

Theorem 3.9.4. *Let u be a solution of (43). Assume $\varphi \in C^{2,1}$ and $\Phi(0) < n+a+4$. Then the free boundary is a $C^{1,\alpha}$ $(n-1)$-dimensional surface around the origin.*

Proof. See the proof of Theorem 2.8.1.

Comments and Further Reading

Sections 3.1 and 3.2 follow [24], to which we refer for the missing proofs. We emphasizes that, in this paper, the optimal regularity of the solution is proved in the case that the contact set is convex.

The rest of the section follows basically the paper [11] except for Sect. 3.4 and part of Sect. 3.5. In particular, the optimal regularity of global solutions follows a different approach, similar to the corresponding proof for the zero obstacle problem.

Lemma 3.8.2 parallels Lemma 2.6.2 and once more is crucial to achieve the monotonicity of tangential derivatives and, consequently, the Lipschitz continuity of the free boundary.

Due to the non homogeneous right hand side in the equation, the boundary Harnack principle in Theorem 3.9.3 is somewhat weaker than the corresponding result in Theorem 2.8.2. Notice the *less than quadratic* decay to zero of the solution at the boundary, necessary to control the effect of the right hand side.

Appendix A: The Fractional Laplacian

Definition and Basic Properties

In this section we briefly recall the definition and some elementary properties of the fractional Laplace operator (see [24]). We denote by $\mathcal{S}$ the Schwartz space of rapidly decreasing C^∞ functions in $\mathbb{R}^n$. Its dual $\mathcal{S}'$ is the space of tempered distributions in $\mathbb{R}^n$.

Definition A1. *For $s \in (-\frac{n}{2}, 1]$ and $f \in S$, we define $(-\Delta)^s f$ through the formula*

$$\widehat{(-\Delta)^s f}(\xi) = |\xi|^{2s} \widehat{f}(\xi).$$

Observe that $(-\Delta)^s f \in C^\infty(\mathbb{R}^n)$ but $(-\Delta)^s f \notin S$ for $s < 1$, because of the singularity at $\xi = 0$ of its Fourier transform. If $s \leq -\frac{n}{2}$, $|\xi|^{2s}$ is not a tempered distribution.

Clearly, $(-\Delta)^0 = Id$ and $(-\Delta)^1 = -\Delta$. We are mainly interested in the case $0 < s < 1$. In this case, we can also compute $(-\Delta)^s f$ using a singular integral:

$$(-\Delta)^s f(x) = c_{n,s} \text{P.V.} \int_{\mathbb{R}^n} \frac{f(x) - f(y)}{|x-y|^{n+2s}} dy.$$

For $s = -\sigma$, $0 < \sigma < \frac{n}{2}$, the operator $(-\Delta)^{-\sigma}$ can be computed as a Riesz Potential:

$$(-\Delta)^{-\sigma} f(x) = c_{n,-\sigma} \int_{\mathbb{R}^n} \frac{f(y)}{|x-y|^{n-2\sigma}} dy.$$

In particular, the last formula indicates that the fundamental solution for $(-\Delta)^s$ is the Riesz kernel

$$\varphi(x) = c_{n,-\sigma} \frac{1}{|x|^{n-2s}}.$$

We can extend by duality the definition of $(-\Delta)^s$ to a large class of distributions.

Definition A2. Let

$$\mathcal{P}_s = \left\{ f \in C^\infty(\mathbb{R}^n) : (1 + |x|^{n+2s}) D^k f \text{ is bounded, } \forall k \geq 0 \right\}$$

In $\mathcal{P}_s$ introduce the topology induced by the seminorms

$$[f]_{k,s} = \sup \left| (1 + |x|^{n+2s}) D^k f \right|.$$

Denote by P'_s the dual of $\mathcal{P}_s$.

If $f \in \mathcal{S}$ then $(-\Delta)^s f \in \mathcal{P}_s$. The symmetry of $(-\Delta)^s$ allows us to define $(-\Delta)^s u$ for $u \in \mathcal{P}'_s$:

$$\langle (-\Delta)^s u, f \rangle = \langle u, (-\Delta)^s f \rangle \qquad \text{for every } f \in \mathcal{S}.$$

Supersolutions and Comparison

The definition of supersolution is defined for $u \in \mathcal{P}'_s$, and the meaning is that it is a nonnegative measure:

Definition A3. Let $u \in \mathcal{P}'_s$. We say that $(-\Delta)^s u \geq 0$ in an open set Ω if for every nonnegative test function $\varphi \in C_0^\infty(\Omega)$ then $\langle (-\Delta)^s u, \varphi \rangle \geq 0$.

Every supersolution shares some properties of superharmonic functions. For instance, u is lower-semicontinuous.

Proposition A1. Let $(-\Delta)^s u \geq 0$ in an open set Ω, then u is lower-semicontinuous in Ω.

Moreover, if the restriction of u on $\text{supp}((-\Delta)^s u)$ is continuous then, u is continuous everywhere. Precisely, we have:

Proposition A2. *Let v be a bounded function in $\mathbb{R}^n$ such that $(-\Delta)^s u \geq 0$. If $E = \text{supp}((-\Delta)^s u)$ and $v_{|E}$ is continuous, then v is continuous in $\mathbb{R}^n$.*

Due to the nonlocal nature of $(-\Delta)^s$, a comparison theorem in a domain Ω must take into account what happens outside Ω. Indeed we have:

Proposition A3. *(Comparison). Let $\Omega \Subset \mathbb{R}^n$ be an open set. Let $(-\Delta)^s u \geq 0$ and $(-\Delta)^s v \leq 0$ in Ω, such that $u \geq v$ in $\mathbb{R}^n \setminus \Omega$ and $u - v$ is lower-semicontinuous in $\overline{\Omega}$. Then $u \geq v$ in $\mathbb{R}^n$. Moreover, if $x \in \Omega$ and $u(x) = v(x)$ then $u = v$ in $\mathbb{R}^n$.*

Also, the set of supersolutions is a directed set, as indicated by the following proposition.

Proposition A4. *Let $\Omega \Subset \mathbb{R}^n$ be an open set. Let $(-\Delta)^s u_1 \geq 0$ and $(-\Delta)^s u_2 \geq 0$ in Ω, such that $u \geq v$ in $\mathbb{R}^n \setminus \Omega$. Then $u = \min\{u_1, u_2\}$ is a supersolution in Ω.*

The proof of the above results are based on a characterization of supersolutions somewhat similar to the super-meanvalue formula for superharmonic functions.

Let $\varphi(x) = c_{n,-\sigma} \frac{1}{|x|^{n-2s}}$ be the fundamental solution of $(-\Delta)^s$. Let $P = P(|x|)$ a paraboloid, tangent from below to φ along ∂B_1. Define $\Phi(x) = P(|x|)$ inside B_1, and $\Phi = \varphi$ outside B_1. Then $\Phi \in C^{1,1}(\mathbb{R}^n)$. For $\lambda > 1$, set

$$\Phi_\lambda(x) = \frac{1}{\lambda^{n-2s}} \Phi\left(\frac{x}{\lambda}\right).$$

Note that $\Phi_{\lambda_1} \geq \Phi_{\lambda_1}$ if $\lambda_1 \geq \lambda_2$. It turns out that $\gamma_\lambda(x) = (-\Delta)^s \Phi_\lambda(x)$ is a continuous and positive function in $L^1(\mathbb{R}^n)$ and that is an approximation of the identity, in the sense that

$$(u * \gamma_\lambda)(x) = \int_{\mathbb{R}^n} u(y) \gamma_\lambda(x - y) \, dy \to u(x) \quad \text{as } \lambda \to 0.$$

In analogy with what happens for the Laplace operator we have:

Proposition A5. *If $(-\Delta)^s u$ is continuous at a point $x \in \mathbb{R}^n$, then*

$$(-\Delta)^s u(x) = \lim_{\lambda \to 0} \frac{C}{\lambda^n} [u(x) - (u * \gamma_\lambda)(x)]$$

where $C = C(s, n)$. In particular, $(-\Delta)^s u \geq 0$ in an open set Ω if and only if u is lower-semicontinuous and

$$(u * \gamma_\lambda)(x) \leq u(x) \quad \text{in } \Omega.$$

Appendix B: The Operator L_a

Definition and Preliminary Facts

For $y > 0$ and $-1 < a < 1$, we define the operator

$$L_a u = \operatorname{div}(y^a \nabla u).$$

The equation $L_a u = 0$ can be written in the form

$$\Delta_x u + \frac{a}{y} u_y + u_{yy} = 0 \tag{57}$$

that induces some analogy with partially symmetric harmonic functions. Indeed, for a integer, let $u \colon \mathbb{R}^n \times \mathbb{R}^{1+a}$ be radially symmetric with respect to the y variable, that is $u(x, y) = U(x, |y|)$. Letting $|y| = r$, the Laplace operator in the variables x and r is given by

$$\Delta u = \Delta_x U + \frac{a}{r} U_r + U_{rr}$$

which is identical to the left hand-side of (57). If we allow non integer values of a then we may think to the operator L_a as a sort of Laplace operator in dimension $n + 1 + a$. Although this is clearly only an analogy, as we shall see later, the solutions of (57) share many properties with harmonic functions.

- *Fundamental solution.* The first issue of this analogy is given by the *fundamental solution* of L_a, given by

$$\Gamma_a(X) = C_{n,a} \frac{1}{|X|^{n+a-1}}$$

if $n + a - 1 > 1$, for an appropriate constant $C_{n,a}$.

  In fact, a direct computation gives that Γ_a is a solution of (57) for $y \neq 0$.

  Note that, for an appropriate constant C_{n+a-1}, $\Gamma_a(x, 0)$ is the fundamental solution of $(-\Delta)^s$, $s = (1 - a)/2$, so that

$$\lim_{y \to 0+} y^a \partial_y \Gamma_a(X) = (-\Delta)^s \Gamma_a(x, 0) = -\delta_0. \tag{58}$$

- *Reflection.* If u is a solution of $L_a u = 0$ in the upper half space and its "conormal derivative" $\lim_{y \to 0+} y^a \partial_y u(x, y)$ vanishes in a ball $B'_R \subset \mathbb{R}^n$ then, reflecting u evenly with respect to y, we obtain a solution in the $n + 1$ dimensional ball B_R.

We first consider the case in which u is smooth.

Lemma B1. Let $u : \mathbb{R}^n \times [0, +\infty)$ be a smooth solution of $L_a u = 0$ in the upper half space, such that

$$\lim_{y \to 0+} y^a \partial_y u (x, y) = 0$$

uniformly in B'_R. Then

$$\tilde{u} (x, y) = \begin{cases} u (x, y) & y \geq 0 \\ u (x, -y) & y < 0 \end{cases}$$

is a weak solution of $\mathrm{div}(|y|^a \nabla u) = 0$ in B_R.

Proof. We show that

$$\int_{B_R} |y|^a \, \nabla \tilde{u} \cdot \nabla h \, dX = 0 \tag{59}$$

for every $h \in C_0^\infty (B_R)$. Let $\varepsilon > 0$ and write:

$$\int_{B_R} |y|^a \, \nabla \tilde{u} \cdot \nabla h \, dX = \int_{B_R \backslash \{|y| < \varepsilon\}} \cdots + \int_{B_R \cap \{|y| < \varepsilon\}} \cdots$$

$$= \int_{B_R \backslash \{|y| < \varepsilon\}} \mathrm{div} \left(|y|^a \, h \nabla \tilde{u} \right) dX$$

$$+ \int_{B_R \cap \{|y| < \varepsilon\}} |y|^a \, \nabla \tilde{u} \cdot \nabla h \, dX$$

$$= - \int_{B_R \cap \{|y| = \varepsilon\}} \varepsilon^a h (x, \varepsilon) \, \tilde{u}_y (x, \varepsilon) \, dx$$

$$+ \int_{B_R \cap \{|y| < \varepsilon\}} |y|^a \, \nabla \tilde{u} \cdot \nabla h \, dX.$$

When $\varepsilon \to 0$, the integral over $B_R \cap \{|y| < \varepsilon\}$ goes to zero, since $|y|^a \, \nabla \tilde{u} \in L_{loc}^1 (\mathbb{R}^{n+1})$. The integral over $B_R \cap \{|y| = \varepsilon\}$ goes to zero since $\varepsilon^a h (x, \varepsilon)$ $\tilde{u}_y (x, \varepsilon) \to 0$. □

If u is not smooth then the meaning of $\lim_{y \to 0+} y^a \partial_y u (x, y) = 0$ is given by (59). Precisely:

Definition B1. We say that $\lim_{y \to 0+} y^a \partial_y u (x, y) = 0$ (resp. ≤ 0) in B_R, in a weak sense, if

$$\int_{B_R} |y|^a \, \nabla \tilde{u} \cdot \nabla h \, dX = 0 \tag{60}$$

for every $h \in C_0^\infty (B_R)$, (resp. $h \geq 0$).

From the proof of Lemma B1, it follows that if $\lim_{y \to 0+} y^a \partial_y u(x, y) = 0$ in a weak sense, then

$$\lim_{y \to 0+} \int_{B'_R} y^a u_y(x, y)\, \varphi(x)\; dx = 0 \qquad \forall \varphi \in C_0^\infty(B'_R). \tag{61}$$

On the other hand, if $y^a \partial_y u(x, y) \in L^1_{loc}(\mathbb{R}^{n+1})$ and (61) holds then

$$\lim_{y \to 0+} y^a \partial_y u(x, y) = 0$$

in B_R, in a weak sense.

- L_a-harmonic extension and conjugate equation. Let $f \in \mathcal{S}$ and $u : \mathbb{R}^n \times [0, +\infty) \to \mathbb{R}$ be the solution of the following problem:

$$\begin{cases} L_a u = 0 & \text{in } \mathbb{R}^n \times (0, +\infty) \\ u(x, 0) = f(x) & \text{in } \mathbb{R}^n \end{cases} \tag{62}$$

We call u the L_a-harmonic extension of f in the upper half space. The function $w(x, y) = y^a u_y(x, y)$ is a solution of the conjugate equation

$$\text{div}\left(y^{-a} \nabla w\right) = 0. \tag{63}$$

Indeed, we have:

$$\text{div}\left(y^{-a} \nabla w\right) = \Delta_x u_y + \partial_y \left(y^{-a} \partial_y (y^a u_y)\right) = \Delta_x u_y + u_{yyy} + a \partial_y (y^{-1} u_y)$$

$$= \partial_y \left[\Delta_x u + u_{yy} + \frac{a}{y} u_y\right] = 0.$$

Harnack Inequality, Liouville Theorem and Mean Value Property

As we have already noted, the operator L_a is a particular case of the class of degenerate elliptic operators considered in [16]. For the following result see Theorems 2.3.8 and 2.3.12 in that paper.

Theorem B2. Assume $L_a v = f(X)$ in B_r. Then:

(1) (Harnack inequality)

$$\sup_{B_{r/2}} v \leq C \left\{ \inf_{B_{r/2}} v + r^2 \sup_{B_r} |f| \right\};$$

(2) $v \in C^{0,\alpha}\left(B_{r/2}\right)$ for some $\alpha \leq 1$ and if $f = 0$

$$\|v\|_{C^{0,\alpha}\left(B_{r/2}\right)} \leq \frac{C}{r^\alpha} \underset{B_r}{\mathrm{osc}}\, v.$$

Using the translational invariance of the equation in the x variable, we obtain the following result.

Lemma B3 (Shauder type estimates). *Assume* $L_a v = 0$ *in* B_r. *Then, for every* $k \geq 1$, *integer:*

$$\left\|D_x^k v\right\|_{L^\infty\left(B_{r/2}\right)} \leq \frac{C}{r^k} \underset{B_r}{\mathrm{osc}}\, v$$

and[6]

$$\left|D_x^k v\right|_{C^{0,\alpha}\left(B_{r/2}\right)} \leq \frac{C}{r^{k+\alpha}} \underset{B_r}{\mathrm{osc}}\, v.$$

Using the above Theorem and the equation in the form

$$\Delta_x v = -v_{yy} - \frac{a}{y} v_y$$

we get:

Lemma B4. *Assume* $L_a v = 0$ *in* B_r. *Then, for every* $r \leq 1$:

$$\left\|v_{yy} + \frac{a}{y} v_y\right\|_{L^\infty\left(B_{r/2}\right)} \leq \frac{C}{r^2} \underset{B_r}{\mathrm{osc}}\, v.$$

The next is a Liouville type result.

Lemma B5. *Let* v *a solution of* $L_a v(X) = 0$ *in* $\mathbb{R}^{n+1}$. *Assume that*

$$v(x,y) = v(x,-y) \qquad \text{and} \qquad |v(X)| \leq C\left(1 + |X|^\gamma\right), \gamma \geq 0.$$

Then v *is a polynomial.*

Proof. We use induction and the following elementary fact: if $\nabla_x v$ is a polynomial and $v(0,y)$ is a polynomial in y, then v is a polynomial.

Assume $\gamma \leq 1$. Taking $r \to \infty$ in Lemma B4 we get

$$v_{yy} + \frac{a}{y} v_y = 0 \quad \text{in } \mathbb{R}^{n+1}.$$

---

[6] $|w|_{C^{0,\alpha}(D)}$ denotes the seminorm

$$\sup_{x,y \in D} \frac{|w(x) - w(y)|}{|x - y|^\alpha}.$$

For x fixed, the solution of this o.d.e. is given by

$$v(x,y) = b(x) y |y|^a + c(x).$$

Since v is symmetric in y we have $b = 0$ and therefore $v(x,y) = c(x)$.

On the other hand, taking $r \to \infty$ in Lemma B3 with $k = 2$, we get $D_x^2 v = 0$ in $\mathbb{R}^{n+1}$. Therefore, for each fixed y, v is a first order polynomial, so that, if $k \leq 1$, $v(x,y) = mx + q$.

Assume now that if $\gamma \leq k - j$, $j = 1, ..., k - 1$, then v is a polynomial of degree $k - j$. Let $k - 1 < \gamma \leq k$. From Lemma B3, $k = 1$ we have

$$|\nabla_x v(X)| \leq C\left(1 + |X|^{\gamma - 1}\right).$$

Moreover, every x-partial derivative is a global solution of the same equation, symmetric in y. Thus, by the inductive hypothesis, we infer that $\nabla_x v$ is a polynomial of degree $k - 1$.

From Lemma B4,

$$\left|v_{yy} + \frac{a}{y}v_y\right| \leq C\left(1 + |X|^{\gamma - 2}\right).$$

Since

$$v_{yy} + \frac{a}{y}v_y = -\Delta_x v$$

and since Δ_x commutes with L_a, we have that $v_{yy} + \frac{a}{y}v_y$ satisfies the same equation as v. Therefore $v_{yy} + \frac{a}{y}v_y$ is a polynomial of degree $k - 2$.

In particular, for $x = 0$,

$$v_{yy}(0,y) + \frac{a}{y}v_y(0,y) = p(y)$$

where p is an even polynomial of degree at most $k - 2$. We can write

$$p(y) = a_0 + a_2 y^2 + ... + a_{2d}y^{2d}.$$

Then, $v(0,y)$ must be

$$v(0,y) = c + by|y|^a + \frac{a_0}{2(1+a)}y^2 + \frac{a_0}{4(3+a)}y^4 + ... + \frac{a_{2d}}{(2d+2)(2d+1+a)}y^{2d+2}.$$

Since v is even in y, we must have $b = 0$ and $v(0,y)$ is a polynomial.

Since $\nabla_x v$ is a polynomial and $v(0,y)$ is a polynomial in y, then v is a polynomial of degree k.   □

Remark B1. Notice that the symmetry hypothesis is necessary. Namely $v(x,y) = y|y|^{-a}$ is a global solution.

We now state a mean value property for supersolution of a nonhomogeneous solution:

Lemma B6 (Mean value property). *Let v be a solution of*

$$L_a v(X) \leq C |y^a| |X|^k \qquad \text{in } B_1.$$

Then, for every $r \leq 1$,

$$v(0) \geq \frac{1}{\omega_{n+a} r^{n+a}} \int_{\partial B_r} v(X) |y|^a \, dS - C r^{k+2}$$

where

$$\omega_{n+a} = \int_{\partial B_1} |y|^a \, dS.$$

Proof. The proof mimic the case of harmonic functions. Consider first $C = 0$.

Define

$$\Gamma(X) = c_{n,a} \max \left\{ \frac{1}{|X|^{n+a+1}} - \frac{1}{r^{n+a+1}}, 0 \right\}$$

where $c_{n,a} = (n+a+1)^{-1} \omega_{n+a}^{-1}$. Note that Γ is supported in B_r, $\Gamma \geq 0$ and a direct calculation yields

$$L_a \Gamma(X) = -\delta_0 + \mu$$

where μ is the measure supported on ∂B_r given by $r^{-n-a} \omega_{n+a}^{-1} |y|^a \, dS$. Thus

$$0 \leq \int_{B_r} -L_a v(X) \Gamma(X) \, dX$$

$$= \int_{B_r} -\nabla v(X) \cdot \nabla \Gamma(X) |y|^a \, dX$$

$$= \int_{B_r} -v(X) \cdot L_a \Gamma(X) \, dX$$

$$= v(0) - \frac{1}{\omega_{n+a} r^{n+a}} \int_{\partial B_r} v(X) |y|^a \, dS$$

which proves the lemma when $C = 0$.

In the case $C > 0$, we apply the above computation to

$$w(X) = v(X) - \frac{C |X|^{k+2}}{(k+2)(k+n+a+1)}$$

since $L_a w(X) \leq 0$. $\qquad \square$

Poincaré Inequalities

Poincaré type inequalities in the context of weighted Sobolev spaces can also be found in [16]. In our case, letting

$$\bar{v} = \frac{1}{\omega_{n+a} r^{n+a}} \int_{\partial B_r} v\left(X\right) |y|^a \, d\sigma,$$

we have[7]:

Lemma B7 (Poincaré inequalities). *Let* $v \in W^{1,2}\left(B_1, |y|^a\right)$. *Then, for* $r \leq 1$:

$$\int_{\partial B_r} \left(v\left(X\right) - \bar{v}\right)^2 |y|^a \, d\sigma \leq C\left(a, n\right) r \int_{B_r} |\nabla v\left(X\right)| \, |y|^a \, dX.$$

and

$$\int_{\partial B_1} \left(v\left(X\right) - v\left(rX\right)\right)^2 |y|^a \, d\sigma \leq C\left(a, n, r\right) \int_{B_1} |\nabla v\left(X\right)| \, |y|^a \, dX. \tag{64}$$

The first inequality is standard, The second one can be proved by integrating ∇v along the lines sX with $s \in (r, 1)$.

Appendix C: Relation Between $(-\Delta)^s$ and L_a

Let $f : \mathbb{R}^n \to \mathbb{R}$ and $u : \mathbb{R}^n \times [0, +\infty) \to \mathbb{R}$ be the L_a-harmonic extension of f in the upper half space that is the solution of the following problem:

$$\begin{cases} L_a u = 0 & \text{in } \mathbb{R}^n \times (0, +\infty) \\ u\left(x, 0\right) = f\left(x\right) & \text{in } \mathbb{R}^n \end{cases}$$

vanishing at infinity.

Lemma C1. *Let* $f \in \mathcal{S}$ *and* u *be its* L_a-harmonic extension and $2s = 1 - a$. *Then, for a suitable constant* κ_a,

$$\kappa_a \lim_{y \to 0+} y^a u_y\left(x, y\right) = (-\Delta)^s f\left(x\right) = \int_{\mathbb{R}^n} \frac{f\left(y\right) - f\left(x\right)}{|y - x|^{n+2s}} dy$$

Proof. We use the Fourier transform in x. Equation (57) transforms into

$$-|\xi|^2 \hat{u}\left(\xi, y\right) + \frac{a}{y} \hat{u}_y\left(\xi, y\right) + \hat{u}_{yy}\left(\xi, y\right) = 0$$

---

[7]A function $v \in W^{1,2}\left(B_1, |y|^a\right)$ has a trace in $L^2\left(\partial B_1, |y|^a\right)$ and the trace operator is compact.

with

$$\hat{u}\left(\xi,0\right) = \hat{f}\left(\xi\right).$$

We claim that

$$\hat{u}\left(\xi,y\right) = \hat{f}\left(\xi\right)\phi\left(|\xi|\,y\right)$$

where $\phi : [0,+\infty) \to \mathbb{R}$ is the minimizer of the functional

$$J\left(v\right) = \int_0^{+\infty} y^a \left[(v')^2 + v^2\right] dy$$

over the class of functions $v \in H^1\left(0,+\infty\right)$, with $v\left(0\right) = 1$. Then ϕ is the solution of the following problem

$$\begin{cases} \phi''\left(y\right) + \frac{a}{y}\phi'\left(y\right) - \phi\left(y\right) = 0 & y > 0 \\ \phi\left(0\right) = 1 \\ \phi\left(+\infty\right) = 0 \end{cases}$$

The scaling $y \longmapsto |\xi|\,y$ shows the claim.
   Then $(X = (x,y))$

$$\int_{\mathbb{R}^n \times [0,+\infty)} y^a \left|\nabla u\right|^2 dX = \int_{\mathbb{R}^n} \int_0^{+\infty} y^a \left[|\xi|^2 |\hat{u}|^2 + |\hat{u}_y|^2\right] dy d\xi$$

$$= \int_{\mathbb{R}^n} \int_0^{+\infty} y^a \, |\xi|^2 \left|\hat{f}\left(\xi\right)\right|^2 \left[|\phi\left(|\xi|\,y\right)|^2 + |\phi'\left(|\xi|\,y\right)|^2\right] dy d\xi$$

$$(|\xi|\,y = z) = \int_{\mathbb{R}^n} \left|\hat{f}\left(\xi\right)\right|^2 |\xi|^{1-a} \int_0^{+\infty} z^a \left[|\phi\left(z\right)|^2 + |\phi'\left(z\right)|^2\right] dz d\xi$$

$$= J\left(\phi\right) \int_{\mathbb{R}^n} \left|\hat{f}\left(\xi\right)\right|^2 |\xi|^{2s} d\xi \equiv \kappa_a \int_{\mathbb{R}^n} \left|\hat{f}\left(\xi\right)\right|^2 |\xi|^{2s} d\xi.$$

We conclude that the Euler equations of the two functionals

$$u \to \int_{\mathbb{R}^n \times [0,+\infty)} y^a \left|\nabla u\right|^2 dX \quad \text{and} \quad f \to \kappa_a \int_{\mathbb{R}^n} \left|\hat{f}\left(\xi\right)\right|^2 |\xi|^{2s} d\xi$$

coincide, that is, for every test function $h : \mathbb{R}^n \times [0,+\infty) \to \mathbb{R}$ with compact support:

$$\int_{\mathbb{R}^n} \int_0^{+\infty} y^a \nabla u \cdot \nabla h \; dx dy = \kappa_a \int_{\mathbb{R}^n} |\xi|^{2s} \hat{f}\left(\xi\right) \overline{\hat{h}\left(\xi,0\right)} d\xi$$

$$= \kappa_a \int_{\mathbb{R}^n} h\left(x,0\right) \left(-\Delta\right)^s u\left(x\right) dx.$$

Since $\operatorname{div}(y^a \nabla u) = 0$,

$$\int_{\mathbb{R}^n} \int_0^{+\infty} y^a \nabla u \cdot \nabla h \, dx dy = -\kappa_a \int_{\mathbb{R}^n} h(x,0) \lim_{y \to 0+} y^a u_y(x,y) \, dx$$

from which $-\kappa_a \lim_{y \to 0+} y^a u_y(x,y) = (-\Delta)^s f(x)$. $\square$

- *Poisson formula.* To find an explicit formula for u, notice that, form the previous paragraphs, the function

$$P(x,y) = -C_{n,a} y^{-a} \partial_y \Gamma_{-a}(x,y)$$

solves $\operatorname{div}(y^a \nabla P) = 0$ for $y > 0$ and, for an appropriate constant $C_{n,a}$, $P(x,y) \to \delta_0$ as $y \to 0+$. Thus

$$P(x,y) = C_{n,a} \frac{y^{1-a}}{\left(|x|^2 + y^2\right)^{(n-a+1)/2}}$$

must be the Poisson kernel for L_a. Therefore, the solution of (62) is given by

$$u(x,y) = \int_{\mathbb{R}^n} P(x-z,y) f(z) \, dz.$$

Remark C1. An alternative way to see that $-\kappa_a \lim_{y \to 0+} y^a u_y(x,y)$ is a realization of $(-\Delta)^s$ is to make the change of variable $z = \left(\frac{y}{1-a}\right)^{1-a}$ representation which reduces $L_a u = 0$ to

$$\Delta_x \tilde{u} + z^\alpha \tilde{u}_{zz} = 0$$

and $\lim_{y \to 0+} y^a u_y(x,y)$ to $\tilde{u}_z(x,0)$ for $\alpha = -2a/(1-a)$.

The Poisson kernel becomes

$$\tilde{P}(x,z) = C_{n,a} \frac{z}{\left(|x|^2 + (1-a)^2 z^{2/(2-a)}\right)^{(n-a+1)/2}}$$

and

$$\tilde{u}(x,z) = \int_{\mathbb{R}^n} \tilde{P}(x-z,y) f(z) \, dz.$$

Using directly this representation, we have

$$\tilde{u}_z(x,0) = \lim_{z \to 0+} \frac{1}{z} \int_{\mathbb{R}^n} \tilde{P}(x-z,y) \left[f(z) - f(x)\right] dz$$

$$= c_{n,a} \text{ p.v.} \int_{\mathbb{R}^n} \frac{f(y) - f(x)}{|y-x|^{n+2s}} dy = -(-\Delta)^s f(x).$$

Comments and Further Reading

Standard references for the properties of the fractional Laplacian are [21] and [47]. Appendix A is taken from [24], to which we refer for the proof of Propositions A1–A5.

Appendix B1 follows [10], while Appendices B2 and B3 are taken from [11].

Appendix C is taken from [10]. In these paper a connection between the operator $L_a u$ and the operator $z^\alpha \Delta_x u + u_{zz}$, where $\alpha = 2a/(1-a)$, is also shown. Operators of this type are known as Baouendi–Grushin operators. For $\alpha > 0$, frequency formulas for these operators appeared before in [17] in the context of unique continuation.

References

1. I. Athanasopoulos, L.A. Caffarelli, Optimal regularity of lower dimensional obstacle problems. Zap. Nauchn. Semin. POMI **310**, 49–66, 226 (2004)
2. I. Athanasopoulos, L.A. Caffarelli, A theorem of real analysis and its application to free boundary problems, C.P.A.M, **38**, 499-502 (1985)
3. I. Athanasopoulos, L.A. Caffarelli, S. Salsa, The structure of the free boundary for lower dimensional obstacle problems. Am. J. Math. **130** (2), 485–498 (2008)
4. W. Alt, L.A. Caffarelli, A. Friedman, Variational problems with two phases and their free boundaries, Trans. A.M.S. **282**(2), 431–461 (1984)
5. J.P. Bouchaud, A. Georges, Anomalous diffusion in disordered media: Statistical mechanics models and physical interpretations. Phys. Rep. **195**(4&5), 1990
6. H. Brézis, D. Kinderlehrer, The smoothness of solutions to nonlinear variational inequalities, Indiana Univ. Math. Journal, **23**(9), 831–844 (1974)
7. L.A. Caffarelli, The obstacle problem revisited, J. Fourier Analysis Appl. **4**, 383–402 (1998)
8. L.A. Caffarelli, Further regularity for the Signorini problem, Comm. P.D.E. **4**(9), 1067–1075 (1979)
9. L.A. Caffarelli, S. Salsa, A geometric approach to free boundary problems. A.M.S. Providence, G.S.M., vol. 68
10. L.A. Caffarelli, L. Silvestre, An extension problem related to the fractional Laplacian. Comm. P.D.E. **32**(8), 1245–1260 (2007)
11. L.A. Caffarelli, S. Salsa, L. Silvestre, Regularity estimates for the solution and the free boundary of the obstacle problem for the fractional Laplacian, Inv. Math. **171**, 425–461 (2008)
12. L.A. Caffarelli, A. Vasseur, Drift diffusion equations with fractional diffusion and the quasi-geostrophic equation, Ann. Math. **171**(3), 1903–1930 (2010)
13. R. Cont, P. Tankov, Financial Modelling with Jump Processes, Chapman & Hall/CRC Press, Financ. Math. ser., Boca raton, FL 2004.
14. G. Duvaut, J.L. Lions, Les inequations en mechanique et en physique, Paris, Dunod, 1972
15. J. Frehse, On Signorini's problem and variational problems with thin obstacles, Ann. Sc. Norm. Sup. Pisa **4**, 343–362 (1977)
16. E. Fabes, C. Kenig, R. Serapioni, The local regularity of solutions of degenerate elliptic equations, Comm. P.D.E., **7**(1), 77–116 (1982)

17. N. Garofalo, Unique continuation for a class of elliptic operators which degenerate on a manifold of arbitrary codimension. J. Diff. Eq. **104**(1), 117–146 (1993)

18. N. Garofalo, A. Petrosian, Some new monotonicity formulas and the singular set in the lower dimensional obstacle problem, Inv. Math. **177**, 415–464 (2009)

19. N. Guillen, Optimal regularity for the Signorini problem, Calculus of Variations and Partial Differential Equations, **36**(4), 533–546 (2009)

20. T. Kilpeläinen, Smooth approximation in weighted Sobolev spaces. Commentat. Math. Univ. Carol., **38**(1), 29–35 (1997)

21. N.S. Landkof, Foundation of Modern Potential Theory, Die Grundlehren der Mathematischen Wissenschaften, Band 180. *Springer-Verlag, New York-Heidelberg*, 1972

22. R. Monneau, On the number of singularities for the obstacle problem in two dimensions, J. Geom. Anal. **13**(2), 359–389 (2003)

23. E. Milakis, L. Silvestre, Regularity for the non linear Signorini problem, Adv. Math. **217**, 1301–1312 (2008)

24. L. Silvestre, Regularity of the obstacle problem for a fractional power of the Laplace operator, C.P.A.M. **60**(1), 67–112 (2007)

25. G. Weiss A homogeneity improvement approach to the obstacle problem, Inv. Math **138**(1), 23–50 (1999)

26. H. Whitney, Analytic extensions of differentiable functions defined in closed sets, Trans. A.M.S. **36**(1), 63–89 (1934)

27. W.P. Ziemer, Weakly differentiable functions, Graduate Text in Math, Springer-Verlag New York, 1989

List of Participants

1. Argiolas Roberto
   roberarg@unica.it
2. Avelin Benny
   benny.avelin@math.umu.se
3. Bianchini Chiara
   chiara.bianchini@math.unifi.it
4. Bianchini Massimiliano
   massimiliano.bianchini@math.unifi.it
5. Bonfanti Giovanni
   g.bonfanti3@campus.unimib.it
6. Bonforte Matteo
   matteo.bonforte@uam.es
7. Candito Pasquale
   pasquale.candito@unirc.it
8. Cellina Arrigo
   arrigo.cellina@unimib.it
9. Charro Caballero Fernando
   fernando.charro@uam.es
10. Chiricotto Maria
    chiricotto@dmmm.uniroma1.it
11. D'Ambrosio Roberta
    rodambrosio@unisa.it
12. D' Ambrosio Lorenzo
    dambros@dm.uniba.it
13. DiBenedetto Emmanuele (**lecturer**)
    em.diben@vanderbilt.edu
14. Di Castro Agnese
    dicastro@mat.uniroma1.it
15. Esposito Pierpaolo
    esposito@mat.uniroma3.it

J. Lewis et al., *Regularity Estimates for Nonlinear Elliptic and Parabolic Problems*, Lecture Notes in Mathematics 2045, DOI 10.1007/978-3-642-27145-8,
© Springer-Verlag Berlin Heidelberg 2012

16. Ferrari Fausto
    ferrari@dm.unibo.it
17. Fimiani Michele
    michelefimiani@gmail.com
18. Floridia Giuseppe
    floridia@dmi.unict.it
19. Fornaro Simona
    simona.fornaro@unipv.it
20. Fugazzola Andrea
    andrea.fugazzola@unipv.it
21. Gianazza Ugo **(editor)**
    gianazza@imati.cnr.it
22. Grillo Gabriele
    gabriele.grillo@polito.it
23. Iagar Razvan Gabriel
    razvan.iagar@uam.es
24. Kokocki Piotr
    koksi@mat.uni.torun.pl
25. Laleoglu Rojbin Ozlem
    rojbin@mat.uc.pt
26. Leonori Tommaso
    leonori@ugr.es
27. Lewis John **(editor)**
    john@ms.uky.edu
28. Lindqvist Peter **(lecturer)**
    lqvist@math.ntnu.no
29. Livrea Roberto
    roberto.livrea@unirc.it
30. Lundstrom Niklas
    niklas.lundstrom@math.umu.se
31. Manfredi Juan **(lecturer)**
    manfredi@pitt.edu
32. Mannucci Paola
    mannucci@math.unipd.it
33. Marcellini Paolo
    marcellini@math.unifi.it
34. Mascolo Elvira
    mascolo@math.unifi.it
35. Masson Mathias
    mathias.masson@tkk.fi
36. Mazzola Marco
    m.mazzola7@campus.unimib.it
37. Pilarczyk Dominika
    dpilarcz@math.uni.wroc.pl

38. Piro Grimaldi Anna
    grimaldi@unica.it
39. Salsa Sandro (**lecturer**)
    sandro.salsa@polimi.it
40. Siljander Juhana
    juhana.siljander@tkk.fi
41. Sirakov Boyan
    sirakov@ehess.fr
42. Stinga Pablo Raùl
    pablo.stinga@uam.es
43. Surnachev Mikhail
    mamsu@swansea.ac.uk
44. Terracini Susanna
    susanna.terracini@unimib.it
45. Urbano Jos Miguel
    jmurb@mat.uc.pt
46. Vazquez Juan Luis
    juanluis.vazquez@uam.es
47. Verzini Gianmaria
    gianmaria.verzini@polimi.it
48. Vespri Vincenzo
    vespri@math.unifi.it

LECTURE NOTES IN MATHEMATICS

 **Springer**

Edited by J.-M. Morel, B. Teissier; P.K. Maini

Editorial Policy (for Multi-Author Publications: Summer Schools / Intensive Courses)

1. Lecture Notes aim to report new developments in all areas of mathematics and their applications - quickly, informally and at a high level. Mathematical texts analysing new developments in modelling and numerical simulation are welcome. Manuscripts should be reasonably selfcontained and rounded off. Thus they may, and often will, present not only results of the author but also related work by other people. They should provide sufficient motivation, examples and applications. There should also be an introduction making the text comprehensible to a wider audience. This clearly distinguishes Lecture Notes from journal articles or technical reports which normally are very concise. Articles intended for a journal but too long to be accepted by most journals, usually do not have this "lecture notes" character.

2. In general SUMMER SCHOOLS and other similar INTENSIVE COURSES are held to present mathematical topics that are close to the frontiers of recent research to an audience at the beginning or intermediate graduate level, who may want to continue with this area of work, for a thesis or later. This makes demands on the didactic aspects of the presentation. Because the subjects of such schools are advanced, there often exists no textbook, and so ideally, the publication resulting from such a school could be a first approximation to such a textbook. Usually several authors are involved in the writing, so it is not always simple to obtain a unified approach to the presentation.

   For prospective publication in LNM, the resulting manuscript should not be just a collection of course notes, each of which has been developed by an individual author with little or no coordination with the others, and with little or no common concept. The subject matter should dictate the structure of the book, and the authorship of each part or chapter should take secondary importance. Of course the choice of authors is crucial to the quality of the material at the school and in the book, and the intention here is not to belittle their impact, but simply to say that the book should be planned to be written by these authors jointly, and not just assembled as a result of what these authors happen to submit.

   This represents considerable preparatory work (as it is imperative to ensure that the authors know these criteria before they invest work on a manuscript), and also considerable editing work afterwards, to get the book into final shape. Still it is the form that holds the most promise of a successful book that will be used by its intended audience, rather than yet another volume of proceedings for the library shelf.

3. Manuscripts should be submitted either online at www.editorialmanager.com/lnm/ to Springer's mathematics editorial, or to one of the series editors. Volume editors are expected to arrange for the refereeing, to the usual scientific standards, of the individual contributions. If the resulting reports can be forwarded to us (series editors or Springer) this is very helpful. If no reports are forwarded or if other questions remain unclear in respect of homogeneity etc, the series editors may wish to consult external referees for an overall evaluation of the volume. A final decision to publish can be made only on the basis of the complete manuscript; however a preliminary decision can be based on a pre-final or incomplete manuscript. The strict minimum amount of material that will be considered should include a detailed outline describing the planned contents of each chapter.

   Volume editors and authors should be aware that incomplete or insufficiently close to final manuscripts almost always result in longer evaluation times. They should also be aware that parallel submission of their manuscript to another publisher while under consideration for LNM will in general lead to immediate rejection.

4. Manuscripts should in general be submitted in English. Final manuscripts should contain at least 100 pages of mathematical text and should always include

   – a general table of contents;
   – an informative introduction, with adequate motivation and perhaps some historical remarks: it should be accessible to a reader not intimately familiar with the topic treated;
   – a global subject index: as a rule this is genuinely helpful for the reader.

   Lecture Notes volumes are, as a rule, printed digitally from the authors' files. We strongly recommend that all contributions in a volume be written in the same LaTeX version, preferably LaTeX2e. To ensure best results, authors are asked to use the LaTeX2e style files available from Springer's web-server at
   ftp://ftp.springer.de/pub/tex/latex/svmonot1/ (for monographs) and
   ftp://ftp.springer.de/pub/tex/latex/svmultt1/ (for summer schools/tutorials).
   Additional technical instructions, if necessary, are available on request from:
   lnm@springer.com.

5. Careful preparation of the manuscripts will help keep production time short besides ensuring satisfactory appearance of the finished book in print and online. After acceptance of the manuscript authors will be asked to prepare the final LaTeX source files and also the corresponding dvi-, pdf- or zipped ps-file. The LaTeX source files are essential for producing the full-text online version of the book. For the existing online volumes of LNM see:
   http://www.springerlink.com/openurl.asp?genre=journal&issn=0075-8434.
   The actual production of a Lecture Notes volume takes approximately 12 weeks.

6. Volume editors receive a total of 50 free copies of their volume to be shared with the authors, but no royalties. They and the authors are entitled to a discount of 33.3 % on the price of Springer books purchased for their personal use, if ordering directly from Springer.

7. Commitment to publish is made by letter of intent rather than by signing a formal contract. Springer-Verlag secures the copyright for each volume. Authors are free to reuse material contained in their LNM volumes in later publications: a brief written (or e-mail) request for formal permission is sufficient.

Addresses:
Professor J.-M. Morel, CMLA,
École Normale Supérieure de Cachan,
61 Avenue du Président Wilson, 94235 Cachan Cedex, France
E-mail: morel@cmla.ens-cachan.fr

Professor B. Teissier, Institut Mathématique de Jussieu,
UMR 7586 du CNRS, Équipe "Géométrie et Dynamique",
175 rue du Chevaleret,
75013 Paris, France
E-mail: teissier@math.jussieu.fr

For the "Mathematical Biosciences Subseries" of LNM:

Professor P. K. Maini, Center for Mathematical Biology,
Mathematical Institute, 24-29 St Giles,
Oxford OX1 3LP, UK
E-mail : maini@maths.ox.ac.uk

Springer, Mathematics Editorial I,
Tiergartenstr. 17,
69121 Heidelberg, Germany,
Tel.: +49 (6221) 4876-8259
Fax: +49 (6221) 4876-8259
E-mail: lnm@springer.com